HISTORY, PHILOSOPHY AND SOCIOLOGY OF SCIENCE

Classics, Staples and Precursors

HISTORY, PHILOSOPHY AND SOCIOLOGY OF SCIENCE

Classics, Staples and Precursors

Selected By

YEHUDA ELKANA
ROBERT K. MERTON
ARNOLD THACKRAY
HARRIET ZUCKERMAN

THE

HISTORY OF CHEMISTRY

BY

THOMAS THOMSON

TWO VOLUMES IN ONE.

A R N O   P R E S S

A New York Times Company

New York – 1975

Reprint Edition 1975 by Arno Press Inc.

Reprinted from a copy in
  The University of Michigan Library

HISTORY, PHILOSOPHY AND SOCIOLOGY OF SCIENCE: '
Classics, Staples and Precursors
ISBN for complete set: 0-405-06575-2
See last pages of this volume for titles.

Manufactured in the United States of America

————◆————

Library of Congress Cataloging in Publication Data

Thomson, Thomas, 1773-1852.
    The history of chemistry.

    (History, philosophy, and sociology of science)
    Reprint of the 1830-31 ed. published by H. Colburn
and R. Bentley, London, which was issued as no. 3 and
10 of The National library.
    1. Chemistry--History. I. Series. II. Series:
The National library ; no. 3 [etc.]
QD11.T5 1975   540'.9   74-26298
ISBN 0-405-06623-6

THE

HISTORY OF CHEMISTRY.

Raeburn, pinx.ᵗ   Dean, sculp.ᵗ

JOSEPH BLACK, M.D. F.R.S.E.

London, Published by Henry Colburn, & Richard Bentley, 1830.

THE

HISTORY OF CHEMISTRY.

BY

THOMAS THOMSON, M.D. F.R.S.E.

PROFESSOR OF CHEMISTRY IN THE UNIVERSITY OF GLASGOW.

TWO VOLUMES IN ONE.

Second Edition.

LONDON:

HENRY COLBURN, AND RICHARD BENTLEY,

NEW BURLINGTON STREET.

PREFACE.

It may be proper, perhaps, to state here, in a very few words, the objects which the author had in view in drawing up the following History of Chemistry. Alchymy, or the art of making gold, with which the science originated, furnishes too curious a portion of the aberrations of the human intellect to be passed over in silence. The writings of the alchymists are so voluminous and so mystical, that it would have afforded materials for a very long work. But I was prevented

from extending this part of the subject to any greater
length than I have done, by considering the small
quantity of information which could have been gleaned
from the reveries of these fanatics or impostors; I
thought it sufficient to give a general view of the nature of their pursuits : but in order to put it in the
power of those who feel inclined to prosecute such investigations, I have given a catalogue of the most
eminent of the alchymists and a list of their works, so
far as I am acquainted with them. This catalogue
might have been greatly extended. Indeed it would
have been possible to have added several hundred
names. But I think the works which I have quoted
are more than almost any reasonable man would think
it worth his while to peruse ; and I can state, from experience, that the information gained by such a perusal
will very seldom repay the trouble.

The account of the chemical arts, with which the
ancients were acquainted, is necessarily imperfect;
because all arts and trades were held in so much contempt by them that they did not think it worth their
while to make themselves acquainted with the pro-

cesses. My chief guide has been Pliny, but many of his descriptions are unintelligible, obviously from his ignorance of the arts which he attempts to describe. Thus circumstanced, I thought it better to be short than to waste a great deal of paper, as some have done, on hypothesis and conjecture.

The account of the Chemistry of the Arabians is almost entirely limited to the works of Geber, which I consider to be the first book on Chemistry that ever was published, and to constitute, in every point of view, an exceedingly curious performance. I was much struck with the vast number of facts with which he was acquainted, and which have generally been supposed to have been discovered long after his time. I have, therefore, been at some pains in endeavouring to convey a notion of Geber's opinions to the readers of this history; but am not sure that I have succeeded. I have generally given his own words, as literally as possible, and, wherever it would answer the purpose, have employed the English translation of 1678.

Paracelsus gave origin to so great a revolution in medicine and the sciences connected with it, that it would

have been unpardonable not to have attempted to lay
his opinions and views before the reader; but, after perusing several of his most important treatises, I found
it almost impossible to form accurate notions on the
subject. I have, therefore, endeavoured to make use
of his own words as much as possible, that the want
of consistency and the mysticism of his opinions may
fall upon his own head. Should the reader find any
difficulty in understanding the philosophy of Paracelsus, he will be in no worse a situation than every one
has been who has attempted to delineate the principles of this prince of quacks and impostors. Van
Helmont's merits were of a much higher kind, and I
have endeavoured to do him justice; though his weak-
nesses are so visible that it requires much candour
and patience to discriminate accurately between his
excellencies and his foibles.

The history of Iatro-chemistry forms a branch of
our subject scarcely less extraordinary than Alchymy
itself. It might have been extended to a much greater
length than I have done. The reason why I did not
enter into longer details was, that I thought the subject

more intimately connected with the history of medicine
than of chemistry : it undoubtedly contributed to the
improvement of chemistry; not, however, by the
opinions or the physiology of the iatro-chemists, but by
inducing their contemporaries and successors to apply
themselves to the discovery of chemical medicines.

The History of Chemistry, after a theory of combustion had been introduced by Beccher and Stahl, becomes much more important. It now shook off the
trammels of alchymy, and ventured to claim its station
among the physical sciences. I have found it necessary
to treat of its progress during the eighteenth century
rather succinctly, but I hope so as to be easily intelligible. This made it necessary to omit the names of
many meritorious individuals, who supplied a share of
the contributions which the science was continually
receiving from all quarters. I have confined myself
to those who made the most prominent figure as chemical discoverers. I had no other choice but to follow
this plan, unless I had doubled the size of this little
work, which would have rendered it less agreeable and
less valuable to the general reader.

With respect to the History of Chemistry during that portion of the nineteenth century which is already past, it was beset with several difficulties. Many of the individuals, of whose labours I had occasion to speak, are still actively engaged in the prosecution of their useful works. Others have but just left the arena, and their friends and relations still remain to appreciate their merits. In treating of this branch of the science (by far the most important of all) I have followed the same plan as in the history of the preceding century. I have found it necessary to omit many names that would undoubtedly have found a place in a larger work, but which the limited extent to which I was obliged to confine myself, necessarily compelled me to pass over. I have been anxious not to injure the character of any one, while I have rigidly adhered to truth, so far as I was acquainted with it. Should I have been so unfortunate as to hurt the feelings of any individual by any remarks of mine in the following pages, it will give me great pain; and the only alleviation will be the consciousness of the total absence on my part of any malignant intention. To gratify the wishes of every individual may, perhaps, be impossible; but I can say, with truth, that my uniform object has been to do justice to the merits of all, so far as my own limited knowledge put it in my power to do.

CONTENTS

OF

THE FIRST VOLUME.

CONTENTS

OF

THE SECOND VOLUME.

—————

HISTORY OF CHEMISTRY.

INTRODUCTION.

CHEMISTRY, unlike the other sciences, sprang originally from delusion and superstition, and was at its commencement exactly on a level with magic and astrology. Even after it began to be useful to man, by furnishing him with better and more powerful medicines than the ancient physicians were acquainted with, it was long before it could shake off the trammels of alchymy, which hung upon it like a nightmare, cramping and blunting all its energies, and exposing it to the scorn and contempt of the enlightened part of mankind. It was not till about the middle of the eighteenth century that it was able to free itself from these delusions, and to venture abroad in all the native dignity of a useful science. It was then that its utility and its importance began to attract the attention of the world; that it drew within its vortex some of the greatest and most active men in every country; and that it advanced towards perfection with an accelerated pace. The field which it now presents to our view is vast and imposing. Its paramount utility is universally acknowledged. It has become a necessary part of education. It has contributed as much to the progress of society, and has done as much to augment the comforts and conveniences of life, and to increase the power and the resources of mankind, as all the other sciences put together.

It is natural to feel a desire to be acquainted with the origin and the progress of such a science; and to know something of the history and character of those numerous votaries to whom it is indebted for its progress and improvement. The object of this little work is to gratify these laudable wishes, by taking a rapid view of the progress of Chemistry, from its first rude and disgraceful beginnings till it has reached its present state of importance and dignity. I shall divide the subject into fifteen chapters. In the first I shall treat of Alchymy, which may be considered as the inauspicious commencement of the science, and which, in fact, consists of little else than an account of dupes and impostors; every where so full of fiction and obscurity, that it is a hopeless and almost impossible task to reach the truth. In the second chapter I shall endeavour to point out the few small chemical rills, which were known to the ancients. These I shall follow in their progress, in the succeeding chapters, till at last, augmented by an infinite number of streams flowing at once from a thousand different quarters, they have swelled to the mighty river, which now flows on majestically, wafting wealth and information to the civilized world.

CHAPTER I.

OF ALCHYMY.

THE word *chemistry* (χημεια, *chemeia*) first occurs in Suidas, a Greek writer, who is supposed to have lived in the eleventh century, and to have written his lexicon during the reign of Alexius Comnenus.* Under the word χημεια in his dictionary we find the following passage:

" CHEMISTRY, the preparation of silver and gold. The books on it were sought out by Dioclesian and burnt, on account of the new attempts made by the Egyptians against him. He treated them with cruelty and harshness, as he sought out the books written by the ancients on the chemistry (Περι χημειας) of gold and silver, and burnt them. His object was to prevent the Egyptians from becoming rich by the knowledge of this art, lest, emboldened by abundance of wealth, they might be induced afterwards to resist the Romans."†

---

* The word χημεια is said to occur in several Greek manuscripts of a much earlier date. But of this, as I have never had an opportunity of seeing them, I cannot pretend to judge. So much fiction has been introduced into the history of Alchymy, and so many ancient names have been treacherously dragged into the service, that we may be allowed to hesitate when no evidence is presented sufficient to satisfy a reasonable man.

† Χημεια, ἡ του αργυρου και χρυσου καλασκευη· ἡς τα βιβλια διερευνησαμενος ὁ Διοκλητιανος εκαυσε, δια τα νεωτερισθεντα

Under the word Δερας, *deras* (*a skin*), in the lexicon, occurs the following passage: " Δερας, the golden fleece, which Jason and the Argonauts (after a voyage through the Black Sea to Colchis) took, together with Medea, daughter of Ætes, the king. But this was not what the poets represent, but a treatise written on skins (δερμασι), teaching how gold might be prepared by chemistry. Probably, therefore, it was called by those who lived at that time, *golden,* on account of its great importance."* ·

From these two passages there can be no doubt that the word *chemistry* was known to the Greeks in the eleventh century; and that it signified, at that time, the art of making gold and silver. It appears, further, that in Suidas's opinion, this art was known to the Egyptians in the time of Dioclesian; that Dioclesian was convinced of its reality; and that, to put an end to it, he collected and burnt all the chemical writings to be found in Egypt. Nay, Suidas affirms that a book, describing the art of making gold, existed at the time of the Argonauts: and that the object of Jason and his followers was to get possession of that invaluable treatise, which the poets disguised under the term *golden fleece.*

The first meaning, then, of chemistry, was the *art of making gold.* And this art, in the opinion of Suidas, was understood at least as early as one thou-

αιγυπ7ιοις Διοκληλιανω· τουλοις ανημερως και φονικως εχρησαλο ό7εδη και τα περι χημειας χρυσου και αργυρου τοις παλαιοις γεγραμμενα βιβλια διερευνησαμενος εκαυσε, προς το μηκελι πλουλον αιγυπλιοις εκ της τοιαυλης προσγινεσθαι τεχνης, μηδε χρημαλων αυλοις θαρρονλας περιουσια του λοιπου ρωμαιοις ανλαιρειν.

* Δερας, το χρυσομαλλον δερας, όπερ ό Ιασων δια της πονλικης θαλασσης συν τοις αργοναυλαις εις την κολχιδα παραγενομενοι έλαβον, και την Μηδειαν την Αιητου του βασιλεως θυγατερα. Τουλο δε ουκ ώς ποιηλικως φερεται· άλλα Βιβλιον ήν έν δερμασι γεγραμενον περισχον όπως δειγινεσθαι δια χημειας χρυσον· εικολως ·ουν οί τολε χρουσουν ώνομαζον αυλο δερας δια την ενεργειαν την ·εξ αυλου.

sand two hundred and twenty-five years before the
Christian era : for that is the period at which the Argonautic expedition is commonly fixed by chronolo-
gists.

Though the lexicon of Suidas be the first printed
book in which the word Chemistry occurs, yet it is
said to be found in much earlier tracts, which still
continue in manuscript. Thus Scaliger informs us
that he perused a Greek manuscript of Zosimus, the
Panapolite, written in the fifth century, and deposited
in the King of France's library. Olaus Borrichius
mentions this manuscript; but in such terms that it
is difficult to know whether he had himself read it;
though he seems to insinuate as much.* The title
of this manuscript is said to be " A faithful Description of the sacred and divine Art of making Gold
and Silver, by Zosimus, the Panapolite."† In this
treatise, Zosimus distinguishes the art by the name
χημια, chemia. From a passage in this manuscript,
quoted by Scaliger, and given also by Olaus Borrichius, it appears that Zosimus carries the antiquity of
the art of making gold and silver, much higher than
Suidas has ventured to do. The following is a literal
translation of this curious passage :

" The sacred Scriptures inform us that there exists
a tribe of genii, who make use of women. Hermes
mentions this circumstance in his Physics ; and almost
every writing (λογος), whether sacred (φανερος) or apo-
cryphal, states the same thing. The ancient and
divine Scriptures inform us, that the angels, captivated
by women, taught them all the operations of nature.
Offence being taken at this, they remained out of
heaven, because they had taught mankind all manner

---

* De Ortu et Progressu Chemiæ, p. 12.

† Σωσιμου του παναπολιτου γνησια γραφη, περι της ιερας, και
θειας τεχνης του χρυσου και αργυριου ποιησιος. Παναπολις
was a city in Egypt.

of evil, and things which could not be advantageous
to their souls. The Scriptures inform us that the
giants sprang from these embraces. Chema is the
first of their traditions respecting these arts. The
book itself they called Chema; hence the art is called
Chemia."

Zosimus is not the only Greek writer on Chemistry.
Olaus Borrichius has given us a list of thirty-eight
treatises, which he says exist in the libraries of Rome,
Venice, and Paris: and Dr. Shaw has increased this
list to eighty-nine.* But among these we find the
names of Hermes, Isis, Horus, Democritus, Cleopatra,
Porphyry, Plato, &c.—names which undoubtedly have
been affixed to the writings of comparatively modern
and obscure authors. The style of these authors, as
Borrichius informs us, is barbarous. They are chiefly
the production of ecclesiastics, who lived between the
fifth and twelfth centuries. In these tracts, the art
of which they treat is sometimes called *chemistry*
(χημεια); sometimes the *chemical art* (χημευτικα);
sometimes the *holy art;* and the *philosopher's stone*.

It is evident from this, that between the fifth century and the taking of Constantinople in the fifteenth
century, the Greeks believed in the possibility of making
gold and silver artificially; and that the art which
professed to teach these processes was called by them
Chemistry.

These opinions passed from the Greeks to the Arabians, when, under the califs of the family of Abas-
sides, they began to turn their attention to science,
about the beginning of the ninth century; and when
the enlightened zeal of the Fatimites in Africa, and
the Ommiades in Spain, encouraged the cultivation
of the sciences. From Spain they gradually made
their way into the different Christian kingdoms of Europe. From the eleventh to the sixteenth century, the art

---

* Shaw's Translation of Boerhaave's Chemistry, i. 20.

of making gold and silver was cultivated in Germany, Italy, France, and England, with considerable assiduity. The cultivators of it were called *Alchymists;* a name obviously derived from the Greek word *chemia,* but somewhat altered by the Arabians. Many alchymistical tracts were written during that period. A considerable number of them were collected by Lazarus Zetzner, and published at Strasburg in 1602, under the title of "Theatrum Chemicum, præcipuos selectorum auctorum tractatus de Chemiæ et Lapidis Philosophici Antiquitate, veritate, jure, præstantia, et operationibus continens in gratiam veræ Chemiæ et Medicinæ Chemicæ Studiosorum (ut qui uberrimam unde optimorum remediorum messem facere poterunt) congestum et in quatuor partes seu volumina digestum." This book contains one hundred and five different alchymistical tracts.

In the year 1610 another collection of alchymistical tracts was published at Basil, in three volumes, under the title of "Artis Auriferæ quam Chemiam vocant volumina tria." It contains forty-seven different tracts.

In the year 1702 Mangetus published at Geneva two very large folio volumes, under the name of "Bibliotheca Chemica Curiosa, seu rerum ad Alchymiam pertinentium thesaurus instructissimus, quo non tantum Artis Auriferæ ac scriptorum in ea nobiliorum Historia traditur; lapidis veritas Argumentis et Experimentis innumeris, immo et Juris Consultorum Judiciis evincitur; Termini obscuriores explicantur; Cautiones contra Impostores et Difficultates in Tinctura Universali conficienda occurrentes declarantur: verum etiam Tractatus omnes Virorum Celebriorum, qui in Magno sudarunt Elixyre, quique ab ipso Hermete, ut dicitur, Trismegisto, ad nostra usque tempora de Chrysopoea scripserunt, cum præcipuis suis Commentariis, concinno ordine dispositi exhibentur." This Bibliotheca contains one hundred and twenty-two alchymistical treatises, many of them of considerable length.

Two additional volumes of the Theatrum Chemicum were afterwards published; but these I have never had an opportunity of seeing.

From these collections, which exhibit a pretty complete view of the writings of the alchymists, a tolerably accurate notion may be formed of their opinions. But before attempting to lay open the theories and notions by which the alchymists were guided, it will be proper to state the opinions which were gradually adopted respecting the origin of Alchymy, and the contrivances by which these opinions were supported.

Zosimus, the Panapolite, in a passage quoted above informs us, that the art of making gold and silver was not a human invention; but was communicated to mankind by angels or demons. These angels, he says, fell in love with women, and were induced by their charms to abandon heaven altogether, and take up their abode upon earth. Among other pieces of information which these spiritual beings communicated to their paramours, was the sublime art of Chemistry, or the fabrication of gold and silver.

It is quite unnecessary to refute this extravagant opinion, obviously founded on a misunderstanding of a passage in the sixth chapter of Genesis. "And it came to pass, when men began to multiply on the face of the earth, and daughters were born unto them, that the sons of God saw the daughters of men, that they were fair; and they took them wives of all which they chose.—There were giants in the earth in those days; and also after that, when the sons of God came in unto the daughters of men, and they bare *children* to them; the same became mighty men, which were of old, men of renown."

There is no mention whatever of angels, or of any information on science communicated by them to mankind.

Nor is it necessary to say much about the opinion advanced by some, and rather countenanced by Olaus

Borrichius, that the art of making gold was the invention of Tubal-cain, whom they represent as the same as Vulcan. All the information which we have respecting Tubal-cain, is simply that he was an instructor of every artificer in brass and iron.* No allusion whatever is made to gold. And that in these early ages of the world there was no occasion for making gold artificially, we have the same authority for believing. For in the second chapter of Genesis, where the garden of Eden is described, it is said, " And a river went out of Eden to water the garden ; and from thence it was parted, and came into four heads : the name of the first is Pison, that is it which encompasseth the whole land of Havilah, where there is gold. And the gold of that land is good : there is bdellium and onyx-stone."

But the most generally-received opinion is, that alchymy originated in Egypt ; and the honour of the invention has been unanimously conferred upon Hermes Trismegistus. He is by some supposed to be the same person with Chanaan, the son of Ham, whose son Mizraim first occupied and peopled Egypt. Plutarch informs us, that Egypt was sometimes called *Chemia*.† This name is supposed to be derived from Chanaan (כנען); thence it was believed that Chanaan was the true inventor of alchymy, to which he affixed his own name. Whether the Hermes ('Ερμῆς) of the Greeks was the same person with Chanaan or his son Mizraim, it is impossible at this distance of time to decide; but to Hermes is assigned the invention of alchymy, or the art of making gold, by almost the unanimous consent of the adepts.

Albertus Magnus informs us, that " Alexander the Great discovered the sepulchre of Hermes, in one of his journeys, full of all treasures, not metallic, but golden, written on a table of *zatadi*, which others call

* Genesis iv. 22.   † De Iside and Osiride, c. 5.

emerald." This passage occurs in a tract of Albertus *de secretis chemicis*, which is considered as supposititious. Nothing is said of the source whence the information contained in this passage was drawn : but, from the quotations produced by Kriegsmann, it would appear that the existence of this emerald table was alluded to by Avicenna and other Arabian writers. According to them, a woman called Sarah took it from the hands of the dead body of Hermes, some ages after the flood, in a cave near Hebron. The inscription on it was in the Phœnician language. The following is a literal translation of this famous inscription, from the Latin version of Kriegsmann :*

1. I speak not fictitious things, but what is true and most certain.

* There are two Latin translations of these tables (unless we are rather to consider them as originals, for no Phœnician nor Greek original exists). I shall insert them both here.

I.—VERBA SECRETORUM HERMETIS TRISMEGISTI.

1. Verum sine mendacio certum et verissimum.

2. Quod est inferius, est sicut quod est superius, et quod est superius est sicut quod est inferius ad perpetranda miracula rei unius.

3. Et sicut omnes res fuerant ab uno meditatione unius : sic omnes res natæ fuerunt ab hac una re adaptatione.

4. Pater ejus est Sol, mater ejus Luna, portavit illud ventus in ventre suo, nutrix ejus terra est.

5. Pater omnis thelesmi totius mundi est hic.

6. Vis ejus integra est, si versa fuerit in terram.

7. Separabis terram ab igne, subtile a spisso suaviter cum magno ingenio.

8. Ascendit a terra in cœlum, iterumque descendit in terram, et recipit vim superiorum et inferiorum, sic habebis gloriam totius mundi. Ideo fugiat a te omnis obscuritas.

9. Hic est totius fortitudinis fortitudo fortis ; quia vincit omnem rem subtilem, omnemque solidam penetrabit.

10. Sic mundus creatus est.

11. Hinc adaptationes erunt mirabiles, quarum modus est hic.

12. Itaque vocatus sum Hermes Trismegistus, habens tres partes philosophiæ totius mundi.

13. Completum est quod dixi de operatione solis.

2. What is below is like that which is above, and what is above is similar to that which is below, to accomplish the miracles of one thing.

3. And as all things were produced by the meditation of one Being, so all things were produced from this one thing by adaptation.

4. Its father is *Sol*, its mother *Luna;* the wind carried it in its belly, the earth is its nurse.

5. It is the cause of all perfection throughout the whole world.

6. Its power is perfect, if it be changed into earth.

7. Separate the earth from the fire, the subtile from the gross, acting prudently and with judgment.

8. Ascend with the greatest sagacity from the earth

II.—Descriptio Arcanorum Hermetis Trismegisti.

1. Vere non ficte, certo verissime aio.

2. Inferiora hæc cum superioribus illis, istaque cum iis vicissim vires sociant, ut producant rem unam omnium mirificissimam.

3. Ac quemadmodum cuncta educta ex uno fuere verbo Dei unius : sic omnes quoque res perpetuo ex hac una re generantur dispositione Naturæ.

4. Patrem ea habet Solem, matrem Lunam : ab aëre in utero quasi gestatur, nutritur a terra.

5. Causa omnis perfectionis rerum ea est per univerum hoc.

6. Ad summam ipsa perfectionem virium pervenit si redierit in humum.

7. In partes tribuite humum ignem passam, attenuans densitatem ejus re omnium suavissima.

8. Summa ascende ingenii sagacitate a terra in cœlum, indeque rursum in terram descende, ac vires superiorum inferiorumque coge in unum : sic potiere gloria totius mundi atque ita abjectæ sortis homo amplius non habere.

9. Isthæc jam res ipsa fortitudine fortior existet ; corpora quippe tam tenuia quam solida penetrando subige.

10. Atque sic quidem quæcunque mundus continet creata fuere.

11. Hinc admiranda evadunt opera, quæ ad eundum modum instituantur.

12. Mihi vero ideo nomen Hermetis Trismegisti impositum fuit, quod trium mundi sapientiæ partium doctor deprehensus sum.

13. Hæc sunt quæ de chemicæ artis prestantissimo opere consignanda esse duxi.

to heaven, and then again descend to the earth, and
unite together the powers of things superior and things
inferior.   Thus you will possess the glory of the whole
world; and all obscurity will fly far away from you.

9. This thing has more fortitude than fortitude itself; because it will overcome every subtile thing, and
penetrate every solid thing.

10. By it this world was formed.

11. Hence proceed wonderful things, which in this
wise were established.

12. For this reason I am called Hermes Trismegistus, because I possess three parts of the philosophy of
the whole world.

13. What I had to say about the operation of *Sol*
is completed.

Such is a literal translation of the celebrated inscription of Hermes Trismegistus upon the emerald
tablet.   It is sufficiently obscure to put it in the power
of commentators to affix almost any explanation to it
that they choose.   The two individuals who have devoted most time to illustrate this tablet, are Kriegsmann and Gerard Dorneus, whose commentaries may
be seen in the first volume of Mangetus's Bibliotheca
Chemica.   They both agree that it refers to the *universal medicine*, which began to acquire celebrity
about the time of Paracelsus, or a little earlier.

This exposition, which appears as probable as any
other, betrays the time when this celebrated inscription seems to have been really written.   Had it been
taken out of the hands of the dead body of Hermes by
Sarah (obviously intended for the wife of Abraham) as
is affirmed by Avicenna, it is not possible that Herodotus, and all the writers of antiquity, both Pagan and
Christian, should have entirely overlooked it; or how
could Avicenna have learned what was unknown to all
those who lived nearest the time when the discovery
was supposed to have been made?   Had it been discovered in Egypt by Alexander the Great, would it

have been unknown to Aristotle, and to all the numerous tribe of writers whom the Alexandrian school produced, not one of whom, however, make the least allusion to it ?   In short, it bears all the marks of a forgery
of the fifteenth century.   And even the tract ascribed
to Albertus Magnus, in which the tablet of Hermes is
mentioned, and the discovery related, is probably also
a forgery; and doubtless a forgery of the same individual who fabricated the tablet itself, in order to
throw a greater air of probability upon a story which
he wished to palm upon the world as true.   His object was in some measure accomplished; for the authenticity of the tablet was supported with much zeal
by Kriegsmann, and afterwards by Olaus Borrichius.

There is another tract of Hermes Trismegistus, entitled "Tractatus Aureus de Lapidis Physici Secreto;"
on which no less elaborate commentaries have been
written.   It professes to teach the process of making
the *philosopher's stone;* and, from the allusions in it,
to the use of this stone, as a universal medicine, was
probably a forgery of the same date as the emerald
tablet.   It would be in vain to attempt to extract any
thing intelligible out of this Tractatus Aureus: it may
be worth while to give a single specimen, that the reader
may be able to form some idea of the nature of the style.

"Take of moisture an ounce and a half; of meri-
dional redness, that is the soul of the sun, a fourth
part, that is half an ounce; of yellow seyr, likewise
half an ounce; and of auripigmentum, a half ounce,
making in all three ounces.   Know that the vine of
wise men is extracted in threes, and its wine at last is
completed in thirty."*

* " Accipe de humore unciam unam et mediam, et de rubore
meridionali, id est anima solis, quartam partem, id est, unciam mediam, et de Seyre citrino, similiter unciam mediam,
et de auripigmenti dimidium, quæ sunt octo, id est unciæ tres.
Scitote quod vitis sapientum in tribus extrahitur, ejusque vinum
in fine triginta peragitur."

Had the opinion, that gold and silver could be artificially formed originated with Hermes Trismegistus, or had it prevailed among the ancient Egyptians, it would certainly have been alluded to by Herodotus, who spent so many years in Egypt, and was instructed by the priests in all the science of the Egyptians. Had *chemistry* been the name of a science real or fictitious, which existed as early as the expedition of the Argonauts, and had so many treatises on it, as Suidas alleges existed in Egypt before the reign of Dioclesian, it could hardly have escaped the notice of Pliny, who was so curious and so indefatigable in his researches, and who has collected in his natural history a kind of digest of all the knowledge of the ancients in every department of practical science. The fact that the term chemistry (χημεια) never occurs in any Greek or Roman writer prior to Suidas, who wrote so late as the eleventh century, seems to overturn all idea of the existence of that pretended science among the ancients, notwithstanding the elaborate attempts of Olaus Borrichius to prove the contrary.

I am disposed to believe, that chemistry or alchymy, understanding by the term the *art of making gold and silver*, originated among the Arabians, when they began to turn their attention to medicine, after the establishment of the caliphs ; or if it had previously been cultivated by Greeks (as the writings of Zosimus, the Panapolite, if genuine, would lead us to suppose), that it was taken up by the Arabians, and reduced by them into regular form and order. If the works of Geber be genuine, they leave little doubt on this point. Geber is supposed to have been a physician, and to have written in the seventh century. He admits, as a first principle, that metals are compounds of mercury and sulphur. He talks of the philosopher's stone ; professes to give the mode of preparing it ; and teaches the way of converting the different metals, known in his time, into medicines, on

whose efficacy he bestows the most ample panegyrics.
Thus the principles which lie at the bottom of alchymy
were implicitly adopted by him. Yet I can nowhere
find in him any attempt to make gold artificially. His
chemistry was entirely devoted to the improvement of
medicine. The subsequent pretensions of the alchymists to convert the baser metals into gold are no
where avowed by him. I am disposed from this to
suspect, that the theory of gold-making was started
after Geber's time, or at least that it was after the
seventh century, before any alchymist ventured to
affirm that he himself was in possession of the secret,
and could fabricate gold artificially at pleasure. For
there is a wide distance between the opinion that gold
may be made artificially and the affirmation that we
are in possession of a method by which this transmutation of the baser metals into gold can be accomplished. The first may be adopted and defended with
much plausibility and perfect honesty; but the second
would require a degree of skill far exceeding that of
the most scientific votary of chemistry at present
existing.

The opinion of the alchymists was, that all the metals are compounds; that the baser metals contain
the same constituents as gold, contaminated, indeed,
with various impurities, but capable, when their impurities are removed or remedied, of assuming all the
properties and characters of gold. The substance
possessing this wonderful power they distinguish by
the name of *lapis philosophorum*, or, philosopher's
stone, and they usually describe it as a red powder,
having a peculiar smell. Few of the alchymists who
have left writings behind them boast of being possessed of the philosopher's stone. Paracelsus, indeed,
affirms, that he was acquainted with the method of
making it, and gives several processes, which, however, are not intelligible. But many affirm that they

had seen the philosopher's stone; that they had portions of it in their possession ; and that they had seen
several of the inferior metals, especially lead and
quicksilver, converted by means of it into gold. Many
stories of this kind are upon record, and so well authenticated, that we need not be surprised at their
having been generally credited.   It will be sufficient
if we state one or two of those which depend upon
the most unexceptionable evidence.   The following
relation is given by Mangetus, on the authority of
M. Gros, a clergyman of Geneva, of the most unexceptionable character, and at the same time a skilful physician and expert chemist :

   About the year 1650 an unknown Italian came to
Geneva, and took lodgings at the sign of the *Green
Cross*.   After remaining there a day or two, he requested De Luc, the landlord, to procure him a man
acquainted with Italian, to accompany him through
the town and point out those things which deserved to
be examined.   De Luc was acquainted with M. Gros,
at that time about twenty years of age, and a student
in Geneva, and knowing his proficiency in the Italian
language, requested him to accompany the stranger.
To this proposition he willingly acceded, and attended
the Italian every where for the space of a fortnight.
The stranger now began to complain of want of money,
which alarmed M. Gros not a little—for at that
time he was very poor—and he became apprehensive,
from the tenour of the stranger's conversation, that he
intended to ask the loan of money from him.   But
instead of this, the Italian asked him if he was acquainted with any goldsmith, whose bellows and other
utensils they might be permitted to use, and who
would not refuse to supply them with the different
articles requisite for a particular process which he
wanted to perform.   M. Gros named a M. Bureau, to
whom the Italian immediately repaired.   He readily

furnished crucibles, pure tin, quicksilver, and the other things required by the Italian. The goldsmith left his workshop, that the Italian might be under the less restraint, leaving M. Gros, with one of his own workmen, as an attendant. The Italian put a quantity of tin into one crucible, and a quantity of quicksilver into another. The tin was melted in the fire and the mercury heated. It was then poured into the melted tin, and at the same time a red powder enclosed in wax was projected into the amalgam. An agitation took place, and a great deal of smoke was exhaled from the crucible; but this speedily subsided, and the whole being poured out, formed six heavy ingots, having the colour of gold. The goldsmith was called in by the Italian, and requested to make a rigid examination of the smallest of these ingots. The goldsmith, not content with the touchstone and the application of aqua fortis, exposed the metal on the cupel with lead, and fused it with antimony, but it sustained no loss. He found it possessed of the ductility and specific gravity of gold ; and full of admiration, he exclaimed that he had never worked before upon gold so perfectly pure. The Italian made him a present of the smallest ingot as a recompence, and then, accompanied by M. Gros, he repaired to the Mint, where he received from M. Bacuet, the mintmaster, a quantity of Spanish gold coin, equal in weight to the ingots which he had brought. To M. Gros he made a present of twenty pieces, on account of the attention that he had paid to him; and, after paying his bill at the inn, he added fifteen pieces more, to serve to entertain M. Gros and M. Bureau for some days, and in the mean time he ordered a supper, that he might, on his return, have the pleasure of supping with these two gentlemen. He went out, but never returned, leaving behind him the greatest regret and admiration. It is needless to add, that M. Gros and M. Bureau continued to enjoy themselves at the inn till the fifteen pieces, which the stranger had left, were exhausted."*

Mangetus gives also the following relation, which he states upon the authority of an English bishop, who communicated it to him in the year 1685, and at the same time gave him about half an ounce of the gold which the alchymist had made:

A stranger, meanly dressed, went to Mr. Boyle, and after conversing for some time about chemical processes, requested him to furnish him with antimony, and some other common metallic substances, which then fortunately happened to be in Mr. Boyle's laboratory. These were put into a crucible, which was then placed in a melting-furnace. As soon as these metals were fused, the stranger showed a powder to the attendants, which he projected into the crucible, and instantly went out, directing the servants to allow the crucible to remain in the furnace till the fire went out of its own accord, and promising at the same time to return in a few hours. But, as he never fulfilled this promise, Boyle ordered the cover to be taken off the crucible, and found that it contained a yellow-coloured metal, possessing all the properties of pure gold, and only a little lighter than the weight of the materials originally put into the crucible.†

The following strange story is related by Helvetius, physician to the Prince of Orange, in his Vitulus Aureus: Helvetius was a disbeliever of the philosopher's stone, and the universal medicine, and even turned Sir Kenelm Digby's sympathetic powder into ridicule. On the 27th of December, 1666, a stranger called upon him, and after conversing for some time about a universal medicine, showed a yellow powder, which he affirmed to be the philosopher's stone, and at the same time five large plates of gold, which had been made

* Preface to Mangetus's Bibliotheca Chemica Curiosa.
† Ibid.

by means of it. Helvetius earnestly entreated that he
would give him a little of this powder, or at least that
he would make a trial of its power; but the stranger
refused, promising however to return in six weeks. He
returned accordingly, and after much entreaty he gave
to Helvetius a piece of the stone, not larger than the
size of a rape-seed. When Helvetius expressed his
doubt whether so small a portion would be sufficient
to convert four grains of lead into gold, the adept
broke off one half of it, and assured him that what
remained was more than sufficient for the purpose.
Helvetius, during the first conference, had concealed
a little of the stone below his nail. This he threw into
melted lead, but it was almost all driven off in smoke,
leaving only a vitreous earth. When he mentioned
this circumstance, the stranger informed him that the
powder must be enclosed in wax, before it be thrown
into the melted lead, lest it should be injured by the
smoke of the lead. The stranger promised to return
next day, and show him the method of making the
projection; but having failed to make his appearance,
Helvetius, in the presence of his wife and son, put six
drachms of lead into a crucible, and as soon as it was
melted he threw into it the fragment of philosopher's
stone in his possession, previously covered over with
wax. The crucible was now covered with its lid, and
left for a quarter of an hour in the fire, at the end of
which time he found the whole lead converted into
gold. The colour was at first a deep green; being
poured into a conical vessel, it assumed a blood-red
colour; but when cold, it acquired the true tint of
gold. Being examined by a goldsmith, he considered
it as pure gold. He requested Porelius, who had the
charge of the Dutch mint, to try its value. Two
drachms of it being subjected to quartation, and solution in aqua fortis, were found to have increased in
weight by two scruples. This increase was doubtless
owing to the silver, which still remained enveloped in

the gold, after the action of the aqua fortis. To endeavour to separate the silver more completely, the
gold was again fused with seven times its weight of
antimony, and treated in the usual manner; but no
alteration took place in the weight.*

It would be easy to relate many other similar nar-
ratives; but the three which I have given are the best
authenticated of any that I am acquainted with. The
reader will observe, that they are all stated on the
authority, not of the persons who were the actors, but
of others to whom they related them; and some of
these, as the English bishop, perhaps not very familiar
with chemical processes, and therefore liable to leave
out or mistate some essential particulars. The evidence, therefore, though the best that can be got, is
not sufficient to authenticate these wonderful stories.
A little latent vanity might easily induce the narrators
to suppress or alter some particulars, which, if known,
would have stripped the statements of every thing mar-
vellous which they contain, and let us into the secret
of the origin of the gold, which these alchymists
boasted that they had fabricated. Whoever will read
the statements of Paracelsus, respecting his knowledge
of the philosopher's stone, which he applied not to the
formation of gold but to medicine, or whoever will
examine his formulas for making the stone, will easily
satisfy himself that Paracelsus possessed no real knowledge on the subject.†

But to convey as precise ideas on this subject as
possible, it may be worth while to state a few of the
methods by which the alchymists persuaded themselves
that they could convert the baser metals into gold.

In the year 1694 an old gentleman called upon
Mr. Wilson, at that time a chemist in London, and
informed him that at last, after forty years' search, he

* Bergmann, Opusc. iv. 121.
† I allude to his *Manuale sive de Lapide Philosophico Medici-
nali.* Opera Paracelsi, ii. 133. Folio edition. Geneva, 1658.

had met with an ample recompence for all his trouble and expenses. This he confirmed with some oaths and imprecations; but, considering his great weakness and age, he looked upon himself as incapable to undergo the fatigues of the process. " I have here," says he, " a piece of sol (*gold*) that I made from silver, about four years ago, and I cannot trust any man but you with so rare a secret. We will share equally the charges and profit, which will render us wealthy enough to command the world." The nature of the process being stated, Mr. Wilson thought it not unreasonable, especially as he aimed at no peculiar advantage for himself. He accordingly put it to the trial in the following manner :

1. Twelve ounces of Japan copper were beat into thin plates, and laid *stratum super stratum* with three ounces of flowers of sulphur, in a crucible. It was exposed in a melting-furnace to a gentle heat, till the sulphureous flames expired. When cold, the æs ustum (*sulphuret of copper*) was pounded, and stratified again; and this process was repeated five times. Mr. Wilson does not inform us whether the powder was mixed with flowers of sulphur every time that it was heated; but this must have been the case, otherwise the sulphuret would have been again converted into metallic copper, which would have melted into a mass. By this first process, then, bisulphuret of copper was formed, composed of equal weights of sulphur and copper.

2. Six pounds of iron wire were put into a large glass body, and twelve pounds of muriatic acid poured upon it. Six days elapsed (during which it stood in a gentle heat) before the acid was saturated with the iron. The solution was then decanted off, and filtered, and six pounds of new muriatic acid poured on the undissolved iron. This acid, after standing a sufficient time, was decanted off, and filtered. Both liquids were put into a large retort, and distilled by a sand-heat. Towards the end, when the drops from the

retort became yellow, the receiver was changed, and the fire increased to the highest degree, in which the retort was kept between four and six hours. When all was cold, the receiver was taken off, and a quantity of flowers was found in the neck of the retort, variously coloured, like the rainbow. The yellow liquor in the receiver weighed ten ounces and a half; the flowers (*chloride of iron*), two ounces and three drams. The liquid and flowers were put into a clean bottle.

3. Half a pound of sal enixum (*sulphate of potash*) and a pound and a half of nitric acid were put into a retort. When the salt had dissolved in the acid, ten ounces of mercury (previously distilled through quick-lime and salt of tartar) were added. The whole being distilled to dryness, a fine yellow mass (*pernitrate of mercury*) remained in the bottom of the retort. The liquor was returned, with half a pound of fresh nitric acid, and the distillation repeated. The distillation was repeated a third time, urging this last cohobation with the highest degree of fire. When all was cold, a various-coloured mass was found in the bottom of the retort: this mass was doubtless a mixture of sulphate of potash, and pernitrate of mercury, with some oxide of mercury.

4. Four ounces of fine silver were dissolved in a pound of aqua fortis; to the solution was added, of the bisulphuret of copper four ounces; of the mixture of sulphate of potash, pernitrate of mercury, and oxide of mercury one ounce and a half, and of the solution of perchloride of iron two ounces and a half. When these had stood in a retort twenty-four hours, the liquor was decanted off, and four ounces of nitric acid were poured upon the little matter that was not dissolved. Next morning a total dissolution was obtained. The whole of this dissolution was put into a retort and distilled almost to dryness. The liquid was poured back, and the distillation repeated three times; the

last time the retort being urged by a very strong fire till no fumes appeared, and not a drop fell.

5. The matter left in the bottom of the retort was now put into a crucible, all the corrosive fumes were gently evaporated, and the residue melted down with a fluxing powder.

This process was expected to yield five ounces of pure gold; but on examination the silver was the same (except the loss of half a pennyweight) as when dissolved in the aqua fortis: there were indeed some grains among the scoria, which appeared like gold, and would not dissolve in aqua fortis. No doubt they consisted of peroxide of iron, or, perhaps, persulphuret of iron.[*]

Mr. Wilson's alchymistical friend, not satisfied with this first failure, insisted upon a repetition of the process, with some alteration in the method and the addition of a certain quantity of gold. The whole was accordingly gone through again; but it is unnecessary to say that no gold was obtained, or at least, the two drams of gold employed had increased in weight by only two scruples and thirteen grains; this addition was doubtless owing to a little silver from which it had not been freed.[†]

I shall now give a process for making the philosopher's stone, which was considered by Mangetus as of great value, and on that account was given by him in the preface to his Bibliotheca Chemica.

1. Prepare a quantity of spirit of wine, so free from water that it is wholly combustible, and so volatile that when a drop of it is let fall it evaporates before it reaches the ground;—this constitutes the first menstruum.

2. Take pure mercury, revived in the usual manner from cinnabar, put it into a glass vessel with common salt and distilled vinegar; agitate violently, and when the vinegar acquires a black colour pour it off and add

---

[*] Wilson's Chemistry, p. 375.   [†] Ibid., p. 379.

new vinegar; agitate again, and continue these repeated agitations and additions till the vinegar ceases
to acquire a black colour from the mercury : the mer
cury is now quite pure and very brilliant.

2. Take of this mercury four parts; of sublimed
mercury* (*mercurii meteoresati*), prepared with your
own hands, eight parts; triturate them together in a
wooden mortar with a wooden pestle, till all the grains
of running mercury disappear. This process is tedious
and rather difficult.

4. The mixture thus prepared is to be put into an
aludel, or a sand-bath, and exposed to a subliming
heat, which is to be gradually raised till the whole
sublimes. Collect the sublimed matter, put it again
into the aludel, and sublime a second time; this process must be repeated five times. Thus a very sweet
and crystallized sublimate is obtained : it constitutes
the salt of wise men (*sal sapientum*), and possesses
wonderful properties.†

5. Grind it in a wooden mortar, and reduce it to
powder; put it into a glass retort, and pour upon it
the spirit of wine (No. 1) till it stands about three
finger-breadths above the powder; seal the retort
hermetically, and expose it to a very gentle heat for
seventy-four hours, shaking it several times a-day;
then distil with a gentle heat and the spirit of wine
will pass over, together with spirit of mercury. Keep
this liquid in a well-stopped bottle, lest it should
evaporate. More spirit of wine is to be poured upon
the residual salt, and after digestion it must be distilled off as before ; and this process must be repeated
till the whole salt is dissolved, and distilled over with
the spirit of wine. You have now performed a great
work. The mercury is now rendered in some measure
volatile, and it will gradually become fit to receive the
tincture of gold and silver. Now return thanks to

* Probably corrosive sublimate.   † Probably calomel.

God, who has hitherto crowned your wonderful work with success; nor is this great work involved in Cimmerian darkness, but clearer than the sun; though preceding writers have imposed upon us with parables, hieroglyphics, fables, and enigmas.

6. Take this mercurial spirit, which contains our magical steel in its belly, put it into a glass retort, to which a receiver must be well and carefully luted: draw off the spirit by a very gentle heat, there will remain in the bottom of the retort the quintessence or soul of mercury; this is to be sublimed by applying a stronger heat to the retort that it may become volatile, as all the philosophers express themselves—

> Si fixum solvas faciesque volare solutum,
> Et volucrum figas faciet te vivere tutum.

This is our luna, our fountain, in which the king and queen may bathe. Preserve this precious quintessence of mercury, which is very volatile, in a well-shut vessel for further use.

8. Let us now proceed to the operation of common gold, which we shall communicate clearly and distinctly, without digression or obscurity; that from vulgar gold we may obtain our philosophical gold, just as from common mercury we obtained, by the preceding processes, philosophical mercury.

In the name of God, then, take common gold, purified in the usual way by antimony, convert it into small grains, which must be washed with salt and vinegar, till it be quite pure. Take one part of this gold, and pour on it three parts of the quintessence of mercury; as philosophers reckon from seven to ten, so we also reckon our number as philosophical, and we begin with three and one; let them be married together like husband and wife, to produce children of their own kind, and you will see the common gold sink and plainly dissolve. Now the marriage is consummated; now two things are converted into one: thus the philosophical sulphur is at hand, as the philosophers say, *the sulphur being dissolved the stone is at hand.* Take then, in the name of God, our philosophical vessel, in which the king and queen embrace each other as in a bedchamber, and leave it till the water is converted into earth, then peace is concluded between the water and fire, then the elements have no longer any thing contrary to each other; because, when the elements are converted into earth they no longer oppose each other; for in earth all elements are at rest. For the philosophers say, " When you shall have seen the water coagulate itself, think that your knowledge is true, and that your operations are truely philosophical." The gold is now no longer common, but ours is philosophical, on account of our processes: at first exceedingly fixed; then exceedingly volatile, and finally exceedingly fixed; and the whole science depends upon the change of the elements. The gold at first was a metal, now it is a sulphur, capable of converting all metals into its own sulphur. Now our tincture is wholly converted into sulphur, which possesses the energy of curing all diseases: this is our universal medicine against all the most deplorable diseases of the human body; therefore, return infinite thanks to Almighty God for all the good things which he has bestowed upon us.

9. In this great work of ours, two modes of fermenting and projecting are wanting, without which the uninitiated will not easily follow our process. The mode of fermenting is as follows: Take of our sulphur above described one part, and project it upon three parts of very pure gold fused in a furnace; in a moment you will see the gold, by the force of the sulphur, converted into a red sulphur of an inferior quality to the first sulphur; take one part of this, and project it upon three parts of fused gold, the whole will be again converted into a sulphur, or a friable mass; mixing one part of this with three parts of gold, you will have

a malleable and extensible metal. If you find it so, well ; if not add other sulphur and it will again pass into sulphur. Now the sulphur will be sufficiently fermented, or our medicine will be brought into a metallic nature.

10. The mode of projecting is this : Take of the fermented sulphur one part, and project it upon ten parts of mercury, heated in a crucible, and you will have a perfect metal ; if its colour is not sufficiently deep, fuse it again, and add more fermented sulphur, and thus it will acquire colour. If it becomes frangible, add a sufficient quantity of mercury and it will be perfect.

Thus, friend, you have a description of the universal medicine, not only for curing diseases and prolonging life, but also for transmuting all metáls into gold. Give therefore thanks to Almighty God, who, taking pity on human calamities, has at last revealed this inestimable treasure, and made it known for the common benefit of all.*

Such is the formula (slightly abridged) of Carolus Musitanus, by which the philosopher's stone, according to him, may be formed. Compared with the formulas of most of the alchymists, it is sufficiently plain. What the *sublimed mercury* is does not appear ; from the process described we should be apt to consider it as *corrosive sublimate* ; on that supposition, the sal sapientum formed in No. 5, would be calomel : the only objection to this supposition is the process described in No. 5 ; for calomel is not soluble in alcohol. The philosopher's stone prepared by this elaborate process could hardly have been any thing else than an *amalgam of gold ;* it could not have contained chloride of gold, because such a preparation, instead of acting medicinally, would have proved a most virulent poison. There is no doubt that amalgam of gold, if

* Mangeti Bibliothecæ Chemicæ Præfatio.

projected into melted lead or tin, and afterwards cupellated, would leave a portion of gold—all the gold of
course that existed previously in the amalgam. It
might therefore have been employed by impostors to
persuade the ignorant that it was really the philosopher's stone; but the alchymists who prepared the
amalgam could not be ignorant that it contained gold.

There is another process given in the same preface
of a very different nature, but too long to be tran-
scribed here, and the nature of the process is not sufficiently intelligible to render an account of it of much
consequence.*

The preceding observations will give the reader some
notion of the nature of the pursuits which occupied the
alchymists: their sole object was the preparation of a
substance to which they gave the name of the philosopher's stone, which possessed the double property of
converting the baser metals into gold, and of curing all
diseases, and of preserving human life to an indefinite
extent. The experiments of Wilson, and the formula
of Musitanus, which have been just inserted, will give
the reader some notion of the way in which they attempted to manufacture this most precious substance.
Being quite ignorant of the properties of bodies, and
of their action on each other, their processes were
guided by no scientific analogies, and one part of the
labour not unfrequently counteracted another; it would
be a waste of time, therefore, to attempt to analyze their
numerous processes, even though such an attempt
could be attended with success. But in most cases,
from the unintelligible terms in which their books are

---

* Whoever wishes to enter more particularly into the processes for making the philosopher's stone contrived by the alchymists, will find a good deal of information on the subject in
Stahl's Fundamenta Chemiæ, vol. i. p. 219, in his chapter *De
lapide philosophorum:* and Junker's Conspectus Chemiæ, vol.
i. p. 604, in his tabula 28, *De transmutatione metallorum universali;* and tabula 29, *De transmutatione metallorum particulari.*

written, it is impossible to divine the nature of the processes by which they endeavoured to manufacture the philosopher's stone, or the nature of the substances which they obtained.*

In consequence of the universality of the opinion that gold could be made by art, there was a set of impostors who went about pretending that they were in possession of the philosopher's stone, and offering to communicate the secret of making it for a suitable reward. Nothing is more astonishing than that persons should be found credulous enough to be the dupes of such impostors. The very circumstance of their claiming a reward was a sufficient proof that they were ignorant of the secret which they pretended to reveal; for what motive could a man have for asking a reward who was in possession of a method of creating gold at pleasure? To such a person money could be no object, as he could procure it in any quantity. Yet, strange as it may appear, they met with abundance of dupes credulous enough to believe their asseverations, and to supply them with money to enable them to perform the wished-for processes. The object of these impostors was either to pocket the money thus furnished, or they made use of it to purchase various substances from which they extracted oils, acids, or similar products, which they were enabled to sell at a profit. To keep the dupes, who thus supplied them with the means of carrying on these processes, in good spirits, it was necessary to show them occasionally small quantities of the baser metals converted into gold; this they performed in various ways. M. Geoffroy, senior, who had an opportunity of witnessing many of their performances,

* Kircher, in his Mundus Subterraneus, has an article on the philosopher's stone, in which he examines the processes of the alchymists, points out their absurdity, and proves by irrefragable arguments that no such substance had ever been obtained. Those who are curious about alchymistical processes may consult that work.

has given us an account of a number of their tricks. It may be worth while to state a few by way of specimen.

Sometimes they made use of crucibles with a false bottom; at the real bottom they put a quantity of oxide of gold or silver, this was covered with a portion of powdered crucible, glued together by a little gummed water or a little wax; the materials being put into this crucible, and heat applied, the false bottom disappears, the oxide of gold or silver is reduced, and at the end of the process is found at the bottom of the crucible, and considered as the product of the operation.

Sometimes they make a hole in a piece of charcoal and fill it with oxide of gold or silver, and stop up the mouth with a little wax; or they soak charcoal in solutions of these metals; or they stir the mixtures in the crucible with hollow rods containing oxide of gold or silver within, and the bottom shut with wax: by these means the gold or silver wanted is introduced during the process, and considered as a product of the operation.

Sometimes they have a solution of silver in nitric acid, or of gold in aqua regia, or an amalgam of gold or silver, which being adroitly introduced, furnishes the requisite quantity of metal. A common exhibition was to dip nails into a liquid, and take them out half converted into gold. The nails consisted of one-half gold, neatly soldered to the iron, and covered with something to conceal the colour, which the liquid removed. Sometimes they had metals one-half gold the other half silver, soldered together, and the gold side whitened with mercury; the gold half was dipped into the transmuting liquid and then the metal heated; the mercury was dissipated, and the gold half of the metal appeared.[*]

As the alchymists were assiduous workmen— as they mixed all the metals, salts, &c. with which they were acquainted, in various ways with each other, and sub-jected such mixtures to the action of heat in close vessels, their labours were occasionally repaid by the discovery of new substances, possessed of much greater activity than any with which they were previously acquainted. In this way they were led to the dis-covery of sulphuric, nitric, and muriatic acids. These, when known, were made to act upon the metals; solu-tions of the metals were obtained, and this gradually led to the knowledge of various metalline salts and preparations, which were introduced with considerable advantage into medicine. Thus the alchymists, by their absurd pursuits, gradually formed a collection of facts, which led ultimately to the establishment of scientific chemistry. On this account it will be proper to notice, in this place, such of them as appeared in Europe during the darker ages, and acquired the highest reputation either on account of their skill as physicians, or their celebrity as chemists. *

1. The first alchymist who deserves notice is Alber-tus Magnus, or Albert Groot, a German, who was born, it is supposed, in the year 1193, at Bollstaedt, and died in the year 1282.† When very young he is said to have been so remarkable for his dulness, that he became the jest of his acquaintances. He studied the sciences at Padua, and afterwards taught at Cologne, and finally in Paris. He travelled through all Germany as Provincial of the order of Dominican Monks, visited Rome, and was made bishop of Ratis-bon: but his passion for science induced him to give up his bishopric, and return to a cloister at Cologne, where he continued till his death.

Albertus was acquainted with all the sciences cul-

* The original author, whom all who have given any account of the alchymists have followed, is Olaus Borrichius, in his Conspectus Scriptorum Chemicorum Celebriorum. He does not inform us from what sources his information was derived.

† Sprengel's History of Medicine, iv. 368.

tivated in his time.   He was at once a theologian, a
physician, and a man of the world : he was an astronomer and an alchymist, and even dipped into magic
and necromancy.   His works are very voluminous.
They were collected by Petr. Jammy, and published
at Leyden in twenty-one folio volumes, in 1651.   His
principal alchymistical tracts are the following :

1. De Rebus Metallicis et Mineralibus.
2. De Alchymia.
3. Secretorum Tractatus.
4. Breve Compendium de Ortu Metallorum.
5. Concordantia Philosophorum de Lapide.
6. Compositum de Compositis.
7 Liber octo Capitum de Philosophorum Lapide.

Most of these tracts have been inserted in the
Theatrum Chemicum.   They are in general plain and
intelligible. In his treatise De Alchymia, for example,
he gives a distinct account of all the chemical substances known in his time, and of the manner of
obtaining them. He mentions also the apparatus then
employed by chemists, and the various processes which
they had occasion to perform.   I may notice the most
remarkable facts and opinions which I have observed
in turning over these treatises.

He was of opinion that all metals are composed of
sulphur and mercury; and endeavoured to account
for the diversity of metals partly by the difference in
the purity, and partly by the difference in the proportions of the sulphur and mercury of which they are
composed.   He thought that water existed also as a
constituent of all metals.

He was acquainted with the water-bath, employed
alembics for distillation, and aludels for sublimation ;
and he was in the habit of employing various lutes,
the composition of which he describes.

He mentions alum and caustic alkali, and seems
to have known the alkaline basis of cream of tartar.
He knew the method of purifying the precious metals

by means of lead and of gold, by cementation; and likewise the method of trying the purity of gold, and of distinguishing pure from impure gold.

He mentions red lead, metallic arsenic, and liver of sulphur. He was acquainted with green vitriol and iron pyrites. He knew that arsenic renders copper white, and that sulphur attacks all the metals except gold.

It is said by some that he was acquainted with gunpowder; but nothing indicating any such knowledge occurs in any of his writings that I have had an opportunity of perusing.*

2. Albertus is said to have had for a pupil, while he taught in Paris, the celebrated Thomas Aquinas, a Dominican, who studied at Bologna, Rome, and Naples, and distinguished himself still more in divinity and scholastic philosophy than in alchymy. He wrote,

1. Thesaurum Alchymiæ Secretissimum.
2. Secreta Alchymiæ Magnalia.
3. De Esse et Essentia Mineralium;

and perhaps some other works, which I have not seen.

These works, so far as I have perused them, are exceedingly obscure, and in various places unintelligible. Some of the terms still employed by modern chemists occur, for the first time, in the writings of Thomas Aquinas. Thus the term *amalgam*, still employed to denote a compound of mercury with another metal, occurs in them, and I have not observed it in any earlier author.

3. Soon after Albertus Magnus, flourished Roger Bacon, by far the most illustrious, the best informed, and the most philosophical of all the alchymists. He was born in 1214, in the county of Somerset. After studying in Oxford, and afterwards in Paris, he became a cordelier friar; and, devoting himself to philosophical

* It is curious that Olaus Borrichius omits Albertus Magnus in the list of alchymistical writers that he has given.

investigations, his discoveries, notwithstanding the pains which he took to conceal them, made such a noise, that he was accused of magic, and his brethren in consequence threw him into prison. He died, it is said, in the year 1284, though Sprengel fixes the year of his death to be 1285.

His writings display a degree of knowledge and extent of thought scarcely credible, if we consider the time when he wrote, the darkest period of the dark ages. In his small treatise De Mirabili Potestate Artis et Naturæ, he begins by pointing out the absurdity of believing in magic, necromancy, charms, or any of those similar opinions which were at that time universally prevalent. He points out the various ways in which mankind are deceived by jugglers, ventriloquists, &c.; mentions the advantages which physicians may derive from acting on the imaginations of their patients by means of charms, amulets, and infallible remedies: he affirms that many of those things which are considered as supernatural, are merely so because mankind in general are unacquainted with natural philosophy. To illustrate this he mentions a great number of natural phenomena, which had been reckoned miraculous; and concludes with several secrets of his own, which he affirms to be still more extraordinary imitations of some of the most singular processes of nature. These he delivers in the enigmatical style of the times; induced, as he tells us, partly by the conduct of other philosophers, partly by the propriety of the thing, and partly by the danger of speaking too plainly.

From an attentive perusal of his works, many of which have been printed, it will be seen that Bacon was a great linguist, being familiar with Latin, Greek, Hebrew, and Arabic; and that he had perused the most important books at that time existing in all these languages. He was also a grammarian; he was well versed in the theory and practice of perspective; he understood the use of convex and concave glasses, and

the art of making them. The camera obscura, burning-glasses, and the powers of the telescope, were
known to him. He was well versed in geography and
astronomy. He knew the great error in the Julian
calendar, assigned the cause, and proposed the remedy.
He understood chronology well; he was a skilful physician, and an able mathematician, logician, meta-
physician, and theologist; but it is as a chemist that
he claims our attention here. The following is a list
of his chemical writings, as given by Gmelin, the
whole of which I have never had an opportunity of
seeing:

1. Speculum Alchymiæ.*
2. Epistola de Secretis Operibus Artis et Naturæ et
de Nullitate Magiæ.
3. De Mirabili Potestate Artis et Naturæ.
4. Medulla Alchymiæ.
5. De Arte Chemiæ.
6. Breviorium Alchymiæ.
7. Documenta Alchymiæ.
8. De Alchymistarum Artibus.
9. De Secretis.
10. De Rebus Metallicis.
11. De Sculpturis Lapidum.
12. De Philosophorum Lapide.
13. Opus Majus, or Alchymia Major.
14. Breviarium de Dono Dei.
15. Verbum abbreviatum de Leone Viridi.
16. Secretum Secretorum.
17. Tractatus Trium Verborum.
18. Speculum Secretorum.

A number of these were collected together, and published at Frankfort in 1603, under the title of " Rogeri
Baconis Angli de Arte Chemiæ Scripta," in a small
duodecimo volume. The Opus Majus was published
in London in 1733, by Dr. Jebb, in a folio volume.

* This tract and the next, which is of considerable length,
will be found in Mangetus's Bibliotheca Chemica Curiosa, i. 613.

Several of his tracts still continue in manuscript in the Harleian and Bodleian libraries at Oxford. He considered the metals as compound of mercury and sulphur. Gmelin affirms that he was aware of the peculiar nature of manganese, and that he was acquainted with bismuth; but after perusing the whole of the Speculum Alchymiæ, the third chapter of which he quotes as containing the facts on which he founds his opinion, I cannot find any certain allusion either to manganese or bismuth. The term *magnesia* indeed occurs, but nothing is said respecting its nature : and long after the time of Paracelsus, bismuth (*bisematum*) was considered as an impure kind of *lead*. That he was acquainted with the composition and properties of *gunpowder* admits of no doubt. In the sixth chapter of his epistle De Secretis Operibus Artis et Naturæ et de Nullitate Magiæ, the following passage occurs :

" For sounds like thunder, and coruscations like lightning, may be made in the air, and they may be rendered even more horrible than those of nature herself. A small quantity of matter, properly manufactured, not larger than the human thumb, may be made to produce a horrible noise and coruscation. And this may be done many ways, by which a city or an army may be destroyed, as was the case when Gideon and his men broke their pitchers and exhibited their lamps, fire issuing out of them with inestimable noise, destroyed an infinite number of the army of the Midianites." And in the eleventh chapter of the same epistle occurs the following passage : " Mix together saltpetre, luru vopo vir con utriet, and sulphur, and you will make thunder and lightning, if you know the method of mixing them." Here all the ingredients of gunpowder are mentioned except charcoal, which is doubtless concealed under the barbarous terms *luru vopo vir con utriet.*

But though Bacon was acquainted with gunpowder, we have no evidence that he was the inventor. How far the celebrated Greek fire, concerning which so much has been written, was connected with gunpowder, it is impossible to say; but there is good evidence to prove that gunpowder was known and used in China before the commencement of the Christian era; and Lord Bacon is of opinion that the thunder and lightning and magic stated by the Macedonians to have been exhibited in Oxydrakes, when it was besieged by Alexander the Great, was nothing else than gunpowder. Now as there is pretty good evidence that the use of gunpowder had been introduced into Spain by the Moors, at least as early as the year 1343, and as Roger Bacon was acquainted with Arabic, it is by no means unlikely that he might have become acquainted with the mode of making the composition, and with its most remarkable properties, by perusing some Arabian writer, with whom we are at present unacquainted. Barbour, in his life of Bruce, informs us that guns were first employed by the English at the battle of Werewater, which was fought in 1327, about forty years after the death of Bacon.

> Two novelties that day they saw,
> That forouth in Scotland had been nene;
> Timbers for helmes was the ane
> That they thought then of great beautie,
> And also wonder for to see.
> The other *crakys* were of war
> That they before heard never air.

In another part of the same book we have the phrase *gynnys for crakys*, showing that the term crakys was used to denote a gun or musket of some form or other. It is curious that the English would seem to have been the first European nation that employed gunpowder in war; they used it in the battle of Crecy, fought in 1346, when it was unknown to the French, and it is supposed to have contributed materially to the brilliant victory which was obtained.

4. Raymond Lully is said to have been a scholar and a friend of Roger Bacon. He was a most voluminous writer, and acquired as high a reputation as any of the alchymists. According to Mutius he was born in Majorca in the year 1235. His father was seneschal to King James the First of Arragon. In his younger days he went into the army; but afterwards held a situation in the court of his sovereign. Devoting himself to science he soon acquired a competent knowledge of Latin and Arabic. After studying in Paris he got the degree of doctor conferred upon him. He entered into the order of Minorites, and induced King James to establish a cloister of that order in Minorca. He afterwards travelled through Italy, Germany, England, Portugal, Cyprus, Armenia and Palestine. He is said by Mutius to have died in the year 1315, and to have been buried in Majorca. The following epitaph is given by Olaus Borrichius as engraven on his tomb:

> Raymundus Lulli, cujus pia dogmata nulli
> Sunt odiosa viro, jacet hic in marmore miro
> Hic M. et CC. Cum P. cœpit sine sensibus esse.

M C C C in these lines denote 1300, and P which is the 15th letter of the alphabet denotes 15, so that if this epitaph be genuine it follows that his death took place in the year 1315.

It seems scarcely necessary to notice the story that Raymond Lully made a present to Edward, King of England, of six millions of pieces of gold, to enable him to make war on the Saracens, which sum that monarch employed, contrary to the intentions of the donor, in his French wars. This story cannot apply to Edward III., because in 1315, at the time of Raymond's death, that monarch was only three years of age. It can scarcely apply to Edward II., who ascended the throne in 1305 : but who had no opportunity of making war, either on the Saracens or French, being totally occupied in opposing the intrigues of his queen and rebellious subjects, to whom he ultimately fell a sacrifice. Edward the First made war both upon the Saracens and the French, and lived during the time of Raymond: but his wars with the Saracens were finished before he ascended the throne, and during the whole of his reign he was too much occupied with his projected conquest of Scotland, to pay much serious attention to any French war whatever. The story, therefore, cannot apply to any of the three Edwards, and cannot be true. Raymond Lully is said to have been stoned to death in Africa for preaching Christianity in the year 1315. Others will have it that he was alive in England in the year 1332, at which time his age would have been 97.

The following table exhibits a list of his numerous writings, most of which are to be found in the Theatrum Chemicum, the Artis Auriferæ, or the Biblotheca Chemica.

1. Praxis Universalis Magni Operis.
2. Clavicula.
3. Theoria et Practica.
4. Compendium Animæ Transmutationis Artis Metallorum.
5. Ultimum Testamentum. Of this work, which professes to give the whole doctrine of alchymy, there is an English translation.
6. Elucidatio Testamenti.
7. Potestas Divitiorum cum Expositione Testamenti Hermetis.
8. Compendium Artis Magicæ, quoad Compositionem Lapidis.
9. De Lapide et Oleo Philosophorum.
10. Modus accipiendi Aurum Potabile.
11. Compendium Alchymiæ et Naturalis Philosophiæ.
12. Lapidarium.
13. Lux Mercuriorum.
14. Experimenta.

15. Ars Compendiosa vel Vademecum.

16. De Accurtatione Lapidis.

Several other tracts besides these are named by Gmelin; but I have never seen any of them. I have attempted several times to read over the works of Raymond Lully, particularly his Last Will and Testament, which is considered the most important of them all. But they are all so obscure, and filled with such unintelligible jargon, that I have found it impossible to understand them. In this respect they form a wonderful contrast with the works of Albertus Magnus and Roger Bacon, which are comparatively plain and intelligible. For an account, therefore, of the chemical substances with which he was acquainted, I am obliged to depend on Gmelin; though I put no great confidence in his accuracy.

Like his predecessors, he was of opinion that all the metals are compounds of sulphur and mercury. But he seems first to have introduced those hieroglyphical figures or symbols, which appear in such profusion in the English translation of his Last Will and Testament, and which he doubtless intended to illustrate his positions. Though what other purpose they could serve, than to induce the reader to consider his statements as allegorical, it is not easy to conjecture. Perhaps they may have been designed to impose upon his contemporaries by an air of something very profound and inexplicable. For that he possessed a good deal of charlatanry is pretty evident, from the slightest glance at his performances.

He was acquainted with cream of tartar, which he distilled: the residue he burnt, and observed that the alkali extracted deliquesced when exposed to the air. He was acquainted with nitric acid, which he obtained by distilling a mixture of saltpetre and green vitriol. He mentions its power of dissolving, not merely mercury, but likewise other metals. He could form aqua regia by adding sal ammoniac or common salt to nitric acid, and he was aware of the property which it had of dissolving gold.

Spirit of wine was well known to him, and distinguished by him by the names of aqua vitæ ardens and argentum vivum vegetabile. He knew the method of rendering it stronger by an admixture of dry carbonate of potash, and of preparing vegetable tinctures by means of it. He mentions alum from Rocca, marcasite, white and red mercurial precipitate. He knew the volatile alkali and its coagulations by means of alcohol. He was acquainted with cupellated silver, and first obtained rosemary oil by distilling the plant with water. He employed a mixture of flour and white of egg spread upon a linen cloth to cement cracked glass vessels, and used other lutes for similar purposes.*

5. Arnoldus de Villa Nova is said to have been born at Villeneuve, a village of Provence, about the year 1240. Olaus Borrichius assures us, that in his time his posterity lived in the neighbourhood of Avignon; that he was acquainted with them, and that they were by no means destitute of chemical knowledge. He is said to have been educated at Barcelona, under John Casamila, a celebrated professor of medicine. This place he was obliged to leave, in consequence of foretelling the death of Peter of Arragon. He went to Paris, and likewise travelled through Italy. He afterwards taught publicly in the University of Montpelier. His reputation as a physician became so great, that his attendance was solicited in dangerous cases by several kings, and even by the pope himself. He was skilled in all the sciences of his time, and was besides a proficient in Greek, Hebrew, and Arabic. When at Paris he studied astrology, and calculating the age of the world, he found that it was to terminate in the year 1335. The theologians of Paris ex-

* Gmelin's Geschitte der Chemie, i. 74

claimed against this and several other of his opinions, and condemned our astrologer as a heretic. This obliged him to leave France; but the pope protected him. He died in the year 1313, on his way to visit Pope Clement V. who lay sick at Avignon. The following table exhibits a pretty full list of his works :

1. Antidotorium.
2. De Vinis.
3. De Aquis Laxativis.
4. Rosarius Philosophorum.
5. Lumen Novum.
6. De Sigillis.
7. Flos Florum.
8. Epistolæ super Alchymia ad Regem Neapolitanum.
9. Liber Perfectionis Magisterii.
10. Succosa Carmina.
11. Questiones de Arte Transmutationis Metallorum.
12. Testamentum.
13. Lumen Luminum.
14. Practica.
15. Speculum Alchymiæ.
16. Carmen.
17. Questiones ad Bonifacium.
18. Semita Semitæ.
19. De Lapide Philosophorum.
20. De Sanguine Humano.
21. De Spiritu Vini, Vino Antimonii et Gemmorum Viribus.

Perhaps the most curious of all these works is the *Rosarium*, which is intended as a complete compend of all the alchymy of his time. The first part of it on the theory of the art is plain enough; but the second part on the practice, which is subdivided into thirty-two chapters, and which professes to teach the art of making the philosopher's stone, is in many places quite unintelligible to me.

He considered, like his predecessors, mercury as a constituent of metals, and he professed a knowledge of the philosopher's stone, which he could increase at pleasure. Gold and gold-water was, in his opinion, one of the most precious of medicines. He employed mercury in medicine. He seems to designate bismuth under the name *marcasite*. He was in the habit of preparing oil of turpentine, oil of rosemary, and spirit of rosemary, which afterwards became famous under the name of Hungary-water. These distillations were made in a glazed earthen vessel with a glass top and helm.

His works were published at Venice in a single folio volume, in the year 1505. There were seven subsequent editions, the last of which appeared at Strasburg in 1613.

6. John Isaac Hollandus and his countryman of the same name, were either two brothers or a father and son; it is uncertain which. For very few circumstances respecting these two laborious and meritorious men have been handed down to posterity. They were born in the village of Stolk in Holland, it is supposed in the 13th century. They certainly were after Arnoldus de Villa Nova, because they refer to him in their writings. They wrote many treatises on chemistry, remarkable, considering the time when they wrote, for clearness and precision, describing their processes with accuracy, and even giving figures of the instruments which they employed. This makes their books intelligible, and they deserve attention because they show that various processes, generally supposed of a more modern date were known to them. Their treatises are written partly in Latin and partly in German. The following list contains the names of most of them:

1. Opera Vegetabilia ad ejus alia Opera Intelligenda Necessaria.

2. Opera Mineralia seu de Lapide Philosophico Libri duo.

3. Tractat vom stein der Weisen.

4. Fragmenta Quædam Chemica.

5. De Triplice Ordine Elixiris et Lapidis Theorea.

6. Tractatus de Salibus et Oleis Metallorum.

7. Fragmentum de Opere Philosophorum.

8. Rariores Chemiæ Operationes.

9. Opus Saturni.

10. De Spiritu Urinæ.

11. Hand der Philosopher.

Olaus Borrichius complains that their *opera mineralia* aboumd with processes ; but that they are ambiguous, and such that nothing certain can be deduced from them even after much labour. Hence they draw on the unwary tyro from labour to labour. I am disposed myself to draw a different conclusion, from what I have read of that elaborate work. It is true that the processes which profess to make the philosopher's stone, are fallacious, and do not lead to the manufacture of gold, as the author intended, and expected : but it is a great deal when alchymistical processes are delivered in such intelligible language that you know the substances employed. This enables us easily to see the results in almost every case, and to know the new compounds which were formed during a vain search for the philosopher's stone. Had the other alchymists written as plainly, the absurdity of their researches would have been sooner discovered, and thus a useless or pernicious investigation would have sooner terminated.

7. Basil Valentine is said to have been born about the year 1394, and is, perhaps, the most celebrated of all the alchymists, if we except Paracelsus. He was a Benedictine monk, at Erford, in Saxony. If we believe Olaus Borrichius, his writings were enclosed in the wall of a church at Erford, and were discovered

long after his death, in consequence of the wall having been driven down by a thunderbolt. But this story is not well authenticated, and is utterly improbable. Much of his time seems to have been taken up in the preparation of chemical medicines. It was he that first introduced antimony into medicine; and it is said, though on no good authority, that he first tried the effects of antimonial medicines upon the monks of his convent, upon whom it acted with such violence that he was induced to distinguish the mineral from which these medicines had been extracted, by the name of *antimoine* (hostile to monks). What shows the improbability of this story is, that the works of Basil Valentine, and in particular his Currus triumphalis Antimonii, were written in the German language. Now the German name for antimony is not *antimoine*, but *speissglass*. The Currus triumphalis Antimonii was translated into Latin by Kerkringius, who published it, with an excellent commentary, at Amsterdam, in 1671.

Basil Valentine writes with almost as much virulence against the physicians of his time, as Paracelsus himself did afterwards. As no particulars of his life have been handed down to posterity, I shall satisfy myself with giving a catalogue of his writings, and then pointing out the most striking chemical substances with which he was acquainted.

The books which have appeared under the name of Basil Valentine, are very numerous; but how many of them were really written by him, and how many are supposititious, is extremely doubtful. The following are the principal:

1. Philosophia Occulta.

2. Tractat von naturlichen und ubernaturlichen Dingen; auch von der ersten tinctur, Wurzel und Geiste der Metallen.

3. Von dern grossen stein der Uhralten.

4. Vier tractatlein vom stein der Weisen.

5. Kurzer anhang und klare repetition oderWieder-
holunge vom grosen stein der Uhralten.

6. De prima Materia Lapidis Philosophici.

7. Azoth Philosophorum seu Aureliæ occultæ de
Materia Lapidis Philosophorum.

8. Apocalypsis Chemica.

9. Claves 12 Philosophiæ.

10. Practica.

11. Opus præclarum ad utrumque, quod pro Testa-
mento dedit Filio suo adoptivo.

12. Letztes Testament.

13. De Microcosmo.

14. Von der grosen Heimlichkeit der Welt und ihrer
Arzney.

15. Von der Wissenschaft der sieben Planeten.

16. Offenbahrung der verborgenen Handgriffe.

17. Conclusiones or Schlussreden.

18. Dialogus Fratris Alberti cum Spiritu.

19. De Sulphure et fermento Philosophorum.

20. Haliographia.

21. Triumph wagen Antimonii.

22. Einiger Weg zur Wahrheit.

23. Licht der Natur.

The only one of these works that I have read with
care, is Kerkringius's translation and commentary on
the Currus triumphalis Antimonii. It is an excellent
book, written with clearness and precision, and contains every thing respecting antimony that was known
before the commencement of the 19th century. How
much of this is owing to Kerkringius I cannot say, as
I have never had an opportunity of seeing a copy of
the original German work of Basil Valentine.

Basil Valentine, like Isaac Hollandus, was of opinion that the metals are compounds of salt, sulphur,
and mercury. The philosopher's stone was composed
of the same ingredients. He affirmed, that there exists

a great similarity between the mode of purifying gold and curing the diseases of men, and that antimony answers best for both. He was acquainted with arsenic, knew many of its properties, and mentions the red compound which it forms with sulphur. Zinc seems to have been known to him, and he mentions bismuth, both under its own name, and under that of *marcasite.* He was aware that manganese was employed to render glass colourless. He mentions nitrate of mercury, alludes to corrosive sublimate, and seems to have known the red oxide of mercury. It would be needless to specify the preparations of antimony with which he was acquainted ; scarcely one was unknown to him which, even at present, exists in the European Pharmacopœias. Many of the preparations of lead were also familiar to him. He was aware that lead gives a sweet taste to vinegar. He knew sugar of lead, litharge, yellow oxide of lead, white carbonate of lead ; and mentions that this last preparation was often adulterated in his time. He knew the method of making green vitriol, and the double chloride of iron and ammonia. He was aware that iron could be precipitated from its solution by potash, and that iron has the property of throwing down copper. He was aware that tin sometimes contains iron, and ascribed the brittleness of Hungarian iron to copper. He knew that oxides of copper gave a green colour to glass; that Hungarian silver contained gold; ·that gold is precipitated from aqua regia by mercury, in the state of an amalgam. He mentions fulminating gold. But the important facts contained in his works are so numerous, while we are so uncertain about the genuineness of the writings themselves, that it will scarcely be worth while to proceed further with the catalogue.

Thus I have brought the history of alchymy to the time of Paracelsus, when it was doomed to undergo a new and important change. It will be better, therefore, not to pursue the history of alchymy further, but to take up the history of true chemistry ; and in the first place to endeavour to determine what chemical facts were known to the Ancients, and how far the science had proceeded to develop itself before the time of Paracelsus.

CHAPTER II.

OF THE CHEMICAL KNOWLEDGE POSSESSED BY THE ANCIENTS.

NOTWITHSTANDING the assertions of Olaus Borrichius, and various other writers who followed him on the same side, nothing is more certain than that the ancients have left no chemical writings behind them, and that no evidence whatever exists to prove that the science of chemistry was known to them. Scientific chemistry, on the contrary, took its origin from the collection and comparison of the chemical facts, made known by the practice and improvement of those branches of manufactures which can only be conducted by chemical processes. Thus the smelting of ores, and the reduction of the metals which they contain, is a chemical process; because it requires, for its success, the separation of certain bodies which exist in the ore chemically combined with the metals; and it cannot be done, except by the application or mixture of a new substance, having an affinity for these substances, and capable, in consequence, of separating them from the metal, and thus reducing the metal to a state of purity. The manufacture of glass, of soap, of leather, are all chemical, because they consist of processes, by means of which bodies, having an affinity for each other, are made to unite in chemical combination. Now I shall in this chapter point out the principal chemical manufactures that were known to the ancients, that we may see how much they contributed towards laying the foundation of the science. The chief sources of our information on this subject are the writings of the Greeks and Romans. Unfortunately the arts and manufactures stood in a very different degree of estimation among the ancients from what they do among the moderns. Their artists and manufacturers were chiefly slaves. The citizens of Greece and Rome devoted themselves to politics or war. Such of them as turned their attention to learning confined themselves to *oratory*, which was the most fashionable and the most important study, or to history, or poetry. The only scientific pursuits which ever engaged their attention, were politics, ethics, and mathematics. For, unless Archimedes is to be considered as an exception, scarcely any of the numerous branches of physics and mechanical philosophy, which constitute so great a portion of modern science, even attracted the attention of the ancients.

In consequence of the contemptible light in which all mechanical employments were viewed by the ancients, we look in vain in any of their writings for accurate details respecting the processes which they followed. The only exception to this general neglect and contempt for all the arts and trades, is Pliny the Elder, whose object, in his natural history, was to collect into one focus, every thing that was known at the period when he lived. His work displays prodigious reading, and a vast fund of erudition. It is to him that we are chiefly indebted for the knowledge of the chemical arts which were practised by the ancients. But the low estimation in which these arts were held, appears evident from the wonderful want of information which Pliny so frequently displays, and the erroneous statements which he has recorded respecting these processes. Still a great deal may be drawn from the information which has been collected and transmitted to us by this indefatigable natural historian.

I.—The ancients were acquainted with SEVEN METALS; namely, gold, silver, mercury, copper, iron, tin, and lead. They knew and employed various preparations of zinc, and antimony, and arsenic; though we have no evidence that these bodies were known to them in the metallic state.

1. Gold is spoken of in the second chapter of Genesis as existing and familiarly known before the flood.

"The name of the first is Pison; that is it which encompasseth the whole land of Havilah, where there is gold. And the gold of that land is good: there is bdellium and the onyx-stone." The Hebrew word for gold, זהב (zeb) signifies to be clear, to shine; alluding, doubtless, to the brilliancy of that metal. The term *gold* occurs frequently in the writings of Moses, and the metal must have been in common use among the Egyptians, when that legislator led the children of Israel out of Egypt.* Gold is found in the earth almost always in a native state. There can be no doubt that it was much more abundant on the surface of the earth, and in the beds of rivers in the early periods of society, than it is at present: indeed this is obvious, from the account which Pliny gives of the numerous places in Asia and Greece, and other European countries, where gold was found in his time.

Gold, therefore, could hardly fail to attract the attention of the very first inhabitants of the globe; its beauty, its malleability, its indestructibility, would give it value: accident would soon discover the possibility of melting it by heat, and thus of reducing the grains or small pieces of it found on the surface of the earth into one large mass. It would be speedily made into ornaments and utensils of various kinds, and thus gradually would come into common use. This we find to have occurred in America, when it was dis-

---

* Exodus xi. 2—xxv. 11, 12, 13, 17, 18, 24, 25, 26—xxviii. 8—xxxii. 2, &c.

covered by Columbus. The inhabitants of the tropical
parts of that vast continent were familiarly acquainted
with gold; and in Mexico and Peru it existed in great
abundance; indeed the natives of these countries
seem to have been acquainted with no other metal, or
at least no other metal was brought into such general
use, except silver, which in Peru was, it is true, still
more common than gold.

Gold, then, was probably the first metal with which
man became acquainted; and that knowledge must
have preceded the commencement of history, since it
is mentioned as a common and familiar substance in
the Book of Genesis, the oldest book in existence, of
the authenticity of which we possess sufficient evidence.
The period of leading the children of Israel out of
Egypt by Moses, is generally fixed to have been one
thousand six hundred and forty-eight years before
the commencement of the Christian era. So early,
then, we are certain, that not only gold, but the
other six malleable metals known to the ancients, were
familiar to the inhabitants of Egypt. The Greeks
ascribe the discovery of gold to the earliest of their
heroes. According to Pliny, it was discovered on
Mount Pangæus by Cadmus, the Phœnician: but
Cadmus's voyage into Greece was nearly coeval with
the exit of the Israelites out of Egypt, at which time
we learn from Moses that gold was in common use
in Egypt. All that can be meant, then, is, that Cadmus first discovered gold in Greece; not that he made
mankind first acquainted with it. Others say that
Thoas and Eaclis, or Sol, the son of Oceanus, first
found gold in Panchaia. Thoas was a contemporary
of the heroes of the Trojan war, or at least was posterior
to the Argonautic expedition, and consequently long
posterior to Moses and the departure of the children
of Israel from Egypt.

2. Silver also was not only familiarly known to the
Egyptians in the time of Moses, but, as we learn from

Genesis, was coined into money before Joseph was set over the land of Egypt by Pharaoh, which happened one thousand eight hundred and seventy-two years before the commencement of the Christian era, and consequently two hundred and twenty-four years before the departure of the children of Israel out of Egypt.

" And Joseph gathered up all the money that was found in the land of Egypt, and in the land of Canaan, for the corn which they bought; and Joseph brought the money into Pharaoh's house.* The Hebrew word כסף (kemep), translated *money*, signifies silver, and was so called from its pale colour. Silver occurs in many other passages of the writings of Moses.† The Greeks inform us, that Erichthonius the Athenian, or Ceacus, were the discoverers of silver; but both of these individuals were long posterior to the time of Joseph.

Silver, like gold, occurs very frequently in the metallic state. This, no doubt, was a still more frequent occurrence in the early ages of the world; it would therefore attract the attention of mankind as early as gold, and for the same reason. It is very ductile, very beautiful, and much more easily fused than gold : it would be therefore more easily reduced into masses, and formed into different utensils and ornaments than even gold itself. The ores of it which occur in the earth are heavy, and would therefore draw the attention of even rude men to them : they have, most of them at least, the appearance of being metallic, and the most common of them may be reduced to the state of metallic silver, simply by keeping them a sufficient time in fusion. Accordingly we find that the Peruvians, before they were overrun by the Spaniards, had made themselves acquainted with the mode of digging out and smelting the ores of silver which occur in

* Genesis xlvii. 14.
† For example, Exodus xi. 2—xxvi. 19, 21 — xxvii. 10, 11, 17, &c.

their country, and that many of their most common utensils were made of that metal.

Silver and gold approached each other nearer in value among the ancients than at present : an ounce of fine gold was worth from ten to twelve ounces of fine silver, the variation depending upon the accidental relation of the supply of both metals. But after the discovery of America, the quantity of silver found in that continent, especially in Mexico, was so great, compared with that of the gold found, that silver became considerably cheaper; so that an ounce of fine gold came to be equivalent to about fourteen ounces and a half of fine silver. Of course these relative values have fluctuated a little according to the abundance of the supply of silver. Though the revolution in the Spanish American colonies has con siderably diminished the supply of silver from the mines, that deficiency seems to have been supplied by other ways, and thus the relative proportion between the value of gold and silver has continued nearly un-altered.

3. That copper must have been known in the earliest ages of society, is sufficiently evident. It occurs frequently native, and could not fail to attract the attention of mankind, from its colour, weight, and malleability. It would not be difficult to fuse it even in the rudest ages : and when melted into masses, as it is malleable and ductile, it would not require much skill to convert it into useful and ornamental utensils. The Hebrew word נחשת (*necheshet*) translated *brass*, obviously means *copper*. We have the authority of the Book of Genesis to satisfy us that copper was known before the flood, and probably as early as either silver or gold.

" And Zillah, she also bore Tubal-cain, an instructor of every artificer in brass *(copper)* and iron."*

* Genesis iv. 22.

The word *copper* occurs in many other passages of the writings of Moses.* That the Hebrew word translated *brass* must have meant copper is obvious, from the following passage : " Out of whose hills thou mayest dig brass."† Brass does not exist in the earth, nor any ore of it, it is always made artificially ; it must therefore have been copper, or an ore of copper, that was alluded to by Moses.

Copper must have been discovered and brought into common use long before iron or steel ; for Homer represents his heroes of the Trojan war as armed with swords, &c. of copper. Copper itself is too soft to be made into cutting instruments ; but the addition of a little tin gives it the requisite hardness. Now we learn from the analyses of Klaproth, that the copper swords of the ancients were actually hardened by the addition of tin.‡

Copper was the metal in common use in the early part of the Roman commonwealth. Romulus coined copper money alone. Numa established a college of workers in copper (*ærariorum fabrúm*).§

The Latin word *æs* sometimes signifies copper, and sometimes brass. It is plain from what Pliny says on the subject, that he did not know the difference between copper and brass ; he says, that an ore of *æs* occurs in Cyprus, called *chalcitis*, where *æs* was first discovered. Here *æs* obviously means copper. In another place he says, that *æs* is obtained from a mineral called *cadmia*. Now from the account of cadmia by Pliny and Dioscorides, there cannot be a doubt that it is the ore to which the moderns have given the name of *calamine*, by means of which brass is made. It is sometimes a silicate and sometimes a carbonate of of zinc ; for both of these ores are confounded together

---

* For example, Exodus xxvii. 2, 3, 4, 6, 10, 11, 17, 18, 19—
xxx. 18, &c. Numbers xxi. 9.
† Deut. viii. 9. ‡ Beitrage, vi. 81. § Plinii Hist. Nat. xxxiv. 1.

under the name of cadmia, and both are employed in the manufacture of brass.

Solinus says, that *æs* was first made at Chalcis, a town in Eubœa. Hence the Greek name, χαλκος (*chalkos*), by which copper was distinguished.

The proper name for brass, by which is meant an alloy of copper and zinc, was *aurichalcum*, or golden, or yellow copper. Pliny says, that long before his time, the ore of aurichalcum was exhausted, so that no more of that beautiful alloy was made. Are we to conclude from this, that there once existed an ore consisting of calamine and ore of copper, mixed or united together? After the exhaustion of the aurichalcum mine, the *salustianum* became the most famous; but it soon gave place to the *livianum*, a copper-mine in Gaul, named after Livia, the wife of Augustus. Both these mines were exhausted in the time of Pliny. The *æs marianum*, or copper of Cordova, was the most celebrated in his time. This last *æs*, he says, absorbs most cadmia, and acquires the greatest resemblance to aurichalcum. We see from this, that in Pliny's time brass was made artificially, and by a process similar to that still followed by the moderns.

The most celebrated alloy of copper among the ancients, was the *æs corinthium*, or Corinthian copper, formed accidentally, as Pliny informs us, during the burning of Corinth by Mummius in the year 608, after the building of Rome, or one hundred and forty-five years before the commencement of the Christian era. There were four kinds of it, of which Pliny gives the following description; not, however, very intelligible:

1. White. It resembled silver much in its lustre, and contained an excess of that metal.

2. Red. In this kind there is an excess of gold.

3. In the third kind, gold, silver, and copper are mixed in equal proportions.

4. The fourth kind is called *hepatizon*, from its having a liver colour. It is this colour which gives it its value.[*]

Copper was put by the ancients to almost all the uses to which it is put by the moderns. One of the great sources of consumption was bronze statues, which were first introduced into Rome after the conquest of Asia Minor. Before that time, the statues of the Romans were made of wood or stoneware. Pliny gives various formulas for making bronze for statues. Of these it may be worth while to put down the most material.

1. To new copper add a third part of old copper. To every hundred pounds of this mixture, twelve pounds and a half of tin[†] are added, and the whole melted together.

2. Another kind of bronze for statues was formed, by melting together

> 100lbs. copper,
> 10lbs. lead,
> 5lbs. tin.

3. Their copper-pots for boiling consisted of 100lbs. of copper, melted with three or four pounds of tin.

The four celebrated statues of horses which, during the reign of Theodosius II. were transported from Chio to Constantinople; and, when Constantinople was taken and plundered by the Crusaders and Venetians in 1204, were sent by Martin Zeno and set up by the doge, Peter Ziani, in the portal of St. Mark; were in 1798, transported by the French to Paris; and finally, after the overthrow of Buonaparte, and the restoration of the Bourbons in 1815, returned to

---

[*] Plinii Hist. Nat. xxxiv. 2.

[†] Pliny's phrase is *plumbum argentorium*. But that the addition was tin, and consequently that plumbum argentorium meant tin, we have the evidence of Klaproth, who analyzed several of these bronze statues, and found them composed of copper, lead, and tin.

Venice and placed upon their ancient pedestals.   The
metal of which these horses had been made was examined by Klaproth, and found by him composed of

Copper, 993
Tin,   7
                ———
            1000*

Klaproth also analyzed an ancient bronze statue in
one of the German cabinets, and found it composed of

Copper, 916
Tin,   75
Lead,   9
                ———
            1000†

Several other old brass and bronze pieces of metal,
very ancient, but found in Germany, were also analyzed by Klaproth.   The result of his analyses was as
follows :

The metal of which the altar of Krodo was made
consisted of

Copper,   69
Zinc,   18
Lead,   13
                ———
            100‡

The emperor's chair, which had in the eleventh century been transported from Harzburg to Goslar, where
it still remains, was found to be composed of

Copper,   92·5
Tin,   5
Lead,   2·5
                ———
            100§

Another piece of metal, which enclosed the high altar
in a church in Germany, was composed of

* Beitrage, vi. 89.
† Beitrage, vi. 118. The statue in question was known by the
name of "The Statue of Püstrichs," at Sondershausen.
‡ Ibid., p. 127.   § Ibid., p. 132.

Copper,   75
Tin,   12·5
Lead,   12·5
            ———
            100*

These analyses, though none of them corresponds exactly with the proportions given by Pliny, confirms sufficiently his general statement, that the bronze of the ancients employed for statues was copper, alloyed with lead and tin.

Some of the bronze statues cast by the ancients were of enormous dimensions, and show decisively the great progress which had been made by them in the art of working and casting metals. The addition of the lead and tin would not only add greatly to the hardness of the alloy, but would at the same time render it more easily fusible. The bronze statue of Apollo, placed in the capitol at the time of Pliny, was forty-five feet high, and cost 500 talents, equivalent to about £50,000 of our money. It was brought from Apollonia, in Pontus, by Lucullus. The famous statue of the sun at Rhodes was the work of Chares, a disciple of Lysippus; it was ninety feet high, was twelve years in making, and cost 300 talents (about £30,000). It was made out of the engines of war left by Demetrius when he raised the siege of Rhodes. After standing fifty-six years, it was overthrown by an earthquake. It lay on the ground 900 years, and was sold by Mauvia, king of the Saracens, to a merchant, who loaded 900 camels with the fragments of it.

Copper was introduced into medicine at rather an early period of society, and various medicinal preparations of it are described by Dioscorides and Pliny. It remains for us to notice the most remarkable of these. Pliny mentions an institution, to which he gives the name of *Seplasia;* the object of which was,

* Ibid., p. 134.

to prepare medicines for the use of medical men. It
seems, therefore, to have been similar to our apothecaries' shops of the present day. Pliny reprobates the
conduct of the persons who had the charge of these
Seplasiæ in his time. They were in the habit of adul-
terating medicines to such a degree, that nothing good
or genuine could be procured from them.*

Both the oxides of copper were known to the ancients, though they were not very accurately distinguished from each other : they were known by the
names *flos æris* and *scoria æris*, or *squama æris*.
They were obtained by heating bars of copper red-hot
and letting them cool, exposed to the air. What fell
off during the cooling was the *flos*, what was driven
off by blows of a hammer was the *squama* or *scoria
æris*. It is obvious, that all these substances were
nearly of the same nature, and that they were in
reality mixtures of the black and red oxides of copper.

Stomoma seems also to have been an oxide of copper, which was gradually formed upon the surface of
the metal, when it was kept in a state of fusion.

These oxides of copper were used as external applications in cases of polypi of the nose, diseases of
the anus, ear, mouth, &c., seemingly as escharotics.

Ærugo, verdigris, was a subacetate of copper,
doubtless often mixed with subacetate of zinc, as not
only copper but brass also was used for preparing it.
The mode of preparing this substance was similar to
the process still followed. Whether verdigris was
employed as a paint by the ancients does not appear ;
for Pliny takes no·notice of any such use of it.

Chalcantum, called also *atramentum sutorium*,
was probably a mixture of sulphate of copper and
sulphate of iron. Pliny's account of the mode of procuring it is too imperfect to enable us to form precise
ideas concerning it ; but it was crystallized on strings,

* Plinii Hist. Nat. xxxiv. 11.

which were extended for the purpose in the solution: its colour was blue, and it was transparent like glass. This description might apply to sulphate of copper; but as the substance was used for blackening leather, and on that account was called *atramentum sutorium*, it is obvious that it must have contained also *sulphate of iron*.

Chalcitis was the name for an ore of copper. The account given of it by Pliny agrees best with copper pyrites, which is now known to be a *sulphur salt*, composed of one atom of sulphide of copper (the acid) united to one atom of sulphide of iron (the base). Pliny informs us, that it is a mixture of *copper*, *misy*, and *sory* : its colour is that of honey. By age, he says, it changes into sory. I think it most probable that native sory, of which Pliny speaks, was sulphuret of copper, and artificial sory sulphate of copper. The native sory is said to constitute black veins in chalcitis. Pliny's description of misy (μισυ) best agrees with copper pyrites. Dioscorides describes it as hard, as having the colour of gold, and as shining like a star.* All this agrees pretty well with copper pyrites.

Scoleca (so called because it assumed the shape of a worm) was formed by triturating alumen, carbonate of soda, and white vinegar, till the matter became green. It was probably a mixture of sulphate of soda, acetate of soda, acetate of alumina, and acetate of copper, probably with more or less oxide of copper, &c., depending upon the proportions of the respective constituents employed.

Such are the preparations of copper, employed by the ancients. They were only used as external applications, partly as escharotics, and partly to induce ulcers to put on a healthy appearance. It does not appear that copper was ever used by the ancients as an internal remedy.

* Lib. v. c. 117.

4. Though *zinc* in the metallic state was unknown to the ancients, yet as they knew some of its ores, and employed preparations of it in medicine, and were in the habit of alloying copper with it, and converting it into brass, it will be proper to state here what was known to them concerning it.

Pliny nowhere makes us acquainted with the process by which copper was converted into brass, nor does he seem to have been acquainted with it; but from several facts incidentally mentioned by him, it is obvious that their process was similar to that which is followed at present by modern brass-makers. The copper in grains is mixed with a certain quantity of calamine (cadmia) and charcoal, and exposed for some time to a moderate heat in a covered crucible. The calamine is reduced to the metallic state, and imbibed by the copper grains. When the copper is thus converted into brass, the temperature is raised sufficiently high to melt the whole : it is then poured out and cast into a slab or ingot.

The cadmia employed by the ancients in medicine was not calamine, but oxide of zinc, which sublimed during the fusion of brass in an open vessel. It was distinguished by a variety of names, according to the state in which it was obtained : the lighter portion was called *capnitis*. *Botryitis* was the name of the portion in the interior of the chimney : the name was derived from some resemblance which it was supposed to have to a bunch of grapes. It had two colours, ash and red. The red variety was reckoned best. This red colour it might derive from some copper mixed with it, but more probably from iron ; for a small quantity of oxide of iron is sufficient to give oxide of zinc a rather beautiful red colour. The portion collected on the sides of the furnace was called *placitis :* it constituted a crust, and was distinguished by different names, according to its colour ; *onychitis* when it was blue externally, but spotted internally : *ostra-*

citis, when it was black and dirty-looking. This last variety was considered as an excellent application to wounds. The best cadmia in Pliny's time was furnished by the furnaces of the Isle of Cyprus: it was used as an external application in ulcers, inflammations, eruptions, &c., so that its use in medicine was pretty much the same as at present. Sulphate and acetate of zinc were unknown to the ancients. No attempt seems to have been made by them to introduce any preparations of zinc as internal medicines.

Pompholyx was the name given to oxide of zinc, sublimed by the combustion of the zinc which exists in brass. *Spodos* seems to have been a mixture of oxides of zinc and copper. There were different varieties of it distinguished by various names.*

5. Iron exists very rarely in the earth in a metallic state, but most commonly in the state of an oxide; and the processes necessary to extract metallic iron from these ores are much more complicated, and require much greater skill, than the reduction of gold, silver, or copper from their respective ores. This would lead us to expect that iron would have been much longer in being discovered than the three metals whose names have been just given. But we learn from the Book of Genesis that iron, like copper and gold, was known before the flood, Tubal-cain being represented as an artificer in copper and iron.† The Hebrew word for iron, ברזל (*berezel*), is said to be derived from בר (*ber*) bright, נזל (*nezel*), to melt; and would lead one to the suspicion, that it referred to *cast* iron rather than *malleable* iron. It is possible that in these early times native iron may have existed as well as native gold, silver, and copper; and in this way Tubal-cain may have become acquainted with the existence and properties of this metal. In the time of Moses, who was learned in all the wisdom of the

---

* See Plinii Hist. Nat. xxxiv. 13. ,   † Genesis iv. 22.

Egyptians, iron must have been in common use in Egypt: for he mentions furnaces for working iron;* ores from which it was extracted;† and tells us that swords‡, knives,‖ axes,§ and tools for cutting stones,¶ were then made of that metal. Now iron in its pure metallic state is too soft to be applied to these uses: it is obvious, therefore, that in Moses's time, not only iron but steel also must have been in common use in Egypt. From this we see how much further advanced the Egyptians were than the Greeks in the knowledge of the manufacture of this most important metal: for during the Trojan war, which was several centuries after the time of Moses, Homer represents his heroes as armed with swords of copper, hardened by tin, and as never using any weapons of iron whatever. Nay, in such estimation was it held, that Achilles, when he celebrated games in honour of Patrocles, proposes a ball of iron as one of his most valuable prizes.**

> " Then hurl'd the hero, thundering on the ground,
> A mass of iron (an enormous round),
> Whose weight and size the circling Greeks admire,
> Rude from the furnace and but shaped by fire.
> This mighty quoit Ætion wont to rear,
> And from his whirling arm dismiss'd in air;
> The giant by Achilles slain, he stow'd
> Among his spoils this memorable load.
> For this he bids those nervous artists vie
> That teach the disk to sound along the sky.
> Let him whose might can hurl this bowl, arise ;
> Who farthest hurls it, takes it as his prize :
> If he be one enrich'd with large domain
> Of downs for flocks and arable for grain,
> Small stock of iron needs that man provide,
> His hinds and swains whole years shall be supplied
> From hence : nor ask the neighbouring city's aid
> For ploughshares, wheels, and all the rural trade."

* Deut. iv. 20.   † Deut. viii. 9.   ‡ Numbers xxxv. 16.
‖ Levit. i. 17.   § Deut. xviii. 5.   ¶ Deut. xxvii. 5.
** Iliad, lib. xxiii. l. 826.

The mass of iron was large enough to supply a shepherd or a ploughman with iron for five years. This circumstance is a sufficient proof of the high estimation in which iron was held during the time of Homer. Were a modern poet to represent his hero as holding out a large lump of iron as a prize, and were he to represent this prize as eagerly contended for by kings and princes, it would appear to us perfectly ridiculous.

Hesiod informs us, that the knowledge of iron was brought over from Phrygia to Greece by the Dactyli, who settled in Crete during the reign of Minos I., about 1431 years before the commencement of the Christian era, and consequently about sixty years before the departure of the children of Israel from Egypt: and it does not appear, that in Homer's time, which was about five hundred years later, the art of smelting iron had been so much improved, as to enable men to apply it to the common purposes of life, as had long before been done by the Egyptians. The general opinion of the ancients was, that the method of smelting iron ore had been brought to perfection by the Chalybes, a small nation situated near the Black Sea,* and that the name *chalybs*, occasionally used for steel, was derived from that people.

Pliny informs us, that the ores of iron are scattered very profusely almost every where: that they exist in Elba; that there was a mountain in Cantabria composed entirely of iron ore; and that the earth in Cappadocia, when watered from a certain river, is converted into iron.† He gives no account of the mode of smelting iron ores; nor does he appear to have been acquainted with the processes; for he says that iron is reduced from its ore precisely in the same way as copper is. Now we know, that the processes for smelting copper and iron are quite different, and

---

* Xenophon's Anabasis, v. 5.   † Plinii Hist. Nat. xxxiv. 14.

founded upon different principles. He says, that in his time many different kinds of iron existed, and they were *stricturæ*, in Latin *a stringenda acie*.

That steel was well known and in common use when Pliny wrote is obvious from many considerations; but he seems to have had no notion of what constituted the difference between iron and steel, or of the method employed to convert iron into steel. In his opinion it depended upon the nature of the water, and consisted in heating iron red-hot, and plunging it, while in that state, into certain waters. The waters at Bilbilis and Turiasso, in Spain, and at Comum, in Italy, possessed this extraordinary virtue. The best steel in Pliny's time came from China; the next best, in point of quality, was manufactured in Parthia.

It would appear, that at Noricum steel was manufactured directly from the ore of iron. This process was perfecly practicable, and it is said still to be practised in certain cases.

The ancients were acquainted with the method of rendering iron, or rather steel, magnetic; as appears from a passage in the fourteenth chapter of the thirty-fourth book of Pliny. Magnetic iron was distinguished by the name of *ferrum vivum*.

When iron is dabbed over with alumen and vinegar it becomes like copper, according to Pliny. Cerussa, gypsum, and liquid pitch, keep it from rusting. Pliny was of opinion that a method of preventing iron from rusting had been once known, but had been lost before his time. The iron chains of an old bridge over the Euphrates had not rusted in Pliny's time; but a few new links, which had been added to supply the place of some that had decayed, were become rusty.

It would appear from Pliny, that the ancients made use of something very like *tractors ;* for he says that pain in the side is relieved by holding near it the point of a dagger that has wounded a man. Water in which red-hot iron had been plunged was recommended as a cure for the dysentery; and the actual cautery with red-hot iron, Pliny informs us, prevents hydrophobia, when a person has been bitten by a mad dog.

Rust of iron and scales of iron were used by the ancients as astringent medicines.

6. Tin, also, must have been in common use in the time of Moses; for it is mentioned without any observation as one of the common metals.* And from the way in which it is spoken of by Isaiah and Ezekiel, it is obvious that it was considered as of far inferior value to silver and gold. Now tin, though the ores of it where it does occur are usually abundant, is rather a scarce metal: that is to say, there are but few spots on the face of the earth where it is known to exist. Cornwall, Spain, in the mountains of Gallicia, and the mountains which separate Saxony and Bohemia, are the only countries in Europe where tin occurs abundantly. The last of these localities has not been known for five centuries. It was from Spain and from Britain that the ancients were supplied with tin; for no mines of tin exist, or have ever been known to exist, in Africa or Asia, except in the East Indies. The Phœnicians were the first nation which caried on a great trade by sea. There is evidence that at a very early period they traded with Spain and with Britain, and that from these countries they drew their supplies of tin. It was doubtless the Phœnicians that supplied the Egyptians with this metal. They had imbibed strongly a spirit of monopoly; and to secure the whole trade of tin they carefully concealed the source from which they drew that metal. Hence, doubtless, the reason why the Grecian geographers, who derived their information from the Phœnicians, represented the Insulæ Cassiterides, or tin

* Numbers xxxi. 22.

islands, as a set of islands lying off the north coast
of Spain. We know that in fact the Scilly islands,
in these early ages, yielded tin, though doubtless the
great supply was drawn from the neighbouring province of Cornwall. It was probably from these islands
that the Greek name for *tin* was derived (κασσιτερος).
Even Pliny informs us, that in his time tin was obtained from the Cassiterides, and from Lusitania
and Gallicia. It occurs, he says, in grains in alluvial
soil, from which it is obtained by washing. It is in
black grains, the metallic nature of which is only re-
cognisable by the great weight. This is a pretty accurate description of *stream tin*, which we know formerly constituted the only ore of that metal wrought
in Cornwall. He says that the ore occurs also along
with grains of gold; that it is separated from the soil
by washing along with the grains of gold, and afterwards smelted separately.

Pliny gives no particulars about the mode of reducing the ore of tin to the metallic state; nor is it at
all likely that he was acquainted with the process.

The Latin term for tin was *plumbum album*. *Stannum* is also used by Pliny; but it is impossible to
understand the account which he gives of it. There
is, he says, an ore consisting of lead, united to silver.
When this ore is smelted, the first metal that flows
out is *stannum*. What flows next is *silver*. What
remains in the furnace is *galena*. This being smelted,
yields *lead*.

Were we to admit the existence of an ore composed
of lead and silver, it is obvious that no such products
could be obtained by simply smelting it.

Cassiteros, or tin, is mentioned by Homer; and,
from the way in which the metal is said by him to
have been used, it is obvious that in his time it bore a
much higher price, and, consequently, was more valued
than at present. In his description of the breastplate
of Agamemnon, he says that it contained ten bands

of steel, twelve of gold, and twenty of tin (κασσιτεροιο).* And in the twenty-third book of the Iliad (line 561), Achilles describes a copper breastplate surrounded with shining tin (φαεινον κασσιτεροιο). Pliny informs us, that in his time tin was adulterated by adding to it about one-third of white copper. A pound of tin, when Pliny lived, cost ten denarii. Now, if we reckon a denarius at $7\frac{3}{4}d.$, with Dr. Arbuthnot, this would make a Roman pound of tin to cost $6s.$ $5\frac{1}{2}d.$ But, as the Roman pound was only equal to three-fourths of our avoirdupois pound, it is plain that in the time of Pliny an avoirdupois pound of tin was worth $8s.$ $7\frac{1}{4}d.$, which is almost seven times the price of tin in the present day.

Tin, in the time of Pliny, was used for covering the inside of copper vessels, as it is at this day. And, no doubt, the process still followed is of the same nature as the process used by the ancients for tinning copper. Pliny remarks, with surprise, that copper thus tinned does not increase in weight. Now Bayen ascertained that a copper pan, nine inches in diameter, and three inches three lines in depth, when tinned, only acquired an additional weight of twenty-one grains. These measures and weights are French. When we convert them into English, we have a copper pan 9·59 inches in diameter, and 3·46 inches deep, which, when tinned, increased in weight 17·23 troy grains. Now the surface of the copper pan, thus tinned, was 176·468 square inches. Hence it follows, that a square inch of copper, when tinned, increases in weight only 0·097 grains. This increase is so small, that we may excuse Pliny, who probably had never seen the increase of weight determined, except by means of a rude Roman statera, for concluding that there was no increase of weight whatever.

Tin was employed by the ancients for mirrors : but

* Iliad xi. 25.

mirrors of silver were gradually substituted; and these in Pliny's time had become so common, that they were even employed by female servants or slaves.

That Pliny's knowledge of the properties of tin was very limited, and far from accurate, is obvious from his assertion that *tin* is less fusible than silver.* It is true that the ancients had no measure to determine the different degrees of heat; but as tin melts at a heat under redness, while silver requires a bright red heat to bring it into fusion, a single comparative trial would have shown him which was most fusible. This trial, it is obvious, had never been made by him.

The ancients seem to have been ignorant of the method of tinning iron. At least, no reference to *tin plate* is made by Pliny, or by any other ancient author, that I have had an opportunity of consulting.

It would appear from Pliny, that both copper and brass were tinned by the Gauls at an early period. Tinned brass was called *æra coctilia*, and was so beautiful that it almost passed for silver. *Plating* (or covering the metal with plates of silver), was gradually substituted for tinning; and finally *gilding* took the place of plating. The trappings of horses, chariots, &c., were thus ornamented. Pliny nowhere gives a description of the process of plating; but there can be little doubt that it was similar to that at present practised. Gilding was accomplished by laying an amalgam of gold on the copper or brass, as at present.

7. Lead appears also to have been in common use among the Egyptians, at the time of Moses.† It was distinguished among the Romans by the name of *plumbum nigrum*. In Pliny's time the lead-mines existed chiefly in Spain and Britain. In Britain lead was so

* Lib. xxxiv. c. 17.   † Numbers xxxi. 22.

abundant, that it was prohibited to extract above a certain quantity in a year. The mines lay on the surface of the earth. Derbyshire was the county in which lead ores were chiefly wrought by the Romans. The rich mines in the north of England seem to have been unknown to them.

Pliny was of opinion that if a lead-mine, after being exhausted, be shut up for some time, the ore will be again renewed.

In the time of Pliny leaden pipes were commonly used for conveying water. The vulgar notion that the ancients did not know that water will always rise in pipes as high as the source from which it proceeds, and that it was this ignorance which led to the formation of aqueducts, is quite unfounded. Nobody can read Pliny without seeing that this important fact was well known in his time.

Sheet lead was also used in the time of Pliny, and applied to the same purposes as at present. But lead was much higher priced among the ancients than it is at present. Pliny informs us that its price was to that of tin as 7 to 10. Hence it must have sold at the rate of 6s. 0¼d. per pound. The present price of lead does not much exceed three halfpence the pound. It is therefore only 1-48th part of the price which it bore in the time of Pliny. This difference must be chiefly owing to the improvements made by the moderns in working the mines and smelting the ores of lead.

Tin, in Pliny's time, was used as a solder for lead. For this purpose it is well adapted, as it is so much easier smelted than lead. But when he says that lead is used also as a solder for tin, his meaning is not so clear. Probably he means an alloy of lead and tin, which, fusing at a lower point than tin, may be used to solder that metal. The addition of some bismuth reduces the fusing point materially; but that metal was unknown to the ancients.

Argentarium is an alloy of equal parts of lead and tin. *Tertiarium*, of two parts lead and one part tin. It was used as a solder.

Some preparations of lead were used by the ancients in medicine, as we know from the description of them given us by Dioscorides and Pliny. These preparations consisted chiefly of protoxide of lead and lead reduced to powder, and partially oxidized by triturating it with water in a mortar. They were applied to ulcers, and employed externally as astringents.

Molybdena was also employed in medicine. Pliny says it was the same as galena. From his description it is obvious that it was *litharge;* for it was in scales, and was more valued the nearer its colour approached to that of gold. It was employed, as it still is, for making plasters. Pliny gives us the process for making the plaster employed by the Roman surgeons. It was made by heating together

3 lbs. molybdena or litharge,
1 lb. wax,
3 heminæ, or 1½ pint, of olive oil.

This process is very nearly the same as the one at present followed by apothecaries for making adhesive plaster.

Psimmythium, or *cerussa*, was the same as our *white lead.* It was made by exposing lead in sheets to the fumes of vinegar. It would seem probable from Pliny's account, though it is confused and inaccurate, that the ancients were in the habit of dissolving cerussa in vinegar, and thus making an impure acetate of lead.

Cerussa was used in medicine. It constituted also a common white paint. At one time, Pliny says, it was found native; but in his time all that was used was prepared artificially.

Cerussa usta seems to have been nearly the same as our *red lead.* It was formed accidentally from cerussa during the burning of the Pyræus. The colour was purple. It was imitated at Rome by burning *silis*

marmarosus, which was probably a variety of some of our ochres.

8. Besides the metals above enumerated, the ancients were also acquainted with quicksilver. Nothing is known about the first discovery of this metal; though it obviously precedes the commencement of history. I am not aware that the term occurs in the writings of Moses. We have therefore no evidence that it was known to the Egyptians at that early period; nor do I find any allusion to it in the works of Herodotus. But this is not surprising, as that author confines himself chiefly to subjects connected with history. Dioscorides and Pliny both mention it as common in their time. Dioscorides gives a method of obtaining it by sublimation from cinnabar. It is remarkable, because it constitutes the first example of a process which ultimately led to distillation.*

Cinnabar is also described by Theophrastus. The term *minium* was applied to it also, till in consequence of the adulteration of cinnabar with *red lead*, the term minium came at last to be restricted to that preparation of lead. Theophrastus describes an artificial cinnabar, which came from the country above Ephesus. It was a shining red-coloured sand, which was collected and reduced to a fine powder by pounding it in vessels of stone. We do not know what it was. The native cinnabar was found in Spain, and was used chiefly as a paint. Dioscorides employs *minium* as the name for what we at present call cinnabar, or bisulphuret of mercury. His cinnabar was a red paint from Africa, produced in such small quantity that painters could scarcely procure enough of it to answer their purposes.

Mercury is described by Pliny as existing native in the mines of Spain, and Dioscorides gives the process for extracting it from cinnabar. It was employed in

* Dioscorides, lib. v. c. 110.

gilding precisely as it is by the moderns. Pliny was aware of its great specific gravity, and of the readiness with which it dissolves gold. The amalgam was squeezed through leather, which separated most of the quicksilver. When the solid amalgam remaining was heated, the mercury was driven off and pure gold remained.

It is obvious from what Dioscorides says, that the properties of mercury were very imperfectly known to him. He says that it may be kept in vessels of glass, or of lead, or of tin, or of silver.* Now it is well known that it dissolves lead, tin, and silver with so much rapidity, that vessels of these metals, were mercury put into them, would be speedily destroyed. Pliny's account of quicksilver is rather obscure. It seems doubtful whether he was aware that native *argentum vivum* and the *hydrargyrum* extracted from cinnabar were the same.

Cinnabar was occasionally used as an external medicine; but Pliny disapproves of it, assuring his readers that quicksilver and all its preparations are virulent poisons. No other mercurial preparations except cinnabar and the amalgam of mercury seem to have been known to the ancients.†

9. The ancients were unacquainted with the metal to which we at present give the name of *antimóny;* but several of the ores of that metal, and of the products of these ores were not altogether unknown to them. From the account of stimmi and stibium, by Dioscorides‡ and Pliny,§ there can be little doubt that these names were applied to the mineral now called *sulphuret of antimony* or crude antimony. It is found most commonly, Pliny says, among the ores of silver,

---

* Lib. v. c. 110.

† The ancients were in the habit of extracting mercury from cinnabar, by a kind of imperfect distillation. The native mercury they called *argentum vivum,* that from cinnabar *hydrargyrus.* See Plinii Hist. Nat. xxxiii. 8.

‡ Lib. v. c. 99.   § Lib. xxxiii. c. 6.

and consists of two kinds, the male and the female; the latter of which is most valued.

This pigment was known at a very early period, and employed by the Asiatic ladies in painting their eyelashes, or rather the insides of their eyelashes, black. Thus it is said of Jezebel, that when Jehu came to Jezreel she painted her face. The original is, *she put her eyes in sulphuret of antimony.*[*] A similar expression occurs in Ezekiel, " For whom thou didst wash thyself, paintedst thy eyes"—literally, put thy eyes in sulphuret of antimony.[†] This custom of painting the eyes black with antimony was transferred from Asia to Greece, and while the Moors occupied Spain it was employed by the Spanish ladies also. It is curious that the term *alcohol*, at present confined to *spirit of wine*, was originally applied to the powder of sulphuret of antimony.[‡] The ancients were in the habit of roasting sulphuret of antimony, and thus converting it into an impure oxide. This preparation was also called stimmi and stibium. It was employed in medicine as an external application, and was conceived to act chiefly as an astringent; Dioscorides describes the method of preparing it. We see, from Pliny's account of stibium, that he did not distinguish between sulphuret of antimony and oxide of antimony.[§]

9. Some of the compounds of arsenic were also known to the ancients; though they were neither acquainted with this substance in the metallic state, nor with its oxide; the poisonous nature of which is so violent that had it been known to them it could not have been omitted by Dioscorides and Pliny.

---

[*] 2 Kings ix. 30.

[†] Chap. 23. v. 40, the Vulgate has it εστιβιζω τους οφθαλμους σου.

[‡] Hartmanni Praxis Chemiatrica, p. 598

[§] Plinii Hist. Nat. xxxiii. 6.

The word σανδαραχη *(sandarache)* occurs in Aristotle, and the term ἀρρενιχον *(arrenichon)* in Theophrastus.* Dioscorides uses likewise the same name with Aristotle. It was applied to a scarlet-coloured mineral, which occurs native, and is now known by the name of *realgar*. It is a compound of arsenic and sulphur. It was employed in medicine both externally and internally, and is recommended by Dioscorides, as an excellent remedy for an inveterate cough.

Auripigmentum and *arsenicum* were names given to the native yellow sulphuret of arsenic. It was used in the same way, and considered by Dioscorides and Pliny as of the same nature with realgar. But there is no reason for supposing that the ancients were acquainted with the compositions of either of these bodies; far less that they had any suspicion of the existence of the metal to which we at present give the name of arsenic.

Such is a sketch of the facts known to the ancients respecting metals. They knew the six malleable metals which are still in common use, and applied them to most of the purposes to which the moderns apply them. Scarcely any information has been left us of the methods employed by them to reduce these metals from their ores. But unless the ores were of a much simpler nature than the modern ores of these metals, of which we have no evidence, the smelting processes with which the ancients were familiar, could scarcely have been contrived without a knowledge of the substances united with the different metals in their ores, and of the means by which these foreign bodies could be separated, and the metals isolated from all impurities. This doubtless implied a certain quantity of chemical knowledge, which having been handed down to the moderns, served as a foundation upon which the modern science of chemistry was

* Περι των λιθων, c. 71.

gradually reared : at the same time it will be admitted that this foundation was very slender, and would of itself have led to little.   Most of the oxides, sulphurets, &c., and almost all the salts into which these metallic bodies enter, were unknown to the ancients.

Besides the working in metals there were some other branches of industry practised by the ancients, so intimately connected with chemical science, that it would be improper to pass them over in silence.   The most important of these are the following :

II.—COLOURS USED BY PAINTERS.

It is well known that the ancient Grecian artists carried the art of painting to the highest degree of perfection, and that their paintings were admired and sought after by the most eminent and accomplished men of antiquity ; and Pliny gives us a catalogue of a great number of first-rate pictures, and a historical account of a vast many celebrated painters of antiquity.   In his own time, he says, the art of painting had lost its importance, statues and tablets having came in place of pictures.

Two kinds of colours were employed by the ancients ; namely, the florid and the austere.   The florid colours, as enumerated by Pliny, were *minium, armenium, cinnaberis, chrysocolla, purpurissum,* and *indicum purpurissum.*

The word *minium* as used by Pliny means *red lead;* though Dioscorides employs it for bisulphuret of mercury or cinnabar.

Armenium was obviously an ochre, probably of a yellow or orange colour.

Cinnaberis was bisulphuret of mercury, which is known to have a scarlet colour.   Dioscorides employs it to denote a vegetable red colour, probably similar to the resin at presen. called *dragon's blood.*

Chrysocolla was a green-coloured paint, and from

Pliny's description of it, could have been nothing else than carbonate of copper or malachite.

Purpurissum was a *lake*, as is obvious from the account of its formation given by Pliny. The colouring matter is not specified, but from the term used there can be little doubt that it was the liquor from the shellfish that yielded the celebrated purple dye of the Tyrians.

Indicum purpurissum was probably *indigo*. This might be implied from the account of it given by Pliny.

The austere colours used by the ancient painters were of two kinds, native and artificial. The native were *sinopis, rubrica, parœtonium, melinum, eretria, auripigmentum*. The artificial were, *ochra, cerussa usta, sandaracha, sandyx, syricum, atramentum*.

Sinopis is the red substance now known by the name of reddle, and used for marking. On that account it is sometimes called *red chalk*. It was found in Pontus, in the Balearian islands, and in Egypt. The price was three denarii, or 1s. 11¼d. the pound weight. The most famous variety of sinopis was from the isle of Lemnos; it was sold sealed and stamped : hence it was called *sphragis*. It was employed to adulterate minium. In medicine it was used to appease inflammation, and as an antidote to poison.

Ochre is merely sinopis heated in a covered vessel. The higher the temperature to which it has been exposed the better it is.

Leucophorum is a compound of

> 6 lbs. sinopis of Pontus,
> 10 lbs. siris,
> 2 lbs. melinum,

triturated together for thirty days. It was used to make gold adhere to wood.

Rubrica from the name, was probably a red ochre.

Parœtonium was a white colour, so called from a

place in Egypt, where it was found. It was obtained also in the island of Crete, and in Cyrene. It was said to be a combination of the froth of the sea consolidated with mud. It consisted probably of carbonate of lime. Six pounds of it cost only one denarius.

Melinum was also a white-coloured powder found in Melos and Samos in veins. It was most probably a carbonate of lime.

Eretria was named from the place where it was found. Pliny gives its medical properties, but does not inform us of its colour. It is impossible to say what it was.

Auripigmentum was yellow sulphuret of arsenic. It was probably but little used as a pigment by the ancient painters.

Cerussa usta was red lead.

Sandaracha was red sulphuret of arsenic. The pound of sandaracha cost 5 as. : it was imitated by red lead. Both it and *ochra* were found in the island Topazos in the Red Sea.

Sandyx was made by torrefying equal parts of true sandaracha and sinopis. It cost half the price of sandaracha. Virgil mistook this pigment for a plant, as is obvious from the following line :

Sponte sua sandix, pascentes vestiet agnos.*

Siricum is made by mixing sinopis and sandyx.

Atramentum was obviously from Pliny's account of it *lamp-black*. He mentions ivory-black as an invention of Apelles : it was called *elephantinum*. There was a native atramentum, which had the colour of sulphur, and got a black colour artificially. It is not unlikely that it contained sulphate of iron, and that it got its black colour from the admixture of some astringent substance.

* Bucol. iv. l. 45.

The ink of the ancients was lamp-black mixed with water, containing gum or glue dissolved in it. *Atramentum indicum* was the same as our *China ink*.

The *purpurissum* was a high-priced pigment. It was made by putting *creta argentaria* (a species of white clay) into the caldrons containing the ingredients for dying purple. The creta imbibed the purple colour and became *purpurissum*. The first portion of *creta* put in constituted the finest and highest-priced pigment. The portions put in afterwards became successively worse, and were, of consequence lower priced. We see, from this description, that it was a lake similar to our modern cochineal lakes.*

That the purpurissum indicum was indigo is obvious from the statement of Pliny, that when thrown upon hot coals it gives out a beautiful purple flame. This constitutes the character of indigo. Its price in Pliny's time was ten denarii, or six shillings and five-pence halfpenny the Roman pound; which is equivalent to $8s.$ $7\frac{1}{3}d.$ the avoirdupo's.

Though few or none of the ancient pictures have been preserved, yet several specimens of the colours used by them still remain in Rome and in the ruins of Herculaneum. Among others the fresco paintings, in the baths of Titus, still remain ; and as these were made for a Roman emperor, we might expect to find the most beautiful and costly colours employed in them. These paints, and some others, were examined by Sir Humphrey Davy, in 1813, while he was in Rome. From his researches we derive some pretty accurate information respecting the colours employed by the painters of Greece and Rome.

1. *Red paints.* Three different kinds of red were found in a chamber opened in 1811, in the baths of Titus, namely, a bright orange red, a dull red, and a brown red. The bright orange red was *minium,* or

* Plinii Hist. Nat. xxxv. 6.

red lead; the other two were merely two varieties of iron ochres. Another still brighter red was observed on the walls; it proved, on examination, to be *vermilion* or *cinnabar*.

2. *Yellow paints.* All the *yellows* examined by Davy proved to be *iron ochres*, sometimes mixed with a little *red lead*. Orpiment was undoubtedly employed, as is obvious from what Pliny says on the subject: but Davy found no traces of it among the yellow colours which he examined. A very deep yellow, approaching orange, which covered a piece of stucco in the ruins near the monument of Caius Cestius, proved to be protoxide of lead, or massicot, mixed with some red lead. The yellows in the Aldobrandini pictures were all ochres, and so were those in the pictures on the walls of the houses at Pompeii.

3. *Blue paints.* Different shades of blues are used in the different apartments of the baths of Titus, which are darker or lighter, as they contain more or less carbonate of lime with which the blue pigment had been mixed by the painter. This blue pigment turned out, on examination, to be a frit composed of alkali and silica, fused together with a certain quantity of oxide of copper. This was the colour called χυανος *(chuanos)* by the Greeks, and *cæruleum* by the Romans. Vitruvius gives the method of preparing it by heating strongly together sand, carbonate of soda, and filings of copper. Davy found that fifteen parts by weight of anhydrous carbonate of soda, twenty parts of powdered opaque flints, and three parts of copper filings, strongly heated together for two hours, gave a substance exactly similar to the blue pigment of the ancients, and which, when powdered, produced a fine deep blue colour. This cæruleum has the advantage of remaining unaltered even when the painting is exposed to the actions of the air and sun.

There is reason to suspect, from what Vitruvius and Pliny say, that glass rendered blue by means of cobalt constituted the basis of some of the blue pigments of the ancients; but all those examined by Davy consisted of glass tinged blue by copper, without any trace of cobalt whatever.

4. *Green paints.* All the green paints examined by Davy proved to be carbonates of copper, more or less mixed with carbonate of lime. I have already mentioned that verdigris was known to the ancients. It was no doubt employed by them as a pigment, though it is not probable that the acetic acid would be able to withstand the action of the atmosphere for a couple of thousand years.

5. *Purple paints.* Davy ascertained that the colouring matter of the ancient purple was combustible. It did not give out the smell of ammonia, at least perceptibly. There is little doubt that it was the *purpurissum* of the ancients, or a clay coloured by means of the purple of the buccinum employed by the Syrians in the celebrated purple dye.

6. *Black and brown paints.* The black paints were lamp-black: the browns were some of them ochres and some of them oxides of manganese.

7. *White paints.* All the ancient white paints examined by Davy were carbonates of lime.[*] We know from Pliny that white lead was employed by the ancients as a pigment; but it might probably become altered in its nature by long-continued exposure to the weather.

III.—GLASS.

It is admitted by some that the word which in our English Bible is translated crystal, means glass, in the following passage of Job: "The gold and the crystal cannot equal it."[†] Now although the exact time when Job was written is not known, it is admitted on all hands to be one of the oldest of the books con

---

[*] Phil. Trans. 1814, p. 97.　　　　[†] Job xxviii. 17

tained in the Old Testament. There are strong reasons for believing that it existed before the time of
Moses ; and some go so far as to affirm that there are
several allusions to it in the writings of Moses. If
therefore glass were known when the Book of Job was
written, it is obvious that the discovery of it preceded
the commencement of history. But even though the
word used in Job should not refer to glass, there can
be no doubt that it was known at a very early period ;
for glass beads are frequently found on the Egyptian
mummies, and they are known to have been embalmed
at a very remote period. The first Greek author who uses
the word glass (ὑαλος, *hyalos*) is Aristophanes. In his
comedy of The Clouds, act ii. scene 1, in the ridiculous dialogue between Socrates and Strepsiades, the
latter announces a method which had occurred to him
to pay his debts. "You know," says he, "the beautiful
transparent stone used for kindling fire." " Do you
mean glass (τον ὑαλον, *ton hyalon*)?" replied Socrates. " I
do," was the answer. He then describes how he would
destroy the writings by means of it, and thus defraud
his creditors. Now this comedy was acted about four
hundred and twenty-three years before the beginning
of the Christian era. The story related by Pliny, respecting the discovery of this beautiful and important
substance, is well known. Some Phœnician merchants,
in a ship loaded with carbonate of soda from Egypt,
stopped, and went ashore on the banks of the river
Belus: having nothing to support their kettles while
they were dressing their food, they employed lumps of
carbonate of soda for that purpose. The fire was
strong enough to fuse some of this soda, and to unite
it with the fine sand of the river Belus: the consequence of this was the formation of glass.* Whether
this story be entitled to credit or not, it is clear that

* Plinii Hist. Nat. xxxvi. 26,

the discovery must have originated in some such accident. Pliny's account of the manufacture of glass, like his account of every other manufacture, is very imperfect : but we see from it that in his time they were in the habit of making coloured glasses; that colourless glasses were most highly prized, and that glass was rendered colourless then as it is at present, by the addition of a certain quantity of oxide of manganese. Colourless glass was very high priced in Pliny's time. He relates, that for two moderate-sized colourless drinking-glasses the Emperor Nero paid 6000 sistertii, which is equivalent to 25l. of our money.

Pliny relates the story of the man who brought a vessel of malleable glass to the Emperor Tiberius, and who, after dimpling it by dashing it against the floor, restored it to its original shape and beauty by means of a hammer; Tiberius, as a reward for this important discovery, ordered the artist to be executed, in order, as he alleged, to prevent gold and silver from becoming useless. But though Pliny relates this story, it is evident that he does not give credit to it; nor does it deserve credit. We can assign no reason why malleable substances may not be transparent; but all of them hitherto known are opaque. Chloride of silver, chloride of lead and iron constitute no exception, for they are not malleable, though by peculiar contrivances they may be extended; and their transparency is very imperfect.

Many specimens of the coloured glasses made by the ancients still remain, particularly the beads employed as ornaments to the Egyptian mummies. Of these ancient glasses several have been examined chemically by Klaproth, Hatchett, and some other individuals, in order to ascertain the substances employed to give colour to the glass. The following are the facts that have been ascertained :

1. *Red glass.* This glass was opaque, and of a

lively copper-red colour. It was probably the kind of red glass to which Pliny gave the name of hæmatinon. Klaproth analyzed it, and obtained from 100 grains of it the following constituents :

Silica	71
Oxide of lead	10
Oxide of copper	7·5
Oxide of iron	1
Alumina	2·5
Lime	1·5

93·5*

No doubt the deficiency was owing to the presence of an alkali. From this analysis we see that the colouring matter of this glass was *red oxide of copper*.

2. *Green glass.* The colour was light verdigris-green, and the glass, like the preceding, was opaque. The constituents from 100 grains were,

Silica	65
Black oxide of copper . .	10
Oxide of lead	7·5
Oxide of iron	3·5
Lime	6·5
Alumina	5·5

98·0†

Thus it appears that both the red and green glass are composed of the same ingredients, though in different proportions. Both owe their colour to copper. The red glass is coloured by the red oxide of that metal; the green by the black oxide, which forms green-coloured compounds, with various acids, particularly with carbonic acid and with silica.

3. *Blue glass.* The variety analyzed by Klaproth had a sapphire-blue colour, and was only translucent

on the edges.   The constituents from 100 grains of it
were,

Silica	81· 5
Oxide of iron	9· 5
Alumina	1· 5
Oxide of copper	0· 5
Lime	0·25

$$93·25*$$

From this analysis it appears that the colouring matter
of this glass was oxide of iron : it was therefore analogous to the lapis lazuli, or ultramarine, in its nature.

Davy, as has been formerly noticed, found another
blue glass, or frit, coloured by means of copper ; and
he showed that the blue paint of the ancients was
often made from this glass, simply by grinding it to
powder.

Klaproth could find no cobalt in the blue glass
which he examined ; but Davy found the transparent
blue glass vessels, which are along with the vases, in
the tombs of Magna Græcia, tinged with cobalt ; and
he found cobalt in all the transparent ancient blue
glasses with which Mr. Millingen supplied him.   The
mere fusion of these glasses with alkali, and subsequent digestion of the product with muriatic acid, was
sufficient to produce a sympathetic ink from them.[†]
The transparent blue beads which occasionally adorn
the Egyptian mummies have also been examined, and
found coloured by cobalt.   The opaque glass beads
are all tinged by means of oxide of copper.   It is
probable from this that all the transparent blue glasses
of the ancients were coloured by cobalt ; yet we find
no allusion to cobalt in any of the ancient authors.
Theophrastus says that copper (χαλκος, *chalcos*) was used
to give glass a fine colour.   Is it not likely that the im-

* Beitrage, p. 144.   † Phil. Trans. 1815, p. 108.

pure oxide of cobalt, in the state in which they used it, was confounded by them with χαλκος (*chalcos*)?

IV.—VASA MURRHINA.

The Romans obtained from the east, and particularly from Egypt, a set of vessels which they distinguished by the name of *vasa murrhina*, and which were held by them in very high estimation. They were never larger than to be capable of containing from about thirty-six to forty cubic inches. One of the largest size cost, in the time of Pliny, about 7000*l.* Nero actually gave for one 3000*l.* They began to be known in Rome about the latter days of the republic. The first six ever seen in Rome were sent by Pompey from the treasures of Mithridates. They were deposited in the temple of Jupiter in the capitol. Augustus, after the battle of Actium, brought one of these vessels from Egypt, and dedicated it also to the gods. In Nero's time they began to be used by private persons; and were so much coveted that Petronius, the favourite of that tyrant, being ordered for execution, and conceiving that his death was owing to a wish of Nero to get possession of a vessel of this kind which he had, broke the vessel in pieces in order to prevent Nero from gaining his object.

There appear to have been two kinds of these vasa murrhina; those that came from Asia, and those that were made in Egypt. The latter were much more common, and much lower priced than the former, as appears from various passages in Martial and Propertius.

Many attempts have been made, and much learning displayed by the moderns to determine the nature of these celebrated vessels; but in general these attempts were made by individuals too little acquainted with chemistry and with natural history in general to qualify them for researches of so difficult a nature. Some will have it that they consisted of a kind of gum;

others that they were made of glass; others, of a particular kind of shell. Cardan and Scaliger assure us that they were *porcelain* vessels; and this opinion was adopted likewise by Whitaker, who supported it with his usual violence and arrogance. Many conceive them to have been made of some precious stone, some that they were of *obsidian;* Count de Veltheim thinks that they were made of the Chinese *agalmatolite*, or *figure stone;* and Dr. Hager conceives that they were made from the Chinese stone *yu.* Bruckmann was of opinion that these vessels were made of sardonyx, and the Abbé Winckelmann joins him in the same conclusion.

Pliny informs us that these vasa murrhina were formed from a species of stone dug out of the earth in Parthia, and especially in Carimania, and also in other places but little known.* They must have been very abundant at Rome in the time of Nero; for Pliny informs us that a man of consular rank, famous for his collection of vasa murrhina, having died, Nero forcibly deprived his children of these vessels, and they were so numerous that they filled the whole inside of a theatre, which Nero hoped to have seen filled with Romans when he came to it to sing in public.

It is clear that the value of these vessels depended on their size. Small vessels bore but a small price, while that of large vessels was very high; this shows us that it must have been difficult to procure a block of the stone out of which they were cut, of a size sufficiently great to make a large vessel.

These vessels were so soft that an impression might be made upon them with the teeth; for Pliny relates the story of a man of consular rank, who drank out of one, and was so enamoured with it that he bit pieces out of the lip of the cup: " Potavit ex eo ante hos annos consularis, ob amorem abraso ejus margine."

* Plinii Hist. Nat. xxxvii. 2.

And what is singular, the value of the cup, so far from being injured by this abrasure, was augmented: " ut tamen injuria illa pretium augeret; neque est hodie murrhini alterius præstantior indicatura."* It is clear from this that the matter of these vessels was neither rock crystal, agate, nor any precious stone whatever, all of which are too hard to admit of an impression from the teeth of a man.

The lustre was vitreous to such a degree that the name *vitrum murrhinum* was given to the artificial fabric, in Egypt.

The splendour was not very great, for Pliny observes, " Splendor his sine viribus nitorque verius quam splendor."

The colours, from their depth and richness, were what gave these vessels their value and excited admiration. The principal colours were purple and white, disposed in undulating bands, and usually separated by a third band, in which the two colours being mixed, assumed the tint of flame: " Sed in pretio varietas colorum, subinde circumagentibus se maculis in purpuram candoremque, et tertium ex utroque ignescentem, velut per transitum coloris, purpura rubescente, aut lacte candescente."

Perfect transparency was considered as a defect, they were merely translucent; this we learn not merely from Pliny, but from the following epigram of Martial:

Nos bibimus vitro, tu murrâ, Pontice : quare ?
Prodat perspicuus ne.duo vina calix.

Some specimens, and they were the most valued, exhibited a play of colour like the rainbow: Pliny says they were very commonly spotted with "sales, verrucæque non eminentes, sed ut in corpore etiam plerumque sessiles." This, no doubt, refers to foreign bodies, such as grains of pyrites, antimony, galena, &c.,

* Plinii Hist. Nat. xxxvii. 2.

which were often scattered through the substances of which the vessels were made.

Such are all the facts respecting the vasa murrhina to be found in the writings of the ancients; they all apply to fluor spar, and to nothing else; but to it they apply so accurately as to leave little doubt that they were in reality vessels of fluor spar, similar to those at present made in Derbyshire.*

The artificial vasa murrhina made at Thebes, in Egypt, were doubtless of glass, coloured to imitate fluor spar as much as possible, and having the semi-transparency which distinguishes that mineral. The imitations being imperfect, these factitious vessels were not much prized nor sought after by the Romans, they were rather distributed among the Arabians and Ethiopians, who were supplied with glass from Egypt.

Rock crystal is compared by Pliny with the stone from which the vasa murrhina were made; the former, in his opinion, had been coagulated by cold, the latter by heat. Though the ancients, as we have seen, were acquainted with the method of colouring glass, yet they prized colourless glass highest on account of its resemblance to rock crystal; cups of it, in Pliny's time, had supplanted those of silver and gold; Nero gave for a crystal cup 150,000 sistertii, or 625*l.*

V.—DYEING AND CALICO-PRINTING.

Very little has been handed down by the ancients respecting the processes of dyeing. It is evident, from Pliny, that they were acquainted with madder, and that preparations of iron were used in the black dyes. The most celebrated dye of all, the *purple*, was dis-

---

* This opinion was first formed by Baron Born, and stated in his Catalogue of Minerals in M. E. Raab's collection, i. 356. But the evidences in favour of it have been brought forward with great clearness and force by M. Roziere. See Jour. de Min. xxxvi. 193.

covered by the Tyrians about fifteen centuries before the Christian era. This colour was given by various kinds of shellfish which inhabit the Mediterranean. Pliny divides them into two genera; the first, comprehending the smaller species, he called *buccinum*, from their resemblance to a hunting-horn; the second, included those called *purpura* : Fabius Columna thinks that these were distinguished also by the name of *murex*.

These shellfish yielded liquor of different shades of colour; they were often mixed in various proportions to produce particular shades of colour. One, or at most two drops of this liquor were obtained from each fish, by extracting and opening a little reservoir placed in the throat. To avoid this trouble, the smaller species were generally bruised whole, in a mortar; this was also frequently done with the large, though the other liquids of the fish must have in some degree injured the colour. The liquor, when extracted, was mixed with a considerable quantity of salt to keep it from putrifying; it was then diluted with five or six times as much water, and kept moderately hot in leaden or tin vessels, for eight or ten days, during which the liquor was often skimmed to separate all the impurities. After this, the wool to be dyed, being first well washed, was immersed and kept therein for five hours; then taken out, cooled, and again immersed, and continued in the liquor till all the colour was exhausted.[*]

To produce particular shades of colour, carbonate of soda, urine, and a marine plant called *fucus*, were occasionally added : one of these colours was a very dark reddish violet—" Nigrantis rosæ colore sublucens."[†] But the most esteemed, and that in which the Tyrians particularly excelled, resembled coagulat-

[*] Plinii Hist. Nat. ix. 38.   [†] Ibid., ix. 36.

ed blood—" laus ei summa in colore sanguinis concreti, nigricans aspectu, idemque suspectu refulgens."*

Pliny says that the Tyrians first dyed their wool in the liquor of the purpura, and afterwards in that of the buccinum; and it is obvious from Moses that this purple was known to the Egyptians in his time.† Wool which had received this double Tyrian dye (*dia bapha*) was so very costly that, in the reign of Augustus, it sold for about 36*l.* the pound. But lest this should not be sufficient to exclude all from the use of it but those invested with the very highest dignities of the state, laws were made inflicting severe penalties, and even death, upon all who should presume to wear it under the dignity of an emperor. The art of dyeing this colour came at length to be practised by a few individuals only, appointed by the emperors, and having been interrupted about the beginning of the twelfth century all knowledge of it died away, and during several ages this celebrated dye was considered and lamented as an irrecoverable loss.‡ How it was afterwards recovered and made known by Mr. Cole, of Bristol, M. Jussieu, M. Reaumur, and M. Duhamel, would lead us too far from our present object, were we to relate it: those who are interested in the subject will find an historical detail in Bancroft's work on Permanent Colours, just referred to.

There is reason to suspect that the Hebrew word translated *fine linen* in the Old Testament, and so celebrated as a production of Egypt, was in reality *cotton*, and not linen. From a curious passage in Pliny, there is reason to believe that the Egyptians in his time, and probably long before, were acquainted with the method of calico-printing, such as is still practised in India

---

* Plinii Hist. Nat. ix. c. 38.   † Exodus xxv. 4.
‡ See Bancroft on Permanent Colours, i. 79.

and the east.   The following is a literal translation of the passage in question:

"There exists in Egypt a wonderful method of dyeing. The white cloth is stained in various places, not with dye stuffs, but with substances which have the property of absorbing (*fixing*) colours, these appl cations are not visible upon the cloth; but when they are dipped into a hot caldron of the dye they are drawn out an instant after dyed.   The remarkable circumstance is, that though there be only one dye in the vat, yet different colours appear upon the cloth; nor can the colour be afterwards removed "*

It is evident enough that these substances applied were different mordants which served to fix the dye upon the cloth; the nature of these mordants cannot be discovered, as nothing specific seems to have been known to Pliny.   The modern mordants are solutions of alumina; of the oxide of tin, oxide of iron, oxide of lead, &c.: and doubtless these, or something equivalent to these, were the substances employed by the ancients.   The purple dye required no mordant, it fixed itself to the cloth in consequence of the chemical affinity which existed between them.   Whether indigo was used by the ancients as a dye does not appear, but there can be no doubt, at least, that its use was known to the Indians at a very remote period.

From these facts, few as they are, there can be little doubt that dyeing, and even calico-printing, had made considerable progress among the ancients; and this could not have taken place without a considerable knowledge of colouring matters, and of the mordants by which these colouring matters were fixed.   These facts, however, were probably but imperfectly understood, and could not be the means of furnishing the ancients with any accurate chemical knowledge.

* Plinii Hist. Nat. xxxv. 11.

VI.—SOAP.

Soap, which constitutes so important and indispensable an article in the domestic economy of the moderns, was quite unknown to the ancient inhabitants of Asia, and even of Greece. No allusion to it occurs in the Old Testament. In Homer, we find Nausicaa, the daughter of the King of the Phæacians, using nothing but water to wash her nuptial garments:

> They seek the cisterns where Phæacian dames
> Wash their fair garments in the limped streams;
> Where gathering into depth from falling rills,
> The lucid wave a spacious bason fills.
> The mules unharness'd range beside the main,
> Or crop the verdant herbage of the plain.
> Then emulous the royal robes they lave,
> And plunge the vestures in the cleansing wave.
> *Odyssey*, vi. l. 99.

We find, in some of the comic poets, that the Greeks were in the habit of adding wood-ashes to water to make it a better detergent. Wood-ashes contain a certain portion of carbonate of potash, which of course would answer as a detergent; though, from its caustic qualities, it would be injurious to the hands of the washerwomen. There is no evidence that carbonate of soda, the *nitrum* of the ancients, was ever used as a detergent; this is the more surprising, because we know from Pliny that it was employed in dyeing, and one cannot see how a solution of it could be employed by the dyers in their processes without discovering that it acted powerfully as a detergent.

The word *soap* (*sapo*) occurs first in Pliny. He informs us that it was an invention of the Gauls, who employed it to render their hair shining; that it was a compound of wood-ashes and tallow, that there were two kinds of it, *hard* and *soft* (*spissus et liquidus*); and that the best kind was made of the ashes of the beech and the fat of goats. Among the Germans it was more employed by the men than the women.* It is curious that no allusion whatever is made by Pliny to the use of soap as a detergent; shall we conclude from this that the most important of all the uses of soap was unknown to the ancients?

It was employed by the ancients as a pomatum; and, during the early part of the government of the emperors, it was imported into Rome from Germany, as a pomatum for the young Roman beaus. Beckmann is of opinion that the Latin word *sapo* is derived from the old German word *sepe*, a word still employed by the common people of Scotland.†

It is well known that the state of soap depends upon the alkali employed in making it. *Soda* constitutes a *hard* soap, and *potash* a *soft* soap. The ancients being ignorant of the difference between the two alkalies, and using wood-ashes in the preparation of it, doubtless formed soft soap. The addition of some common salt, during the boiling of the soap, would convert the soft into hard soap. As Pliny informs us that the ancients were acquainted both with hard and soft soap, it is clear that they must have followed some such process.

VII.—STARCH.

The manufacture of starch was known to the ancients. Pliny informs us that it was made from wheat and from *siligo*, which was probably a variety or subspecies of wheat. The invention of starch is ascribed by Pliny to the inhabitants of the island of Chio, where in his time the best starch was still made. Pliny's description of the method employed by the ancients of

---

* Plinii Hist. Nat. xxviii. 12. The passage of Pliny is as follows : "Prodest et sapo; Gallorum hoc inventum rutilandis capillis ex sevo et cinere. Optimus fagino et caprino, duobus modis, spissus et liquidus : uterque apud Germanos majore in usu viris quam feminis."

† Hist. of Inventions, iii. 239.

making starch is tolerably exact. Next to the Chian starch that of Crete was most celebrated; and next to it was the Egyptian. The qualities of starch were judged of by the weight; the lightest being always reckoned the best.

VIII.—BEER.

That the ancients were acquainted with wine is universally known. This knowledge must have been nearly coeval with the origin of society; for we are informed in Genesis that Noah, after the flood, planted a vineyard, and made wine, and got intoxicated by drinking the liquid which he had manufactured.* Beer also is a very old manufacture. It was in common use among the Egyptians in the time of Herodus, who informs us that they made use of a kind of wine made from barley, because no vines grew in their country.† Tacitus informs us, that in his time it was the drink of the Germans.‡ Pliny informs us that it was made by the Gauls, and by other nations. He gives it the name of *cerevisia* or *cervisia;* the name obviously alluding to the grain from which it was made.

But though the ancients seem acquainted with both wine and beer, there is no evidence of their having ever subjected these liquids to distillation, and of having collected the products. This would have furnished them with ardent spirits or alcohol, of which there is every reason to believe they were entirely ignorant. Indeed, the method employed by Dioscorides to obtain mercury from cinnabar, is a sufficient proof that the true process of distillation was unknown to them. He mixed cinnabar with iron filings, put the

---

* Genesis ix. 20.

† ″Οινῳ δ᾽ ἐκ κριθεων πεποιημενῳ διαχρεονται· ου γαρ σφι εισι ἐν τῇ χωρῃ ἄμπελοι. Euterpe chap. 77.

‡ De Moribus Germanorum, c. 23. " Potui humor ex hordeo aut frumento in quandam similitudinem vini corruptus."

mixture into a pot, to the top of which a cover of stoneware was luted. Heat was applied to the pot, and when the process was at an end, the mercury was found adhering to the inside of the cover. Had they been aware of the method of distilling the quicksilver ore into a receiver, this imperfect mode of collecting only a small portion of the quicksilver, separated from the cinnabar, would never have been practised. Besides, there is not the smallest allusion to ardent spirits, either in the writings of the poets, historians, naturalists, or medical men of ancient Greece; a circumstance not to be accounted for had ardent spirits been known, and applied even to one-tenth of the uses to which they are put by the moderns.

IX.—STONEWARE.

The manufacture of stoneware vessels was known at a very early period of society. Frequent allusions to the potter's wheel occur in the Old Testament, showing that the manufacture must have been familiar to the Jewish nation. The porcelain of the Chinese boasts of a very high antiquity indeed. We cannot doubt that the processes of the ancients were similar to those of the moderns, though I am not aware of any tolerably accurate account of them in any ancient author whatever.

Moulds of plaster of Paris were used by the ancients to take casts precisely as at present.*

The sand of Puzzoli was used by the Romans, as it is by the moderns, to form a mortar capable of hardening under water.

Pliny gives us some idea of the Roman bricks, which are known to have been of an excellent quality. There were three sizes of bricks used by the Romans.

1. Lydian, which were $1\frac{1}{2}$ foot long and 1 foot broad.

* Plinii Hist. Nat. xxxv. 12.

2. Tetradoron, which was a square of 16 inches each side.

3. Pentadoron, which was a square, each side of which was 20 inches long.

Doron signifies the palm of the hand : of course it was equivalent to 4 inches.

X.——PRECIOUS STONES AND MINERALS.

Pliny has given a pretty detailed description of the precious stones of the ancients; but it is not very easy to determine the specific minerals to which he alludes.

1. The description of the diamond is tolerably precise. It was found in Ethiopia, India, Arabia, and Macedonia. But the Macedonian diamond, as well as the adamas cyprius and siderites, were obviously not diamonds, but soft stones.

2. The *emerald* of the ancients *(smaragdus)* must have varied in its nature. It was a green, transparent, hard stone ; and, as colour was the criterion by which the ancients distinguished minerals and divided them into species, it is obvious that very different minerals must have been confounded together, under the name of emerald. Sapphire, beryl, doubtless fluor spar when green, and probably even serpentine, nephrite, and some ores of copper, seem to have occasionally got the same name. There is no reason to believe that the *emerald* of the moderns was known before the discovery of America. At least it has been only found in modern times in America. Some of the emeralds described by Pliny as losing their colour by exposure to the sun, must have been fluor spars. There is a remarkably deep and beautiful green fluor spar, met with some years ago in the county of Durham, in one of the Weredale mines that possesses this property. The emeralds of the ancients were of such a size ($13\frac{1}{2}$ feet, large enough to be cut into a pillar), that we can

consider them in no other light than as a species of rock.

3. Topaz of the ancients had a green colour, which is never the case with the modern topaz. It was found in the island Topazios, in the Red Sea.* It is generally supposed to have been the *chrysolite* of the moderns. But Pliny mentions a statue of it six feet long. Now chrysolite never occurs in such large masses. Bruce mentions a green substance in an emerald island in the Red Sea, not harder than glass. Might not this be the emerald of the ancients?

4. *Calais,* from the locality and colour was probably the Persian turquoise, as it is generally supposed to be.

5. Whether the *prasius* and *chrysoprasius* of Pliny were the modern stones to which these names are given, we have no means of determining. It is generally supposed that they are, and we have no evidence to the contrary.

6. The *chrysolite* of Pliny is supposed to be our *topaz :* but we have no other evidence of this than the opinion of M. Du Tems.

7. *Asteria* of Pliny is supposed by Saussure to be our sapphire. The lustre described by Pliny agrees with this opinion. The stone is said to have been very hard and colourless.

8. *Opalus* seems to have been our *opal.* It is called, Pliny says, *pæderos* by many, on account of its beauty. The Indians called it *sangenon.*

9. *Obsidian* was the same as the mineral to which we give that name. It was so called because a Roman named Obsidianus first brought it from Egypt. I have a piece of obsidian, which the late Mr. Salt brought from the locality specified by Pliny, and which possesses all the characters of that mineral in its purest state.

---

* The word topazo is said by Pliny to signify, in the language of the Troglodytes, *to seek.*

10. *Sarda* was the name of *carnelian*, so called because it was first found near Sardis. The *sardonyx* was also another name for *carnelian*.

11. Onyx was a name sometimes given to a rock, *gypsum;* sometimes it was a light-coloured *chalcedony*. The Latin name for chalcedony was *carchedonius*, so called because Carthage was the place where this mineral was exposed to sale. The Greek name for Carthage was Καρχηδων (*carchedon*).

12. *Carbunculus* was the garnet; and *anthrax* was a name for another variety of the same mineral.

13. The *oriental amethyst* of Pliny was probably a sapphire. The fourth species of amethyst described by Pliny, seems to have been our amethyst. Pliny derives the name from α and (*a*) μυθη (*mythe*), *wine*, because it has not quite the colour of wine. But the common derivation is from α and μυθνω, *to intoxicate*, because it was used as an amulet to prevent intoxication.

14. The *sapphire* is described by Pliny as always opaque, and as unfit for engraving on. We do not know what it was.

15. The *hyacinth* of Pliny is equally unknown. From its name it was obviously of a blue colour. Our hyacinth has a reddish-brown colour, and a great deal of hardness and lustre.

16. The *cyanus* of Pliny may have been our *cyanite*.

17. *Astrios* agrees very well, as far as the description of Pliny goes, with the variety of felspar called *adularia*.

18. *Belioculus* seems to have been our *catseye*.

19. *Lychnites* was a violet-coloured stone, which became electric by heat. Unless it was a *blue tourmalin*, I do not know what it could be.

20. The *jasper* of the ancients was probably the same as ours.

21. *Molochites* may have been our *malachite*. The name comes from the Greek word μολοχη, *mallow*, or *marshmallow*.

22. Pliny considers *amber* as the juice of a tree concreted into a solid form. The largest piece of it that he had ever seen weighed 13 lbs. Roman weight, which is nearly equivalent to 9¾ lbs. avoirdupois. *Indian amber*, of which he speaks, was probably *copal*, or some transparent resin. It may be dyed, he says, by means of *anchusa* and the *fat of kids*.

23. *Lapis specularis* was foliated sulphate of lime, or selenite.

24. *Pyrites* had the same meaning among the ancients that it has among the moderns ; at least as far as iron pyrites or bisulphuret of iron is concerned. Pliny describes two kind of pyrites ; namely, the *white (arsenical pyrites)*, and the *yellow* (iron pyrites). It was used for striking fire with steel, in order to kindle tinder. Hence the name *pyrites* or *firestone*.

25. *Gagates*, from the account given of it by Pliny, was obviously pit-coal or jet.

26. *Marble* had the same meaning among the ancients that it has among the moderns. It was sawed by the ancients into slabs, and the action of the saw was facilitated by a sand brought for the purpose from Ethiopia and the isle of Naxos. It is obvious that this sand was powdered corundum, or emery.

27. *Creta* was a name applied by the ancients not only to chalk, but to *white clay*.

28. *Melinum* was an *oxide of iron*. Pliny gives a list of one hundred and fifty-one species of stones in the order of the alphabet. Very few of the minerals contained in this list can be made out. He gives also a list of fifty-two species of stones, whose names are derived from a fancied resemblance which the stones are supposed to bear to certain parts of animals. Of these, also, very few can be made out.

XI.—MISCELLANEOUS OBSERVATIONS.

The ancients seem to have been ignorant of the nature and properties of air, and of all gaseous bodies.

Pliny's account of air consists of a single sentence:
" Aër densatur nubibus; furit procellis."   " Air is
condensed in clouds, it rages in storms."   Nor is his
description of water much more complete, since it consists only of the following phrases : " Aquæ subeunt
in imbres, rigescunt in grandines, tumescunt in fluc-
tus, præcipitantur in torrentes."*   "Water falls in
showers, congeals in hail, swells in waves, and rushes
down in torrents." In the thirty-eighth chapter of the
second book, indeed, he professes to treat of *air ;* but
the chapter contains merely an enumeration of meteorological phenomena, without once touching upon
the nature and properties of air.

Pliny, with all the philosophers of antiquity, admitted the existence of the four elements, fire, air, water,
and earth ; but though he enumerates these in the fifth
chapter of his first book, he never attempts to explain
their nature or properties. Earth, among the ancients,
had two meanings, namely, the planet on which we
live, and the soil upon which vegetables grow. These
two meanings still exist in common language. The
meaning afterwards given to the *term,* earth, by the
chemists, did not exist in the days of Pliny, or, at
least, was unknown to him ; a sufficient proof that
chemistry, in his time, had made no progress as a
science ; for some notions respecting the properties and
constituents of those supposed four elements must have
constituted the very foundation of scientific chemistry.

The ancients were acquainted with none of the acids
which at present constitute so numerous a tribe, except *vinegar,* or *acetic acid ;* and even this acid was
not known to them in a state of purity. They knew
none of the saline bases, except lime, soda, and potash,
and these very imperfectly. Of course the whole
tribe of salts was unknown to them, except a very few,
which they found ready formed in the earth, or which

* Plinii Hist. Nat. ii. 63.

they succeeded in forming by the action of vinegar on lead and copper.   Hence all that extensive and most important branch of chemistry, consisting of the combinations of the acids and bases, on which scientific chemistry mainly depends, must have been unknown to them.

Sulphur occurring native in large quantities, and being remarkable for its easy combustibility, and its disagreeable smell when burning, was known in the very earliest ages.   Pliny describes four kinds of sulphur, differing from each other, probably, merely in their purity.   These were

1. Sulphur vivum, or apyron.   It was dug out of the earth solid, and was doubtless pure, or nearly so. It alone was used in medicine.

2. Gleba—used only by fullers.

3. Egula—used also by fullers.

Pliny says, it renders woollen stuffs white and soft. It is obvious from this, that the ancients knew the method of bleaching flannel by the fumes of sulphur, as practised by the moderns.

4. The fourth kind was used only for sulphuring matches.

Sulphur, in Pliny's time, was found native in the Æolian islands, and in Campania.   It is curious that he never mentions Sicily, whence the great supply is drawn for modern manufacture.

In medicine, it seems to have been only used externally by the ancients.   It was considered as excellent for removing eruptions.   It was used also for fumigating.

The word *alumen*, which we translate *alum*, occurs often in Pliny; and is the same substance which the Greeks distinguished by the name of στυπτηρια (*stypteria*). It is described pretty minutely by Dioscorides, and also by Pliny.   It was obviously a natural production, dug out of the earth, and consequently quite different from our alum, with which the ancients were unacquainted.

Dioscorides says that it was found abundantly in
Egypt ; that it was of various kinds, but that the slaty
variety was the best.   He mentions also many other
localities.   He says that, for medical purposes, the
most valued of all the varieties of alumen were the
slaty, the *round*, and the *liquid*.   The slaty alumen
is very white, has an exceedingly astringent taste, a
strong smell, is free from stony concretions, and
gradually cracks and emits long capillary crystals from
these rifts ; on which account it is sometimes called
trichites.   This description obviously applies to a kind
of slate-clay, which probably contained pyrites mixed
with it of the decomposing kind.   The capillary crystals were probably similar to those crystals at present
called *hair-salt* by mineralogists, which exude pretty
abundantly from the shale of the coal-beds, when it
has been long exposed to the air.   *Hair-salt* differs
very much in its nature.   Klaproth ascertained by
analysis, that the *hair-salt* from the quicksilver-mines
in Idria is sulphate of magnesia, mixed with a small
quantity of sulphate of iron.[*]   The *hair-salt* from the
abandoned coal-pits in the neighbourhood of Glasgow
is a double salt, composed of sulphate of alumina, and
sulphate of iron, in definite proportions; the composition being

> 1   atom protosulphate of iron,
> $1\frac{1}{2}$ atom sulphate of alumina,
> 15   atoms water.

I suspect strongly that the capillary crystals from
the schistose alumen of Dioscorides were nearly of the
same nature.

From Pliny's account of the uses to which alumen
was applied, it is quite obvious that it must have
varied very much in its nature.   *Alumen nigrum* was
used to strike a black colour, and must therefore have
contained iron.   It was doubtless an impure native

[*] Beitrage, iii. 104.

sulphate of iron, similar to many native productions of the same nature still met with in various parts of the world, but not employed; their use having been superseded by various artificial salts, more definite in their nature, and consequently more certain in their application, and at the same time cheaper and more abundant than the native.

The alumen employed as a mordant by the dyers, must have been a sulphate of alumina more or less pure; at least it must have been free from all sulphate of iron, which would have affected the colour of the cloth, and prevented the dyer from accomplishing his object.*

What the *alumen rotundum* was, is not easily conjectured. Dioscorides says, that it was sometimes made artificially; but that the artificial alumen rotundum was not much valued. The best, he says, was full of air-bubbles, nearly white, and of a very astringent taste. It had a slaty appearance, and was found in Egypt or the Island of Melos.

The *liquid alumen* was limpid, milky, of an equal colour, free from hard concretions, and having a fiery shade of colour.† In its nature, it was similar to the alumen candidum; it must therefore have consisted chiefly, at least, of sulphate of alumina.

Bitumen and naphtha were known to the ancients, and used by them to give light instead of oil; they were employed also as external applications in cases of disease, and were considered as having the same virtues as sulphur. It is said, that the word translated *salt* in the New Testament—" Ye are the salt of the earth: but if the salt have lost his savour, wherewith shall it be salted? It is henceforth good for nothing, but to be cast out, and to be trodden under foot

---

* " Quoniam inficiendis claro colore lanis candidum liquidumque utilissimum est, contraque fuscis et obscuris nigrum."—*Plinii*, xxxv. 15.

† See Dioscorides, lib. v. c. 123. Plinii Hist. Nat. xxxv. 18.

of men"*—it is said, that the word salt in this passage refers to asphalt, or bitumen, which was used by the Jews in their sacrifices, and called *salt* by them. But I have not been able to find satisfactory evidence of the truth of this opinion. It is obvious from the context, that the word translated *salt* could not have had that meaning among the Jews; because salt never can be supposed to lose its savour. Bitumen, while liquid, has a strong taste and smell, which it loses gradually by exposure to the air, as it approaches more and more to a solid form.

Asphalt was one of the great constituents of the Greek fire. A great bed of it still existing in Albania, supplied the Greeks with this substance. Concerning the nature of the Greek fire, it is clear that many exaggerated and even fabulous statements have been published. The obvious intention of the Greeks being, probably, to make their invention as much dreaded as possible by their enemies. Nitre was undoubtedly one of the most important of its constituents; though no allusion whatever is ever made. We do not know when *nitrate of potash*, the nitre of the moderns, became known in Europe. It was discovered in the east; and was undoubtedly known in China and India before the commencement of the Christian era. The property of nitre, as a supporter of combustion, could not have remained long unknown after the discovery of the salt. The first person who threw a piece of it upon a red-hot coal would observe it. Accordingly we find that its use in fireworks was known very early in China and India; though its prodigious expansive power, by which it propels bullets with so great and destructive velocity, is a European invention, posterior to the time of Roger Bacon.

* Matthew v. 13.—" Ὑμεις εστε το ἅλας της γης· εαν δε το ἅλας μωρανθῃ, ἐν τινι ἁλισθησεται; ἐις ουδεν ισχωει ἐτι ἲι μη βληθηναι εξω, καὶ καταπατεισθαι ὑπο των ἀνθρωπων."

The word *nitre* (נתר) had been applied by the ancients to *carbonate of soda*, a production of Egypt, where it is still formed from sea-water, by some unknown process of nature in the marshes near Alexandria. This is evident, not merely from the account given of it by Dioscorides and Pliny; for the following passage, from the Old Testament, shows that it had the same meaning among the Jews: " As he that taketh away a garment in cold weather, is as vinegar upon nitre: so is he that singeth songs to a heavy heart."* Vinegar poured upon saltpetre produces no sensible effect whatever, but when poured upon carbonate of soda, it occasions an *effervescence*. When saltpetre came to be imported to Europe, it was natural to give it the same name as that applied to carbonate of soda, to which both in taste and appearance it bore some faint resemblance. Saltpetre possessing much more striking properties than carbonate of soda much more attention was drawn to it, and it gradually fixed upon itself the term *nitre*, at first applied to a different salt. When this change of nomenclature took place does not appear; but it was completed before the time of Roger Bacon, who always applies the term *nitrum* to our nitrate of potash and never to carbonate of soda.

In the preceding history of the chemical facts known to the ancients, I have taken no notice of a well-known story related of Cleopatra. This magnificent and profligate queen boasted to Antony that she would herself consume a million of sistertii at a supper. Antony smiled at the proposal, and doubted the possibility of her performing it. Next evening a magnificent entertainment was provided, at which Antony, as usual, was present, and expressed his opinion that the cost of the feast, magnificent as it was, fell far short of the sum specified by the queen. She

---

* Proverbs xxv. 20.

requested him to defer computing till the dessert was finished. A vessel filled with vinegar was placed before her, in which she threw two pearls, the finest in the world, and which were valued at ten millions of sistertii ; these pearls were dissolved by the vinegar,* and the liquid was immediately drunk by the queen. Thus she made good her boast, and destroyed the two finest pearls in the world.† This story, supposing it true, shows that Cleopatra was aware that vinegar has the property of dissolving pearls. But not that she knew the nature of these beautiful productions of nature. We now know that pearls consist essentially of carbonate of lime, and that the beauty is owing to the thin concentric laminæ, of which they are composed.

Nor have I taken any notice of lime with which the ancients were well acquainted, and which they applied to most of the uses to which the moderns put it. Thus it constituted the base of the Roman mortar, which is known to have been excellent. They employed it also as a manure for the fields, as the moderns do. It was known to have a corrosive nature when taken internally ; but was much employed by the ancients externally, and in various ways as an application to ulcers. Whether they knew its solubility in water does not appear ; though, from the circumstance of its being used for making mortar, this fact could hardly escape them. These facts, though of great importance, could scarcely be applied to the rearing of a chemical structure, as the ancients could have no notion of the action of acids upon lime, or of the numerous salts which it is capable of forming. Phenomena which must have remained unknown till the discovery of the acids enabled experimenters to try their effects upon limestone and quicklime. Not even a conjecture appears in any ancient writer that I have looked into,

* " Cujus asperitas visque in tabem margeritas resolvit."
† Plinii Hist. Nat. ix. 35.

about the difference between quicklime and lime-stone. This difference is so great that it must have been remarked by them, yet nobody seems ever to have thought of attempting to account for it. Even the method of burning or calcining lime is not de-scribed by Pliny; though there can be no doubt that the ancients were acquainted with it.

Nor have I taken any notice of leather or the me-thod of tanning it. There are so many allusions to leather and its uses by the ancient poets and histo-rians, that the acquaintance of the ancients with it is put out of doubt. But so far as I know, there is no description of the process of tanning in any ancient author whatever.

CHAPTER III.

CHEMISTRY OF THE ARABIANS.

HITHERTO I have spoken of Alchymy, or of the chemical manufactures of the ancients. The people to whom scientific chemistry owes its origin are the Arabians. Not that they prosecuted scientific chemistry themselves; but they were the first persons who attempted to form chemical medicines. This they did by mixing various bodies with each other, and applying heat to the mixture in various ways. This led to the discovery of some of the mineral acids. These they applied to the metals, &c., and ascertained the effects produced upon that most important class of bodies. Thus the Arabians began those researches which led gradually to the formation of scientific chemistry. We must therefore endeavour to ascertain the chemical facts for which we are indebted to the Arabians.

When Mahomet first delivered his dogmas to his countrymen they were not altogether barbarous. Possessed of a copious and expressive language, and inhabiting a burning climate, their imaginations were lively and their passions violent. Poetry and fiction were cultivated by them with ardour, and with considerable success. But science and inductive philosophy, had made little or no progress among them. The fatalism introduced by Mahomet, and the blind enthusiasm which he inculcated, rendered them furious bigots and determined enemies to every kind of intellectual improvement. The rapidity with which they overran Asia, Africa, and even a portion of Europe, is universally known. At that period the western world, was sunk into extreme barbarism, and the Greeks, with whom the remains of civilization still lingered, were sadly degenerated from those sages who graced the classic ages. Bent to the earth under the most grinding but turbulent despotism that ever disgraced mankind, and having their understandings sealed up by the most subtle and absurd, and un-comprising superstition, all the energy of mind, all the powers of invention, all the industry and talent, which distinguished their ancestors, had completely forsaken them. Their writers aimed at nothing new or great, and were satisfied with repeating the scientific facts determined by their ancestors. The lamp of science fluttered in its socket, and was on the eve of being extinguished.

Nothing good or great could be expected from such a state of society. It was, therefore, wisely deter-mined by Providence that the Mussulman conquerors, should overrun the earth, sweep out those miserable governors, and free the wretched inhabitants from the trammels of despotism and superstition. As a des-potism not less severe, and a superstition still more gloomy and uncompromising, was substituted in their place, it may seem at first sight, that the conquests of the Mahometans brought things into a worse state than they found them. But the listless inactivity, the almost deathlike torpor which had frozen the minds of mankind, were effectually roused. The Mussulmans displayed a degree of energy and activity which have few parallels in the history of the world : and after the conquests of the Mahometans were completed, and the Califs quietly seated upon the greatest and most powerful throne that the world had ever seen ; after Almanzor, about the middle of the eighth century, had

founded the city of Bagdad, and settled a permanent
and flourishing peace, the arts and sciences, which
usually accompany such a state of society, began to
make their appearance.

That calif founded an academy at Bagdad, which
acquired much celebrity, and gradually raised itself
above all the other academies in his dominions. A
medical college was established there with powers to
examine all those persons who intended to devote
themselves to the medical profession. So many professors and pupils flocked to this celebrated college,
from all parts of the world, that at one time their number amounted to no fewer than six thousand. Public
hospitals and laboratories were established to facilitate
a knowledge of diseases, and to make the students
acquainted with the method of preparing medicines.
It was this last establishment which originated with the
califs that gave a first beginning to the science of
chemistry.

In the thirteenth century the calif Mostanser re-
established the academy and the medical college at
Bagdad : for both had fallen into decay, and had
been replaced by an infinite number of Jewish semi-
naries. Mostanser gave large salaries to the professors, collected a magnificent library, and established a
new school of pharmacy. He was himself often present at the public lectures.

The successor of Mostanser was the calif Haroun-
Al-Raschid, the perpetual hero of the Arabian tales.
He not only carried his love for the sciences further
than his predecessors, but displayed a liberality and a
tolerance for religious opinions, which was not quite
consistent with Mahometan bigotry and superstition.
He drew round him the Syrian Christians, who translated the Greek classics, rewarded them liberally, and
appointed them instructors of his Mahometan subjects, especially in medicine and pharmacy. He protected the Christian school of Dschondisabour, founded

by the Nestorian Christians, before the time of Mahomet, and still continuing in a flourishing state : always surrounded by literary men, he frequently condescended to take a part in their discussions, and not unfrequently, as might have been expected from his rank, came off victorious.

The most enlightened of all the califs was Almamon, who has rendered his name immortal by his exertions in favour of the sciences. It was during his reign that the Arabian schools came to be thoroughly acquainted with Greek science; he procured the translation of a great number of important works. This conduct inflamed the religious zeal of the faithful, who devoted him to destruction, and to the divine wrath, for favouring philosophy, and in that way diminishing the authority of the Koran. Almamon purchased the ancient classics, from all quarters, and recommended the care of doing so in a particular manner to his ambassadors at the court of the Greek emperors. To Leo, the philosopher, he made the most advantageous offers, to induce him to come to Bagdad; but that philosopher would not listen to his invitation. It was under the auspices of this enlightened prince, that the celebrated attempt was made to determine the size of the earth by measuring a degree of the meridian. The result of this attempt it does not belong to this work to relate.

Almotassem and Motawakkel, who succeeded Almamon, followed his example, favoured the sciences, and extended their protection to men of science who were Christians. Motawakkel re-established the celebrated academy and library of Alexandria. But he acted with more severity than his predecessors with regard to the Christians, who may perhaps have abused the tolerance which they enjoyed.

The other vicars of the prophet, in the different Mahometan states, followed the fine example set them by Almamon. Already in the eighth century the sovereigns of Mogreb and the western provinces of Africa showed themselves the zealous friends of the sciences. One of them called Abdallah-Ebn-Ibadschab rendered commerce and industry flourishing at Tunis. He himself cultivated poetry and drew numerous artists and men of science into his state. At Fez and in Morocco the sciences flourished, especially during the reign of the Edrisites, the last of whom, Jahiah, a prince possessed of genius, sweetness, and goodness, changed his court into an academy, and paid attention to those only who had distinguished themselves by their scientific knowledge.

But Spain was the most fortunate of all the Mahometan states, and had arrived at such a degree of prosperity both in commerce, manufactures, population, and wealth, as is hardly to be credited. The three Abdalrahmans and Alhakem carried, from the eighth to the tenth century, the country subject to the Calif of Cordova to the highest degree of splendour. They protected the sciences, and governed with so much mildness, that Spain was probably never so happy under the dominion of any Christian prince. Alhakem established at Cordova an academy, which for several ages was the most celebrated in the whole world. All the Christians of Western Europe repaired to this academy in search of information. It contained, in the tenth century, a library of 280,000 volumes. The catalogue of this library filled no less than forty-four volumes. Seville, Toledo, and Murcia, had likewise their schools of science and their libraries, which retained their celebrity as long as the dominion of the Moors lasted. In the twelfth century there were seventy public libraries in that part of Spain which belonged to the Mahometans. Cordova had produced one hundred and fifty authors, Almeria fifty-two, and Murcia sixty-two.

The Mahometan states of the east continued also to favour the sciences. An emir of Irak, Adad-El-

Daula by name, distinguished himself towards the end of the tenth century by the protection which he afforded to men of science. To him almost all the philosophers of the age dedicated their works. Another emir of Irak, Saif-Ed-Daula, established schools at Kufa and at Bussora, which soon acquired great celebrity. Abou-Mansor-Baharam, established a public library at Firuzabad in Curdistan, which at its very commencement contained 7000 volumes. In the thirteenth century there existed a celebrated school of medicine in Damascus. The calif Malek-Adel endowed it richly, and was often present at the lectures with a book under his arm.

Had the progress of the sciences among the Arabians been proportional to the number of those who cultivated them, we might hail the Saracens as the saviours of literature during the dark and benighted ages of Christianity; but we must acknowledge with regret, that notwithstanding the enlightened views of the califs, notwithstanding the multiplicity of academies and libraries, and the prodigious number of writers, the sciences received but little improvement from the Arabians. There are very few Arabian writers in whose works we find either philosophical ideas, successful researches, new facts, or great and new and important truths. How, indeed, could such things be expected from a people naturally hostile to mental exertion; professing a religion which stigmatizes all exercise of the judgment as a crime, and weighed down by the heavy yoke of despotism? It was the religion of the Arabians, and the despotism of their princes, that opposed the greatest obstacles to the progress of the sciences, even during the most flourishing period of their civilization.* Fortunately

* For a fuller account of the progress of science among the Arabians than would be consistent with this work, the reader is referred to Mortucla's Hist. des Mathématiques, i. 351; Sprengel's Hist. de la Médecine, ii. 246.

chemistry was the branch of science least obnoxious to the religious prejudices of the Mahometans. It was in it, therefore, that the greatest improvements were made : of these improvements it will be requisite now to endeavour to give the reader some idea. Astrology and alchymy, they both derived from the Greeks : neither of them were inconsistent with the taste of the nation—neither of them were anathematized by the Mahometan creed, though Islamism prohibited magic and all the arts of divination. Alchymy may have suggested the chemical processes—but the Arabians applied them to the preparation of medicines, and thus opened a new and most copious source of investigation.

The chemical writings of the Arabians which I have had an opportunity of seeing and perusing in a Latin dress, being ignorant of the original language in which they were written, are those of Geber and Avicenna.

Geber, whose real name was Abou-Moussah-Dschafar-Al-Soli, was a Sabean of Harran, in Mesopotamia, and lived during the eighth century. Very little is known respecting the history of this writer, who must be considered as the patriarch of chemistry. Golius, professor of the oriental languages in the University of Leyden, made a present of Geber's work in manuscript to the public library. He translated it into Latin, and published it in the same city in folio, and afterwards in quarto, under the title of " Lapis Philosophorum."* It was translated into English by Richard Russel in 1678, under the title of, " The Works of Geber, the most famous Arabian Prince and Philosopher."† The works of Geber, so far as they

---

* Boerhaave's Chemistry (Shaw's translation), i. 26. *Note.*

† Golius was not, however, the first translator of Geber. A translation of the longest and most important of his tracts into Latin appeared in Strasburg, in 1529. There was another translation published in Italy, from a manuscript in the Vatican. There probably might be other translations. I have

appeared in Latin or English, consist of four tracts.
The first is entitled, "Of the Investigation or Search
of Perfection." The second is entitled, "Of the Sum
of Perfection, or of the perfect Magistery." The
third, "Of the Invention of Verity or Perfection."
And the last, "Of Furnaces, &c.; with a Recapitulation of the Author's Experiments."

The object of Geber's work is to teach the method
of making the philosopher's stone, which he distinguishes usually by the name of *medicine of the third
class.* The whole is in general written with so much
plainness, that we can understand the nature of the
substances which he employed, the processes which
he followed, and the greater number of the products
which he obtained. It is, therefore, a book of some
importance, because it is the oldest chemical treatise
in existence,* and because it makes us acquainted
with the processes followed by the Arabians, and the
progress which they had made in chemical investigations. I shall therefore lay before the reader the most
important facts contained in Geber's work.

1. He considered all the metals as compounds of
mercury and sulphur: this opinion did not originate
with him. It is evident from what he says, that the
same notion had been adopted by his predecessors—
men whom he speaks of under the title of the
ancients.

2. The metals with which he was acquainted were
gold, silver, copper, iron, tin, and *lead.* These are
usually distinguished by him under the names of *Sol,
Luna, Venus, Mars, Jupiter,* and *Saturn.* Whether

compared four different copies of Geber's works, and found
some differences, though not very material. I have followed
Russel's English translation most commonly, as upon the whole
the most accurate that I have seen.

* Of course I exclude the writings of the Greek ecclesiastics
mentioned in a previous part of this work, which still continue in manuscript; because, I am ignorant of what they
contain.

these names of the planets were applied to the metals
by Geber, or only by his translators, I cannot say;
but they were always employed by the alchymists,
who never designated the metals by any other ap-
pellations.

3. Gold and silver he considered as perfect metals;
but the other four were imperfect metals. The difference between them depends, in his opinion, partly
upon the proportions of mercury and sulphur in each,
and partly upon the purity or impurity of the mercury
and sulphur which enters into the composition of each.

Gold, according to him, is created of the most
subtile substance of mercury and of most clear fixture,
and of a small substance of sulphur, clean and of pure
redness, fixed, clear, and changed from its own nature,
tinging that; and because there happens a diversity in
the colours of that sulphur, the yellowness of gold
must needs have a like diversity.* His evidence that
gold consisted chiefly of mercury, is the great ease
with which mercury dissolves gold. For mercury, in
his opinion, dissolves nothing that is not of its own
nature. The lustre and splendour of gold is another
proof of the great proportion of mercury which it contains. That it is a fixed substance, void of all burning sulphur, he thinks evident by every operation in
the fire, for it is neither diminished nor inflamed.
His other reasons are not so intelligible.†

Silver, like gold, is composed of much mercury and
a little sulphur; but in the gold the sulphur is red;
whereas the sulphur that goes to the formation of
silver is white. The sulphur in silver is also clean,
fixed, and clear. Silver has a purity short of that of
gold, and a more gross inspissation. The proof of
this is, that its parts are not so condensed, nor is it
so fixed as gold; for it may be diminished by fire,
which is not the case with gold.‡

* Sum of Perfection, book ii. part i. chap. 5.
† Ibid.   ‡ Ibid., chap. 6.

Iron is composed of earthy mercury and earthy sulphur, highly fixed, the latter in by far the greatest quantity. Sulphur, by the work of fixation, more easily destroys the easiness of liquefaction than mercury. Hence the reason why iron is not fusible, as is the case with the other metals.*

Sulphur not fixed melts sooner than mercury; but fixed sulphur opposes fusion. What contains more fixed sulphur, more slowly admits of fusion than what partakes of burning sulphur, which more easily and sooner flows.†

Copper is composed of sulphur unclean, gross and fixed as to its greater part; but as to its lesser part not fixed, red, and livid, in relation to the whole not overcoming nor overcome and of gross mercury. ‡

When copper is exposed to ignition, you may discern a sulphureous flame to arise from it, which is a sign of sulphur not fixed; and the loss of the quantity of it by exhalation through the frequent combustion of it, shows that it has fixed sulphur. This last being in abundance, occasions the slowness of its fusion and the hardness of its substance. That copper contains red and unclean sulphur, united to unclean mercury, is, he thinks, evident, from its sensible qualities. §

Tin consists of sulphur of small fixation, white with a whiteness not pure, not overcoming but overcome, mixed with mercury partly fixed and partly not fixed, white and impure.‖ That this is the constitution of tin he thinks evident; for when calcined, it emits a sulphureous stench, which is a sign of sulphur not fixed: it yields no flame, not because the sulphur is fixed, but because it contains a great portion of mercury. In tin there is a twofold sulphur and also a twofold mercury. One sulphur is less fixed, because in calcining it gives out a stench as sulphur. The fixed

* Sum of Perfection, book ii. part i. chap. 7.
† Ibid.   ‡ Ibid., chap. 8.   § Ibid.   ‖ Ibid., chap. 9.

sulphur continues in the tin after it is calcined.   He
thinks that the twofold mercury in tin is evident, from
this, that before calcination it makes a crashing
noise when bent, but after it has been thrice calcined, that crashing noise can no longer be perceived.* Geber says, that if lead be washed with
mercury, and after its washing melted in a fire not
exceeding the fire of its fusion, a portion of the mercury will remain combined with the lead, and will
give it the crashing noise and all the qualities of tin.
On the other hand, you may convert tin into lead.
By manifold repetition of its calcination, and the administration of fire convenient for its reduction, it is
turned into lead.†

Lead, in Geber's opinion, differs from tin only in
having a more unclean substance commixed of the
two more gross substances, sulphur and mercury.
The sulphur in it is burning and more adhesive to the
substance of its own mercury, and it has more of the
substance of fixed sulphur in its composition than tin
has.‡

Such are the opinions which Geber entertained respecting the composition of the metals.   I have been
induced to state them as nearly in his own words as
possible, and to give the reasons which he has assigned
for them, even when his facts were not quite correct,
because I thought that this was the most likely way of
conveying to the reader an accurate notion of the sentiments of this father of the alchymists, upon the very
foundation of the whole doctrine of the transmutation
of metals.   He was of opinion that all the imperfect
metals might be transformed into gold and silver, by
altering the proportions of the mercury and sulphur of
which they are composed, and by changing the nature
of the mercury and sulphur so as to make them the
same with the mercury and sulphur which constitute

* Sum of Perfection, book ii. part i. chap. 9.
† Ibid.   ‡ Ibid., chap. 10.

gold and silver. The subtance capable of producing these important changes he calls sometimes the *philosopher's stone*, but generally the *medicine*. He gives the method of preparing this important *magistery*, as he calls it. But it is not worth while to state his process, because he leaves out several particulars, in order to prevent the foolish from reaping any benefit from his writings, while at the same time those readers who possess the proper degree of sagacity will be able, by studying the different parts of his writings, to divine the nature of the steps which he omits, and thus profit by his researches and explanations. But it will be worth while to notice the most important of his processes, because this will enable us to judge of the state of chemistry in his time.

4. In his book on furnaces, he gives a description of a furnace proper for calcining metals, and from the fourteenth chapter of the fourth part of the first book of his Sum of Perfection, it is obvious that the method of calcining or oxidizing iron, copper, tin, and lead, and also mercury and arsenic were familiarly known to him.

He gives a description of a furnace for distilling, and a pretty minute account of the glass or stoneware, or metallic aludel and alembic, by means of which the process was conducted. He was in the habit of distilling by surrounding his aludel with hot ashes, to prevent it from being broken. He was acquainted also with the water-bath. These processes were familiar to him. The description of the distillation of many bodies occurs in his work ; but there is not the least evidence that he was acquainted with ardent spirits. The term *spirit* occurs frequently in his writings, but it was applied to volatile bodies in general, and in particular to sulphur and white arsenic, which he considered as substances very similar in their properties. Mercury also he considered as a spirit.

The method of distilling *per descensum*, as is practised in the smelting of zinc, was also known to him. He describes an apparatus for the purpose, and gives several examples of such distillations in his writings.

He gives also a description of a furnace for melting metals, and mentions the vessels in which such processes were conducted. He was acquainted with crucibles; and even describes the mode of making cupels, nearly similar to those used at present. The process of cupellating gold and silver, and purifying them by means of lead, is given by him pretty minutely and accurately: he calls it *cineritium*, or at least that is the term used by his Latin translator.

He was in the habit of dissolving salts in water and acetic acid, and even the metals in different menstrua. Of these menstrua he nowhere gives any account; but from our knowledge of the properties of the different metals, and from some processes which he notices, it is easy to perceive what his solvents must have been; namely, the mineral acids which were known to him, and to which there is no allusion whatever in any preceding writer that I have had an opportunity of consulting. Whether Geber was the discoverer of these acids cannot be known, as he nowhere claims the discovery: indeed his object was to slur over these acids, as much as possible, that their existence, or at least their remarkable properties, might not be suspected by the uninitiated. It was this affectation of secrecy and mystery that has deprived the earliest chemists of that credit and reputation to which they would have been justly entitled, had their discoveries been made known to the public in a plain and intelligible manner.

The mode of purifying liquids by filtration, and of separating precipitates from liquids by the same means, was known to Geber. He called the process *distillation through a filter*.

Thus the greater number of chemical processes, such as they were practised almost to the end of the eighteenth century, were known to Geber. If we compare his works with those of Dioscorides and Pliny, we shall perceive the great progress which chemistry or rather pharmacy had made. It is more than probable that these improvements were made by the Arabian physicians, or at least by the physicians who filled the chairs in the medical schools, which were under the protection of the califs: for as no notice is taken of these processes by any of the Greek or Roman writers that have come down to us, and as we find them minutely described by the earliest chemical writers among the Arabians, we have no other alternative than to admit that they originated in the east.

I shall now state the different chemical substances or preparations which were known to Geber, or which he describes the method of preparing in his works.

1. Common salt. This substance occurring in such abundance in the earth, and being indispensable as a seasoner of food, was known from the earliest ages. But Geber describes the method which he adopted to free it from impurities. It was exposed to a red heat, then dissolved in water, filtered, crystallized by evaporation, and the crystals being exposed to a red heat, were put into a close vessel, and kept for use.* Whether the identity of sal-gem (*native salt*) and common salt was known to Geber is nowhere said. Probably not, as he gives separate directions for purifying each.

2. Geber gives an account of the two fixed alkalies, *potash* and *soda*, and gives processes for obtaining them. Potash was obtained by burning cream of tartar in a crucible, dissolving the residue in water, filtering the solution, and evaporating to dryness.† This would yield a pure carbonate of potash.

Carbonate of soda he calls *sagimen vitri*, and salt of soda. He mentions plants which yield it when burnt, points out the method of purifying it, and even

* Investigation and Search of Perfection, chap. 3.
† Invention of Verity, chap. 4.

describes the method of rendering it caustic by means of quicklime. *

3. Saltpetre, or nitrate of potash, was known to him; and Geber is the first writer in whom we find an account of this salt. Nothing is said respecting its origin; but there can be little doubt that it came from India, where it was collected, and known long before Europeans were acquainted with it. The knowledge of this salt was probably one great cause of the superiority of the Arabians over Europeans in chemical knowledge; for it enabled them to procure *nitric acid*, by means of which they dissolved all the metals known in their time, and thus acquired a knowledge of various important saline compounds, which were of considerable importance.

There is a process for preparing saltpetre artificially, in several of the Latin copies of Geber, though it does not appear in our English translation. The method was to dissolve sagimen vitri, or carbonate of soda, in aqua fortis, to filter and crystallize by evaporation.† If this process be genuine, it is obvious that Geber must have been acquainted with nitrate of soda; but I have some doubts about the genuineness of the passage, because the term *aqua fortis* occurs in it. Now this term occurs nowhere else in Geber's work: even when he gives the process for procuring nitric acid, he calls it simply water; but observes, that it is a water possessed of much virtue, and that it constitutes a precious instrument in the hands of the man who possesses sagacity to use it aright.

4. Sal ammoniac was known to Geber, and seems to have been quite common in his time. There is no evidence that it was known to the Greeks or Romans, as neither Dioscorides nor Pliny make any allusion to it. The word in old books is sometimes *sal armoniac*, sometimes *sal ammoniac*. It is supposed to

---

* Search of Perfection, chap. 3.
† De Investigatione Perfect. chap. 4.

have been brought originally from the neighbourhood of the temple of Jupiter Ammon : but had this been the case, and had it occurred native, it could scarcely have been unknown to the Romans, under whose dominions that part of Africa fell. In the writings of the alchymists, sal ammoniac is mentioned under the following whimsical names :

> Anima sensibilis,
> Aqua duorum fratrum ex sorore,
> Aquila,
> Lapis aquilinis,
> Cancer,
> Lapis angeli conjungentis,
> Sal lapidum,
> Sal alocoph.

Geber not only knew sal ammoniac, but he was aware of its volatility; and gives various processes for subliming it, and uses it frequently to promote the sublimation of other bodies, as of oxides of iron and copper. He gives also a method of procuring it from urine, a liquid which, when allowed to run into putrefaction, is known to yield it in abundance Sal ammoniac was much used by Geber, in his various processes to bring the inferior metals to a state of greater perfection. By adding it or common salt to aqua fortis, he was enabled to dissolve gold, which certainly could not be accomplished in the time of Dioscorides or Pliny. The description, indeed, of Geber's process for dissolving gold is left on purpose in a defective state; but an attentive reader will find no great difficulty in supplying the defects, and thus understanding the whole of the process.

5. Alum, precisely the same as the alum of the moderns, was familiarly known to Geber, and employed by him in his processes. The manufacture of this salt, therefore, had been discovered between the time when Pliny composed his Natural History and

the eighth century, when Geber wrote; unless we admit that the mode of making it had been known to the Tyrian dyers, but that they had kept the secret so well, that no suspicion of its existence was entertained by the Greeks and Romans. That they employed *alumina* as a mordant in some of their dyes, is evident; but there is no proof whatever that *alum*, in the modern sense of the word, was known to them.

Geber mentions three alums which he was in the habit of using; namely, icy alum, or Rocca alum; Jamenous alum, or alum of Jameni, and feather alum. *Rocca*, or *Edessa*, in Syria, is admitted to have been the place where the first manufactory of alum was established; but at what time, or by whom, is quite unknown : we know only that it must have been posterior to the commencement of the Christian era, and prior to the eighth century, when Geber wrote. Jameni must have been another locality where, at the time of Geber, a manufactory of alum existed. *Feather alum* was undoubtedly one of the native impure varieties of *alum*, known to the Greeks and Romans. Geber was in the habit of distilling alum by a strong heat, and of preserving the water which came over as a valuable menstruum. If alum be exposed to a red heat in glass vessels, it will give out a portion of sulphuric acid : hence water distilled from alum by Geber was probably a weak solution of sulphuric acid, which would undoubtedly act powerfully as a solvent of iron, and of the alkaline carbonates. It was probably in this way that he used it.

6. Sulphate of iron or copperas, as it is called (*cuperosa*), in the state of a crystalline salt, was well known to Geber, and appears in his time to have been manufactured.

7. Baurach, or borax, is mentioned by him, but without any description by which we can know whether or not it was our borax : the probability is that it was.

Both glass and borax were used by him when the oxides of metals were reduced by him to the metallic state.

8. Vinegar was purified by him by distilling it over, and it was used as a solvent in many of his processes.

9. Nitric acid was known to him by the name of *dissolving water*. He prepared it by putting into an alembic one pound of sulphate of iron of Cyprus, half a pound of saltpetre, and a quarter of a pound of alum of Jameni: this mixture was distilled till every thing liquid was driven over. He mentions the red fumes which make their appearance in the alembic during the process. * This process, though not an economical one, would certainly yield nitric acid; and it is remarkable, because it is here that we find the first hint of the knowledge of chemists of this most important acid, without which many chemical processes of the utmost importance could not be performed at all.

10. This acid, thus prepared, he made use of to dissolve silver: the solution was concentrated till the nitrate of silver was obtained by him in a crystallized state. This process is thus described by him: " Dissolve silver calcined in solutive water (*nitric acid*), as before; which being done, coct it in a phial with a long neck, the orifice of which must be left unstopped, for one day only, until a third part of the water be consumed. This being effected, set it with its vessel in a cold place, and then it is converted into small fusible stones, like crystal."†

11. He was in the habit also of dissolving sal ammoniac in this nitric acid, and employing the solution, which was the aqua regia of the old chemists, to dissolve gold. ‡ He assures us that this aqua regia would dissolve likewise sulphur and silver. The latter assertion is erroneous. But sulphur is easily converted

* Invention of Verity, chap. 23.
† Ibid., chap. 21. ‡ Ibid., chap. 23.

into sulphuric acid by the action of aqua regia, and of course it disappears or dissolves.

12. Corrosive sublimate is likewise described by Geber in a very intelligible manner. His method of preparing it was as follows: " Take of mercury one pound, of dried sulphate of iron two pounds, of alum calcined one pound, of common salt half a pound, and of saltpetre a quarter of a pound: incorporate altogether by trituration and sublime; gather the white, dense, and ponderous portions which shall be found about the sides of the vessel. If in the first sublimation you find it turbid or unclean (which may happen by reason of your own negligence), sublime a second time with the same fuses." * Still more minute directions are given in other parts of the work: we have even some imperfect account of the properties of corrosive sublimate.

13. Corrosive sublimate is not the only preparation of mercury mentioned by Geber. He informs us that when mercury is combined with sulphur it assumes a red colour, and becomes cinnabar.† He describes the affinities of mercury. for the different metals. It adheres easily to three metals; namely, lead, tin, and gold; to silver with more difficulty. To copper with still more difficulty than to silver; but to iron it unites in nowise unless by artifice.‡ This is a tolerably accurate account of the matter. He says, that mercury is the heaviest body in nature except gold, which is the only metal that will sink in it.§ Now this was true, applied to all the substances known when Geber lived.

He gives an account of the method of forming the peroxide of mercury by heat; that variety of it formerly distinguished by the name of *red precipitati per se*. " Mercury," he says, " is also coagulated by

* Invention of Verity, chap. 8.
† Sum of Perfection, book i. part iii. chap. 4.
‡ Ibid., chap. 6.   § Ibid.

long and constant retention in fire, in a glass vessel with a very long neck and round belly; the orifice of the neck being kept open, that the humidity may vanish thereby."* He gives another process for preparing this oxide, possible, perhaps, though certainly requiring very cautious regulation of the fire. " Take," says he, " of mercury one pound, of vitriol (sulphate of iron) rubified two pounds, and of saltpetre one pound. Mortify the mercury with these, and then sublime it from rock alum and saltpetre in equal weights."†

14. Geber was acquainted with several of the compounds of metals with sulphur. He remarks that sulphur when fused with metals increases their weight.‡ Copper combined with sulphur becomes yellow, and mercury red.§ He knew the method of dissolving sulphur in caustic potash, and again precipitating it by the addition of an acid. His process is as follows: " Grind clear and gummose sulphur to a most subtile powder, which boil in a lixivium made of ashes of *heartsease* and quicklime, gathering from off the surface its oleaginous combustibility, until it be discerned to be clear. This being done, stir the whole with a stick, and then warily take off that which passeth out with the lixivium, leaving the more gross parts in the bottom. Permit that extract to cool a little, and upon it pour a fourth part of its own quantity of distilled vinegar, and then will the whole suddenly be congealed as milk. Remove as much of the clear lixivium as you can ; but dry the residue with a gentle fire and keep it."‖

15. It would appear from various passages in Geber's works that he was acquainted with arsenic in the metallic state. He frequently mentions its com-

* Sum of Perfection, book i. part iv. chap. 16.
† Invention of Verity, chap. 10.
‡ Sum of Perfection, book i. part iii. chap. 4.
§ Ibid.   ‖ Invention of Verity, chap. 6.

bustibility, and considers it as the *compeer* of sulphur. And in his book on *Furnaces*, chapter 25 (or 28 in some copies), he expressly mentions *metallic arsenic (arsenicum metallinum)*, in a preparation not very intelligible, but which he considered of great importance. The white oxide of arsenic or arsenious acid, was obviously well known to him. He gives more than one process for obtaining it by sublimation.* He observes in his Sum of Perfection, book i. part iv. chap. 2, which treats of sublimation, " Arsenic, which before its sublimation was evil and prone to adustion, after its sublimation, suffers not itself to be inflamed; but only resides without inflammation."

Geber states the fact, that when arsenic is heated with copper that metal becomes white.† He gives also a process by which the white arseniate of iron is obviously made. " Grind one pound of iron filings with half a pound of sublimed arsenic (arsenious acid). Imbibe the mixture with the water of saltpetre, and salt-alkali, repeating this imbibation thrice. Then make it flow with a violent fire, and you will have your iron white. Repeat this labour till it flow sufficiently with peculiar dealbation.‡

16. He mentions oxide of copper under the name of *æs ustum*, the red oxide of iron under the name of *crocus* of iron. He mentions also litharge and red lead.§ But as all these substances were known to the Greeks and Romans, it is needless to enter into any particular details.

17. I am not sure what substance Geber understood by the word *marchasite*. It was a substance which must have been abundant, and in common use, for he refers to it frequently, and uses it in many of his processes; but he nowhere informs us what it is. I sus-

* Invention of Verity, chap. 7.
† Sum of Perfection, book ii. part. ii. chap. 11.
‡ Invention of Verity, chap. 14.   § Ibid., chap. 4 and 12.

pect it may have been sulphuret of antimony, which was certainly in common use in Asia long before the time of Geber. But he also makes mention of antimony by name, or at least the Latin translator has made use of the word *antimonium*. When speaking of the reduction of metals after heating them with sulphur, he says, " The reduction of tin is converted into clear antimony ; but of lead, into a dark-coloured antimony, as we have found by proper experience."* It is not easy to conjecture what meaning the word antimony is intended to convey in this passage. In another passage he says, " Antimony is calcined, dissolved, clarified, congealed, and ground to powder, so it is prepared."†

18. Geber's description of the metals is tolerably accurate, considering the time when he wrote. As an example I shall subjoin his account of gold. " Gold is a metallic body, yellow, ponderous, mute, fulged, equally digested in the bowels of the earth, and very long washed with mineral water; under the hammer extensible, fusible, and sustaining the trial of the cupel and cementation."‡ He gives an example of copper being changed into gold. " In copper-mines," he says, " we see a certain water which flows out, and carries with it thin scales of copper, which (by a continual and long-continued course) it washes and cleanses. But after such water ceases to flow, we find these thin scales with the dry sand, in three years time to be digested with the heat of the sun ; and among these scales the purest gold is found : therefore we judge those scales were cleansed by the benefit of the water, but were equally digested by heat of the sun, in the dryness of the sand, and so brought to equality."§ Here we have an example of plausible reasoning from

* Sum of Perfection, book ii. part iii. chap. 10.
† Invention of Verity, chap. 4.
‡ Sum of Perfection, book i. part iii. chap. 8.
§ Ibid., book i. part iii. chap. 8.

defective premises. The gold grains doubtless existed in the sand before, while the scales of copper in the course of three years would be oxidized and converted into powder, and disappear, or at least lose all their metallic lustre.

Such are the most remarkable chemical facts which I have observed in the works of Geber. They are so numerous and important, as to entitle him with some justice to the appellation of the father and founder of chemistry. Besides the metals, sulphur and salt, with which the Greeks and Romans were acquainted, he knew the method of preparing sulphuric acid, nitric acid, and aqua regia. He knew the method of dissolving the metals by means of these acids, and actually prepared nitrate of silver and corrosive sublimate. He was acquainted with potash and soda, both in the state of carbonates and caustic. He was aware that these alkalies dissolve sulphur, and he employed the process to obtain sulphur in a state of purity.

But notwithstanding the experimental merit of Geber, his spirit of philosophy did not much exceed that of his countrymen. He satisfied himself with accounting for phenomena by occult causes, as was the universal custom of the Arabians; a practice quite inconsistent with real scientific progress. That this was the case will appear from the following passage, in which Geber attempts to give an explanation of the properties of the *great elixir* or *philosopher's stone :* " Therefore, let him attend to the properties and ways of action of the composition of the greater elixir. For we endeavour to make one substance, yet compounded and composed of many, so permanently fixed, that being put upon the fire, the fire cannot injure; and that it may be mixed with metals in flux and flow with them, and enter with that which in them is of an ingressible substance, and be fermented with that which in them is of a permixable substance; and be consolidated with that which in them is of a consolidable substance; and be fixed with that which in them is of a fixable substance; and not be burnt by those things which burn not gold and silver; and take away consolidation and weights with due ignition.*

The next Arabian whose name I shall introduce into this history, is Al-Hassain-Abou-Ali-Ben-Abdallah-Ebn-Sina, surnamed Scheik Reyes, or prince of physicians, vulgarly known by the name of *Avicenna*. Next to Aristotle and Galen, his reputation was the highest, and his authority the greatest of all medical practitioners; and he reigned paramount, or at least shared the medical sceptre till he was hurled from his throne by the rude hands of Paracelsus.

Avicenna was born in the year 978, at Bokhara, to which place his father had retired during the emirate of the calif Nuhh, one of the sons of the celebrated Almansor. Ali, his father, had dwelt in Balkh, in the Chorazan. After the birth of Avicenna he went to Asschena in Bucharia, where he continued to live till his son had reached his fifteenth year. No labour nor expense was spared on the education of Avicenna, whose abilities were so extraordinary that he is said to have been able to repeat the whole Koran by heart at the age of ten years. Ali gave him for a master Abou-Abdallah-Annatholi, who taught him grammar, dialectics, the geometry of Euclid, and the astronomy of Ptolemy. But Avicenna quitted his tuition because he could not give him the solution of a problem in logic. He attached himself to a merchant, who taught him arithmetic, and made him acquainted with the Indian numerals from which our own are derived. He then undertook a journey to Bagdad, where he studied philosophy under the great Peripatician, Abou-Nasr-Alfarabi, a disciple of Mesue the elder. At the same time he applied himself to medicine, under the tuition of the Nestorian, Abou-Sahel-Masichi. He

---

* Investigation of Perfections, chap. 11.

informs us himself that he applied with an extraordinary ardour to the study of the sciences. He was in the habit of drinking great quantities of liquids during the night, to prevent him from sleeping; and he often obtained in a dream a solution of those problems at which he had laboured in vain while he was awake. When the difficulties to be surmounted appeared to him too great, he prayed to God to communicate to him a share of his wisdom; and these prayers, he assures us, were never offered in vain. The metaphysics of Aristotle was the only book which he could not comprehend, and after reading them over forty times, he threw them aside with great anger at himself.

Already, at the age of sixteen, he was a physician of eminence; and at eighteen he performed a brilliant cure on the calif Nuhh, which gave him such celebrity that Mohammed, Calif of Chorazan, invited him to his palace; but Avicenna rather chose to reside at Dschordschan, where he cured the nephew of the calif Kabus of a grievous distemper.

Afterwards he went to Ray, where he was appointed physician to Prince Magd-Oddaula. Here he composed a dictionary of the sciences. Sometime after this he was raised to the dignity of vizier at Hamdan; but he was speedily deprived of his office and thrown into prison for having favoured a sedition. While incarcerated he wrote many works on medicine and philosophy. By-and-by he was set at liberty, and restored to his dignity; but after the death of his protector, Schems-Oddaula, being afraid of a new attempt to deprive him of his liberty, he took refuge in the house of an apothecary, where he remained long concealed and completely occupied with his literary labours. Being at last discovered he was thrown into the castle of Berdawa, where he was confined for four months. At the end of that time a fortunate accident enabled him to make his escape, in the disguise of a monk. He repaired to Ispahan, where he lived much respected at the court of the calif Ola-Oddaula. He did not live to a great age, because he had worn out his constitution by too free an indulgence of women and wine. Having been attacked by a violent colic, he caused eight injections, prepared from long pepper, to be thrown up in one day. This excessive use of so irritating a remedy, occasioned an excoriation of the intestines, which was followed by an attack of epilepsy. A journey to Hamdan, in company with the calif, and the use of mithridate, into which his servant by mistake had put too much opium, contributed still further to put an end to his life. He had scarcely arrived at the town when he died in the fifty-eighth year of his age, in the year 1036.

Avicenna was the author of the immense work entitled " Canon," which was translated into Latin, and for five centuries constituted the great standard, the infallible guide, the confession of faith of the medical world. All medical knowledge was contained in it; and nothing except what was contained in it was considered by medical men as of any importance. When we take a view of the Canon, and compare it with the writings of the Greeks, and even of the Arabians, that preceded it, we shall find some difficulty in accounting for the unbounded authority which he acquired over the medical world, and for the length of time during which that authority continued.

But it must be remembered, that Avicenna's reign occupies the darkest and most dreary period of the history of the human mind. The human race seems to have been asleep, and the mental faculties in a state of complete torpor. Mankind, accustomed in their religious opinions to obey blindly the infallible decisions of the church, and to think precisely as the church enjoined them to think, would naturally look for some means to save them the trouble of thinking on medical subjects; and this means they found fortunately in the canons of Avicenna. These canons,

in their opinion, were equally infallible with the decisions of the holy father, and required to be as implicitly obeyed. The whole science of medicine was
reduced to a simple perusal of Avicenna's Canon, and
an implicit adherence to his rules and directions.

When we compare this celebrated work with the
medical writings of the Greeks, and even of the
Arabians, the predecessors of Avicenna, we shall be
surprised that it contains little or nothing which can
be considered as original ; the whole is borrowed from
the writings of Galen, or Ætius, or Rhazes : scarcely
ever does he venture to trust his own wings, but rests
entirely on the sagacity of his Greek and Arabian
predecessors. Galen is his great guide ; or, if he ever
forsake him, it is to place himself under the direction
of Aristotle.

The Canon contains a collection of most of the
valuable information contained in the writings of the
ancient Greek physicians, arranged, it must be allowed, with great clearness. The Hhawi of Razes is almost as complete ; but it wants the *lucidus ordo* which
distinguishes the Canon of Avicenna. I conceive that
the high reputation which Avicenna acquired, was
owing to the care which he bestowed upon his arrangement. He was undoubtedly a man of abilities, but
not of inventive genius. There is little original matter
in the Canon. But the physicians in the west, while
Avicenna occupied the medical sceptre, had no opportunity of judging of the originality of their oracle,
because they were unacquainted with the Greek language, and could not therefore consult the writings of
Galen or Ætius, except through the corrupt medium
of an Arabian version.

But it is not the medical reputation of Avicenna that
induced me to mention his name here. Like all the
Arabian physicians, he was also a chemist ; and his
chemical tracts having been translated into Latin, and
published in Western Europe, we are enabled to judge

of their merit, and to estimate the effect which they may have had upon the progress of chemistry. The first Latin translation of the chemical writings of Avicenna was published at Basil in 1572; they consist of two separate books; the first, under the name of " Porta Elementorum," consists of a dialogue between a master and his pupil, respecting the mysteries of Alchymy. He gives an account of the four elements, fire, air, water, earth, and gives them their usual qualities of dry, moist, hot, and cold. He then treats of air, which, he says, is the food of fire, of water, of honey, of the mutual conversion of the elements into each other; of milk and cheese, of the mixture of fire and water, and that all things are composed of the four elements. There is nothing in this tract which has any pretension to novelty; he merely retails the opinions of the Greek philosophers.

The other treatise is much larger, and professes to teach the whole art of alchymy; it is divided into ten parts, entitled " Dictiones." The first diction treats of the philosopher's stone in general; the second diction treats of the method of converting light things into heavy, hard things into soft; of the mutation of the elements; and of some other particulars of a nature not very intelligible. The third diction treats of the formation of the elixir; and the same subject is continued in the fourth.

The fifth diction is one of the most important in the whole treatise; it is in general intelligible, which is more than can be said of those that precede it. This diction is divided into twenty-eight chapters: the first chapter treats of copper, which, he says, is of three kinds; permenian copper, natural copper, and Navarre copper. But of these three varieties he gives no account whatever; though he enlarges a good deal on the qualities of copper—not its properties, but its supposed medicinal action. It is hot and dry, he says, but in the calx of it there is humidity. His account of the composition of copper is the same with that of Geber.

The second chapter treats of lead, the third of tin, and in the remaining chapters he treats successively of brass, iron, gold, silver, marcasite, sulphuret of antimony, which is distinguished by the name of alcohol; of soda, which he says is the juice of a plant called *sosa*. And he gives an unintelligible process by which it is extracted from that plant, without mentioning a syllable about the combustion to which it is obvious that it must have been subjected.

In the twelfth chapter he treats of saltpetre, which, he says, is brought from Sicily, from India, from Egypt, and from Herminia. He describes several varieties of it, but mentions nothing about its characteristic property of deflagrating upon burning coals. He then treats successively of common salt, of sal-gem, of vitriol, of sulphur, of orpiment, and of sal ammoniac, which, he says, comes from Egypt, from India, and from Forperia. In the nineteenth and subsequent chapters he treats of aurum vivum, of hair, of urine, of eggs, of blood, of glass, of white linen, of horse-dung, and of vinegar.

The sixth diction, in thirty-three chapters, treats of the calcination of the metals, of sublimation, and of some other processes. I think it unnecessary to be more particular, because I cannot perceive any thing in it that had not been previously treated of by Geber.

The seventh diction treats of the preparation of blood and eggs, and the method of dividing them into their four elements. It treats also of the elixir of silver, and the elixir of gold; but it contains no chemical fact of any importance.

The eighth diction treats of the preparation of the ferment of silver, and of gold. The ninth diction treats of the whole magistery, and of the nuptials of the sun and moon; that is, of gold and silver. The tenth diction treats of weights.

The chemical writings of Avicenna are of little value, and apply chemistry rather to the supposed medical qualities of the different substances treated of, than to the advancement of the science. All the chemical knowledge which he possesses is obviously drawn from Geber. Geber, then, may be looked upon as the only chemist among the Arabians to whom we are indebted for any real improvements and new facts. It is true that the Arabian physicians improved considerably the materia medica of the Greeks, and introduced many valuable medicines into common use which were unknown before their time. It is enough to mention corrosive sublimate, manna, opium, asafœtida. It would be difficult to make out many of the vegetable substances used by the Arabian chemists; because the plants which they designated by particular names, can very seldom be identified. Botany at that time had made so little progress, that no method was known of describing plants so as to enable other persons to determine what they were.

CHAPTER IV.

OF THE PROGRESS OF CHEMISTRY UNDER PARACELSUS AND
HIS DISCIPLES.

HITHERTO we have witnessed only the first rude
beginnings, or, as it were, the early dawn of the chemical day.   It is from the time of Paracelsus that the
true commencement of chemical investigations is to be
dated.   Not that Paracelsus or his followers understood the nature of the science, or undertook any
regular or successful investigation.   But Paracelsus
shook the medical throne of Galen and Avicenna to
its very foundation; he roused the latent energies of
the human mind, which had for so long a period lain
torpid; he freed medical men from those trammels,
and put an end to that despotism which had existed
for five centuries.   He pointed out the importance of
chemical medicines, and of chemical investigations, to
the physician.   This led many laborious men to turn
their attention to the subject.   Those metals which
were considered as likely to afford useful medicines,
mercury for example, and antimony, were exposed to
the action of an infinite number of reagents, and a
prodigious collection of new products obtained and
introduced into medicine.   Some of these were better,
and some worse, than the preparations formerly employed; but all of them led to an increase of the
stock of chemical knowledge, which now began to
accumulate with considerable rapidity.   It will be

proper, therefore, to give a somewhat particular account of the life and opinions of Paracelsus, so far as they can be made out from his writings, because, though he was not himself a scientific chemist, he may be truly considered as the man through whose means the stock of chemical knowledge was accumulated, which was afterwards, by the ingenuity of Beccher, and Stahl, moulded into a scientific form.

Philippus Aureolus Theophrastus Paracelsus Bombast ab Hohenheim (as he denominates himself) was born at Einsideln, two German miles from Zurich. His father was called William Bombast von Hohenheim. He was a very near relation of George Bombast von Hohenheim, who became afterwards grand master of the order of Johannites. William Bombast von Hohenheim practised medicine at Einsideln.* After receiving the first rudiments of his education in his native city, he became a wandering scholastic, as was then the custom with poor scholars. He wandered from province to province, predicting the future by the position of the stars, and the lines on the hand, and exhibiting all the chemical processes which he had learned from founders and alchymists. For his initiation in alchymy, astrology, and medicine, he was indebted to his father, who was much devoted to these three sciences. Paracelsus mentions also the names of several ecclesiastics from whom he received chemical information; among others, Tritheimius, abbot of Spanheim; Bishop Scheit, of Stettbach; Bishop Erhart, of Laventall; Bishop Nicolas, of Hippon; and Bishop Matthew Schacht. He seems also to have served some years as an army surgeon, for he mentions many cures which he performed in the Low Countries, in the States of the Church, in the kingdom of Naples, and during the wars against the Venetians, the Danes, and the Dutch.

* See Testamentum Paracelsi, passim.

There is some uncertainty whether he received a regular college education, as was then the practice with all medical men. He acknowledges himself that his medical antagonists reproached him with never having frequented their schools; and he is perpetually affirming, that a physician should receive all his knowledge from God, and not from man. But if we can trust his own assertions, there can be no doubt that he took a regular medical degree, which implies a regular college education. He tells us, in his preface to his Chirurgia Magna, that he visited the universities of Germany, France, and Italy. He assures his readers, that he was the ornament of the schools where he studied. He even speaks of the oath which he was obliged to take when he received his medical degree; but where he studied, or where and when he received his medical degree, are questions which neither Paracelsus nor his disciples, nor his biographers, have enabled us to solve. If he ever attended a university, he must have neglected his studies, otherwise he could not have been ignorant, as he confessedly was, of the very first elements of the most common kinds of knowledge. But if he neglected the universities, he laboured long and assiduously with the rich Sigismond Fuggerus, of Schwartz, in order to learn the true secret of forming the philosopher's stone.

He gives us some details of the numerous journeys that he made, as was customary with the alchymists of the time, into the mountains of Bohemia, the East, and Sweden, to inspect the mines, to get himself initiated into the mysteries of the eastern adepts, to inspect the wonders of nature, and to view the celebrated diamond mountain, the position of which, however, he unfortunately forgets to specify.

In the preface to his Chirurgia Magna, he informs us that he traversed Spain, Portugal, England, Prussia, Poland, and Transylvania; where he not only profited by the information of the medical men with whom he became acquainted, but that he drew much precious information from old women, gipsies, conjurors, and chemists. * He spent several years in Hungary; and informs us that at Weissenburg, in Croatia, and in Stockholm, he was taught by several old women to prepare drinks capable of curing ulcers. He is said also to have made a voyage into Egypt, and even into Tartary; and he accompanied the son of the Kan of the Tartars to Constantinople, in order to learn the secret of the philosopher's stone from Trismogin, who inhabited that capital. This prodigious activity, this constant motion from place to place, left him but little leisure for reading: accordingly he informs us himself, that during the space of ten years he never opened a book, and that his whole library consisted only of six sheets. The inventory of his books, drawn up after his death, confirms this recital; for they consisted only of the Bible, the Concordance to the Bible, the New Testament, and the Commentaries of St. Jerome on the Evangelists.

We know not at what period he returned back to Germany; but at the age of thirty-three the great number of fortunate cures which he had performed rendered him an object of admiration to the people, and of jealousy to the rival physicians of the time. He assures us that he cured eighteen princes whose diseases had been aggravated by the practitioners devoted to the system of Galen. Among others he cured Philip, Margrave of Baden, of a dysentery, who promised him a great reward, but did not keep his promise, and even treated him in a way unworthy of that

---

* " Hispania, Portugallia, Anglia, Borussia, Lithuania, Polonia, Pannonia, Valachia, Transylvania, Croatia, Illyrico, immo omnibus totius Europæ nationibus peragratis, undeque non solum apud medicos, sed et chirurgos, tonsores, aniculas, magos, chymistas, nobiles ac ignobiles, optima, selectiora ac secretiora, quæ uspiam extarent remedia, inquisivi acriter."—*Præfatio Chirurgiæ Magnæ.* Opera Paracelsi, tom. iii.

prince. This cure, however, and others of a similar nature, added greatly to his celebrity; and in order to raise his reputation to the highest possible pitch, he announced publicly that he was able to cure all the diseases hitherto reckoned incurable; and that he had discovered an elixir, by means of which the life of man might be prolonged at pleasure to any extent whatever. He began the practice, which has since been so successfully followed in this country, of dispensing medicines gratuitously to the poor, in order to induce the rich to apply to him for assistance when they were overtaken with diseases.

In the year 1526 Paracelsus was appointed professor of physic and surgery in the University of Basil. This appointment was given him, it is said, by the recommendation of Œcolampadius. He introduced the custom of lecturing in the common language of the country, as is at present the universal practice: but during the time of Paracelsus, and long after indeed, all lectures were delivered in Latin. The new method which he followed in explaining the theory and practice of the art; the numerous fortunate cures which he stated in confirmation of his method of treatment; the emphasis with which he spoke of his secrets for prolonging life, and for curing every kind of disease without distinction, but still more his lecturing in a language which was understood by the whole population, drew to Bâle an immense crowd of idle, enthusiastic, and credulous hearers.

The lectures which he delivered on Practical Medicine still remain, written in a confused mixture of German and barbarous Latin, and containing little or nothing except a farrago of empirical remedies, advanced with the greatest confidence. They have a much greater resemblance to a collection of quack advertisements than to the sober lectures of a professor in a university. In the month of November, 1526, he wrote to Christopher Clauser, a physician in

Zurich, that as Hippocrates was the first physician among the Greeks, Avicenna among the Arabians, Galen among the Pergamenians, and Marsilius among the Italians, so he was beyond dispute the greatest physician among the Germans. Every country produces an illustrious physician, whose medicines are adapted to the climate in which he lived, but not suited to other countries. The remedies of Hippocrates were good to the Greeks, but not suitable to the Germans; thus it was necessary that an inspired physician should spring up in every country, and that he was the person destined to teach the Germans the art of curing all diseases. *

Paracelsus began his professorial career by burning publicly, in his class-room, and in the presence of his pupils, the works of Galen and Avicenna, assuring his hearers that the strings of his shoes possessed more knowledge than those two celebrated physicians. All the universities united had not, he assured them, as much knowledge as was contained in his own beard, and the hairs upon his neck were better informed than all the writers that ever existed put together. To give the reader an idea of the arrogant absurdity of his pretensions, I shall translate a few sentences of the preface to his tract, entitled " Paragranum," where he indulges in his usual strain of rodomontade : " Me, me you shall follow, you Avicenna, you Galen, you Rhazes, you Montagnana, you Mesue. I shall not follow you, but you shall follow me. You, I say, you inhabitants of Paris, you inhabitants of Montpelier, you Suevi, you Misnians, you inhabitants of Cologne, you inhabitants of Vienna; all you whom the Rhine and the Danube nourish, you who inhabit the islands

---

* See the dedication to his treatise *De Gradibus et Compositionibus Receptorum et Naturalium.* Opera Paracelsi, vol. ii. p. 144. I always refer to the folio edition of Paracelsus's works, in three volumes, published at Geneva in 1658, by M. de Tournes, which is the edition in my possession.

of the sea; you also Italy, you Dalmatia, you Athens, you Greek, you Arabian, you Israelite—I shall not follow you, but you shall follow me. Nor shall any one lurk in the darkest and most remote corner whom the dogs shall not piss upon. I shall be the monarch, the monarchy shall be mine. If I administer, and I bind up your loins, is he with whom you are at present delighted a Cacophrastus? This ordure must be eaten by you."

"What will your opinion be when you see your Cacophrastus constituted the chief of the monarchy? What will you think when you see the sect of Theophrastus leading on a solemn triumph, if I make you pass under the yoke of my philosophy? your Pliny will you call Cacopliny, and your Aristotle, Cacoaristotle? If I plunge them together with your Porphyry, Albertus, &c., and the whole of their compatriots into my *necessary*." But the terms become now so coarse and indelicate, that I cannot bring myself to proceed further with the translation. Enough has been given to show the extreme arrogance and folly of Paracelsus.

So far, however, was this impudence and grossness from injuring the interest of Paracelsus, that we are assured by Ramus and Urstisius that it contributed still further to increase it. The coarseness of his language was well suited to the vulgarity of the age; and his arrogance and boasting were considered, as usual, as a proof of superior merit. The cure which he performed on Frobenius, drew the attention of Erasmus himself, who consulted him about the diseases with which he was afflicted; and the letters that passed between them are still preserved. The epistle of Paracelsus is short, enigmatical, and unintelligible; that of Erasmus is distinguished by that clearness and elegance which characterize his writings.* But Frobenius died

* Opera Paracelsi, i. 485.

in the month of October, 1527, and the antagonists of Paracelsus attributed his death (and probably with justice) to the violent remedies which had been administered to a man whose constitution had been destroyed by the gout.

His death contributed not a little to tarnish the glory of Paracelsus: but he suffered the greatest injury from the habits of intoxication in which he indulged, and from the vulgarity of the way in which he spent his time. He hardly ever went into his class-room to deliver a lecture till he was half intoxicated, and scarcely ever dictated to his secretaries till he had lost the use of his reason by a too liberal indulgence in wine. If he was summoned to visit a patient, he scarcely ever went but in a state of intoxication. Not unfrequently he passed the whole night in the alehouse, in the company of peasants, and when morning came, was quite incapable of performing the duties of his station. On one occasion, after a debauch, which lasted the whole night, he was called next morning to visit a patient; on entering the room, he inquired if the sick person had taken any thing: " Nothing," was the answer, " except the body of our Lord." " Since you have already," says he, " provided yourself with another physician, my presence here is unnecessary," and he left the apartment instantly. When Albertus Basa, physician to the king of Poland, visited Paracelsus in the city of Basle, he carried him to see a patient whose strength was completely exhausted, and which, in his opinion, it was impossible to restore; but Paracelsus, wishing to make a parade of his skill, administered to him three drops of his laudanum, and invited him to dine with him next day. * The invita-

* There were two laudanums of Paracelsus; one was *red oxide of mercury*, the other consisted of the following substances : Chloride of antimony, 1 ounce; hepatic aloes, 1 ounce; rose-water, ½ ounce; saffron, 3 ounces; ambergris, 2 drams. All these well mixed.

tion was accepted, and the sick man dined next day with his physician.

Towards the end of the year 1527 a disgraceful dispute into which he entered brought his career, as a professor, to a sudden termination. The canon Cornelius, of Lichtenfels, who had been long a martyr to the gout, employed him as his physician, and promised him one hundred florins if he could cure him. Paracelsus made him take three pills of laudanum, and having thus freed him from pain, demanded the sum agreed upon; but Lichtenfels refused to pay him the whole of it. Paracelsus summoned him before the court, and the magistrate of Basle decided that the canon was bound to pay only the regular price of the medicine administered. Irritated at this decision, our intoxicated professor uttered a most violent invective against the magistrate, who threatened to punish him for his outrageous conduct. His friends advised him to save himself by flight. He took their advice, and thus abdicated his professorship. But, by this time, his celebrity as a teacher had been so completely destroyed by his foolish and immoral conduct, that he had lost all his hearers. In consequence of this state of things, his flight from Basle produced no sensation whatever in that university.

Paracelsus betook himself, in the first place, to Alsace, and sent for his faithful follower, the bookseller, Operinus, together with the whole of his chemical apparatus. In 1528 we find him at Colmar, where he recommenced his ambulating life of a theosophist, which he had led during his youth. His book upon syphilis, known at that time by the name of Morbus Gallicus, was dedicated at Colmar, to the chief magistrate of Colmar, Hieronymus Bonerus.* In 1531 he was at Saint-Gallen; in 1535, at Pfeffersbade, and in 1536, at Augsburg, where he dedicated his Chirur-

* Opera Paracelsi, iii. 101.

gia Magna to Malhausen. At the request of John de
Leippa, Marshal of Bohemia, he undertook a journey
into Moravia; as that nobleman, having been informed
that Paracelsus understood the method of curing the
gout radically, was anxious to put himself under his
care. Paracelsus lived for a long time at Kroman,
and its environs. John de Leippa, instead of receiving any benefit from the medicines administered to
him, became daily worse, and at last died. This was
the fate also of the lady of Zerotin, in whom the
remedies of Paracelsus produced no fewer than twenty-
four epileptic fits in one day. Paracelsus, instead of
waiting the disgrace with which the death of this lady
would have overwhelmed him, announced his intention
of going to Vienna, that he might see how they would
treat him in that capital.

It is said, that from Vienna he went into Hungary;
but in 1538, we find him in Villach, where he dedicated his Chronica et Origo Carinthiæ to the states of
Carinthia.* His book, De Natura Rerum, had been
dedicated to Winkelstein, and the dedication is dated
also at Villach, in the year 1537.† In 1540 he was
at Mindelheim, and in 1541, at Strasburg, where he
died, in St. Stephen's hospital, in the forty-eighth
year of his age.

To form an accurate idea of this most extraordinary
man, we must attend to his habits, and to the situation in which he was placed. He had acquired such
a habit of moving about, that he assures us himself he
found it impossible for him to continue for any length
of time in one place. He was always surrounded by
a number of followers, whom neither his habits of intoxication, nor the foolish and immoral conduct in
which he was accustomed to indulge, could induce
to forsake him. The most celebrated of these was
Operinus, a printer at Basle, on whom Paracelsus

* Opera Paracelsi, i. 243. † Ibid., ii. 84.

lavishes the most excessive praises, in his book De Morbo Gallico. But Operinus loaded his master with obloquy, being provoked at him because he had not made him acquainted with the secret of the philosopher's stone, as he had promised to do. We must therefore be cautious in believing the stories that he relates to the discredit of his master. We know the names of two others of his followers; Francis, who assures us that Paracelsus was devoted to the transmutation of metals; and George Vetter, who considered him as a magician; as was the opinion also of Operinus. Paracelsus himself, speaks of Dr. Cornelius, whom he calls his secretary, and in honour of whom he wrote several of his libels. Other libels are dedicated to Doctors Peter, Andrew, and Ursinus, to the licentiate Pancrace, and to Mr. Raphael. On this occasion he complains bitterly of the infidelity of his servants, who, he says, had succeeded in stealing from him several of his secrets; and had by this means been enabled to establish their reputation. He accuses equally the barbers and bathers that followed him, and is no less severe upon the physicians of every country through which he travelled.

When we attempt to form an accurate conception of the medical and philosophical opinions of this singular man, we find ourselves beset with almost insurmountable difficulties. His statements are so much at variance with each other, in his different pieces, and so much confusion reigns with respect to the order of publication, that we know not what to fix on as his last and maturest opinions. His style is execrable; filled with new words of his own coining, and of mysticisms either introduced to excite the admiration of the ignorant, or from the fanaticism and credulity of the writer, who was undoubtedly, to a considerable extent, the dupe of his own impostures. That he was in possession of the philosopher's stone, or of a medicine capable of prolonging life to an indefinite length, as he all along asserted, he could not himself believe; but he had boasted so long and so loudly of his wonderful cures, and of the efficacy of his medicines, that there can be no doubt that he ultimately placed implicit faith in them. The blunders of the transcribers whom he employed to copy his works, may perhaps account for some of the contradictions which they contain. But how can we look for a regular system of opinions from a man who generally dictated his works when in a state of intoxication, and thus laboured under an almost constant deprivation of reason.

His obscurity was partly the effect of design, and no doubt was intended to exalt the notions entertained of his profundity. He uses common words in new significations, without giving any indication of the change which he introduced. Thus *anatomy*, in the writings of Paracelsus, signifies not the dissection of dead animals to determine their structure, but it means the nature, force, and magical designation of a thing. And as, according to the Platonic and Cabalistic theory, every earthly body is formed after the model of a heavenly body, Paracelsus calls *anatomy* the knowledge of that model, of that ideal, or of that paradigm after which all things are created. He terms the fundamental force of a thing *a star*, and defines alchymy the art of drawing out the stars of metals. The star is the source of all knowledge. When we eat, we introduce into our bodies *the star*, which is then modified, and favours nutrition.

It is probable that many of his obscure and unintelligible expressions are the fruit of ignorance. Thus he uses the term *pagoyus*, instead of *paganus*. He gives the name of *pagoyæ* to the four *entities*, or causes of diseases, founded on the influence of the stars, to the elementary qualities; to the occult qualities, and to the influence of spirits; because these had been already admitted by the *Pagans*. But the fifth *entity*, or cause of disease, which has God immediately for

its author, is *non pagoya*. The *undimia* of Paracelsus
is our *œdema;* only he applies the name to every kind
of dropsy. The Latin word *tonitru*, we find is declined
by Paracelsus. Thus he says, *lapis tonitrui.* The
well-known line of Ovid,

> Tollere nodosam nescit medicina podagram,

He travestied into

> Nescit tartaream Roades curare podagram.*

Roades, he says, means medicines for horses; and
if any person wishes a more elegant verse, he may
make it for himself.† He employs, also, a great number of words to which no meaning whatever can be
attached; and to which, in all probability, he himself
had affixed none.

As is the case with all fanatics, he treated with contempt every kind of knowledge acquired by labour
and application; and boasted that his wisdom was
communicated to him directly by God Almighty. The
theosophist who is worthy of partaking of the divine
light, has no occasion for adopting a positive religion,
nor of subjecting himself to any kind of religious cere-
mony. The divine light within, which assimilates him
to the Deity, more than compensates for all these vulgar
usages, and raises the illuminated votary far above the
beggarly elements of external worship. Accordingly,
Paracelsus has been accused of treating the public
worship of the Deity with contempt. Not satisfied
with the plain sense of the book, he attempted to explain in a mystical manner the words and syllables of
the Bible. He accused Luther of not going far enough.
" Luther," says he, " is not worthy of untying the
strings of my shoes : should I undertake a reformation,
I would begin by sending the pope and the reformers
themselves to school." God, says Paracelsus, is the

* Opera Paracelsi, i. 328.
† "Qui elegantiorem optat, ille eum condat."—*Ibid,*

first and most excellent of writers. The Holy Scripture conducts us to all truth, and teaches us all things. But medicine, philosophy, and astronomy, are among the number of things. Therefore, when we want to know what magical medicine is, we must consult the Apocalypse. The Bible, with its paraphrases, is the key to the theory of diseases. It puts it in our power to understand St. John, who, like Daniel, Ezekiel, Moses, &c., was a magician, a cabalist, a diviner. The first duty of a physician is to study the Cabala, without which he must every moment commit a thousand blunders. " Learn," says he, " the cabalistic art, which includes under it all the others." " Man invents nothing, the devil invents nothing; it is God alone who unveils to us the light of nature." " God honoured at first with his illumination the blind pagans, Apollo, Æsculapius, Machaon, Podalirius, and Hippocrates, and imparted to them the genius of medicine; their successors were the sophists." One would suppose, from this passage, that Paracelsus had read and studied Hippocrates, and that he held him in high estimation. But the commentaries which he has left on some of the aphorisms, show evidently that he did not even understand the Greek physician. " The compassion of God," says he, " is the only foundation of medical science, and not a knowledge of the great masters, or of the writings which they have left in Greek and Latin." " God often acts in dreams by the light of nature, and points out to man the manner of curing diseases." " This knowledge renders all those objects visible which would otherwise escape the sight; and when faith is joined with it, nothing is then impossible to the theosophist, who may transport the ocean to the top of Mount Ætna, and Olympus into the Red Sea." Paracelsus predicts that by the year 1590 Christian theosophy would be generally spread over the world, and that the Galenical schools would be almost or entirely overthrown.

We find in Paracelsus some traces of the opinions of the Gnostics and Arians, who considered Christ as the first emanation of the Deity. He calls the first man *parens hominis;* and makes all spirits emanate from him. He is the *limbus minor,* or the last creature, into whom enters the great *limbus,* or the seed of all the creatures, the infinite being. All the sciences, and all the arts of man, are derived from this great *limbus;* and he who can sink himself in the little *limbus,* that is to say, in Adam, and who can communicate by faith with Jesus Christ, may invoke all *spirits.* Those who owe their science to this *limbus,* are the best informed ; those who derive it from the stars, occupy the last rank ; and those who owe it to the light of nature, are intermediate between the preceding. Jesus Christ, in his capacity of *limbus minor* and first man, being always an emanation of the Divinity ; and, consequently, a subordinate personage. These ideas explain to us why Paracelsus passed for an Arian, and was supposed not to believe in the Divinity of Jesus Christ. He was of opinion that the faithful performed miracles, and operated magical cures by their simple confidence in God the Father, and not by their faith in Christ; but he adds, however, that we ought to pray to Jesus, in order to obtain his intercession.

From the preceding attempt to explain the opinions of Paracelsus, it will be evident to the reader that he was both a fanatic and impostor, and that his theory (if such a name can be given to the reveries of a drunkard), consisted in uniting medicine with the doctrines of the Cabala. A few more observations will be necessary to develop his dogmas still further.

Every body, in his opinion, and man in particular, is double, consisting of a material and spiritual substance.* The spiritual, which may be called the

* Archidoxorum, lib. i. Opera Paracelsi, ii. 4.

sideric, results from the celestial influences; and we may trace after it a figure capable of producing all kinds of magical effects. When we can act upon the body itself, we act at the same time upon the spiritual form by characters and conjurations.[*] Yet, in another passage, he blames all magical ceremonies, and ascribes them to want of faith. The celestial intelligences impress upon material bodies certain signs, which manifest their influence. The perfection of art consists in understanding the meaning of these signs, and in determining from them the nature, qualities, and essence of a body. Adam, the first man, had a perfect knowledge of the Cabala; he could interpret the signatures of all things. It was this which enabled him to assign to the animals names which suited them best. A man who renounces all sensuality, and is blindly obedient to the will of God, is capable of taking a share in the actions which celestial intelligences perform; and consequently is possessed of the philosopher's stone. Never does he want any thing; all creatures in earth and in heaven are obedient to him; he can cure all diseases, and prolong his life as long as he pleases; because he possesses the tincture which Adam and the patriarch's before the flood employed to prolong the term of their existence.[†] Beelzebub, the chief of the demons, is also subject to the power of magic: and who can blame the theosophist for believing in the devil? He ought, however, to take care to prevent this malignant spirit from commanding him. Paracelsus was often wont to say, " If God does not aid me, the devil will help me."

[*] De longa Vita. Opera Paracelsi, ii. 46.
[†] Archidoxorum, lib. viii. Opera Paracelsi, ii. 29. In this book he gives the method of preparing the elixir of life. It seems to have been nothing else than a solution of *common salt* in water; for the quintessence of gold, with which this solution was to be mixed, was doubtless an imaginary substance.

Pantheism was one of the principal dogmas of the Cabala; and Paracelsus adopts it in all its grossness. He affirms perpetually that every thing is animated in the universe; that every thing which exists, eats, drinks, and voids excrements: even minerals and liquids take food and void the digested remains of their nourishment.* This opinion leads necessarily to the admission of a great number of spiritual substances, intermediate between material and immaterial in every part of the sublunary world, in water, air, earth, and fire; who, as well as man, eat, drink, converse, beget children; but which approach pure spirits in this, that they are more transparent, and infinitely more agile than all other animal bodies. Man possesses a soul, of which these pure spirits are destitute. Hence it happens that these spiritual substances are at once body and spirit without a soul. When they die (for like the human race they are subject to death), no soul remains. Like us they are exposed to diseases. Their names vary according to the places that they occupy. When they inhabit the air, they are called *sylphs;* when the water, *nymphs;* when the earth, *pigmies;* when the fire, *salamanders.*† The inhabitants of the waters are also called *undinæ,* and those of the fire *vulcani.* The sylphs approach nearest to our nature, as they live in the air like us. The sylphs, nymphs, and pigmies, sometimes obtain permission from God to make themselves visible, to converse with men, to indulge in carnal pleasures, and to produce children. But the salamanders have no relation to man. These spiritual beings are acquainted with the future, and capable of revealing it to man. They appear under the form of *ignes fatui.* We have also

---

* Modus Pharmacandi. Opera Paracelsi, i. 811.

† Liber de Nymphis, Sylphis, Pygmæis, et Salamandris, et de ceteris Spiritibus. Opera Paracelsi, ii. 388. If the reader can understand this singular book, his sagacity will be greater than mine.

the history of the fairies and the giants; and are told
how these spiritual beings are the guardians of concealed treasures; and how these sylphs, nymphs, pigmies, and salamanders, may be charmed, and their
treasures taken from them.

This division of man into body and spirit, and of
the things of nature into visible and invisible, has in
all ages of the world, been adopted by fanatics, because it enabled them to explain the history of ghosts,
and a thousand similar prejudices. Hence the distinction between soul and spirit, which is so very ancient; and hence the three following harmonies to
which the successors of Paracelsus paid a particular
attention:

*Soul,   Spirit,   Body,
Mercury,   Sulphur, Salt,
Water,   Air,   Earth.*

The will and the imagination of man acts principally
by means of the spirit. Hence the reason of the
efficacy of sorcery and magic. The *nævi materni* are
the impressions of these *vice-men*, and Paracelsus
calls them *cocomica signa*. The *sideric* body of man
draws to him, by imagination, all that surrounds him,
and particularly the stars, on which it acts like a magnet. In this manner, women with child, and during
the regular period of monthly evacuation, having a
diseased imagination, are not only capable of poisoning a mirror by their breath, but of injuring the infants in their wombs, and even also of poisoning the
moon. But it seems needless to continue this disagreeable detail of the absurd and ridiculous opinions
which Paracelsus has consigned to us in his different
tracts.

The Physiology of Paracelsus (if such a name can
be applied to his reveries) is nothing else than an application of the laws of the Cabala to the explanation
of the functions of the body. There exists, he assures
us, an intimate connexion between the sun and the

heart, the moon and the brain, Jupiter and the liver, Saturn and the spleen, Mercury and the lungs, Mars and the bile, Venus and the kidneys. In another part of his works, he informs us that the sun acts on the umbilicus and the middle parts of the abdomen, the moon on the spine, Mercury on the bowels, Venus on the organs of generation, Mars on the face, Jupiter on the head, and Saturn on the extremities. The pulse is nothing else than the measure of the temperature of the body, according to the space of the six places which are in relation to the planets. Two pulses under the sole of the feet belong to Saturn and Jupiter, two at the elbow to Mars and Venus, two in the temples to the moon and mercury. The pulse of the sun is found under the heart. The *macrocosm* has also seven pulses, which are the revolutions of the seven planets, and the irregularity or intermittence of these pulses, is represented by the eclipses. The moon and Saturn are charged in the macrocosm with thickening the water, which causes it to congeal. In like manner the moon of the microcosm, that is to say the brain, coagulates the blood. Hence *melancholy per-sons*, whom Paracelsus calls *lunatics*, have a thick blood. We ought not to say of a man that he has such and such a complexion; but that it is Mars, Venus, &c., so that a physician ought to know the planets of the microcosm, the arctic and antarctic pole, the meridian, the zodiac, the east and the west, before trying to explain the functions or cure the diseases.* This knowledge is acquired by a continual comparison of the macrocosm with the microcosm. What must have been the state of medicine at the time when Paracelsus wrote, when the propagator of

---

* Paragrani Alterius, tract. ii. Opera Paracelsi, i. 235. The reader who has the curiosity to consult this tract, will find abundance of similar stuff, which I did not think worth translating.

such opinions could be reckoned one of the greatest of its reformers ?

The system of Galen had for its principal basis the doctrine of the four elements, *fire, air, water,* and *earth.* Paracelsus neglected these elements, and multiplied the substances of the disease itself. He admits, strictly speaking, three or four elements ; namely, the *star,* the *root,* the *element,* the *sperm,* which he distinguishes by the name of the *true seed.* All these elements were originally confounded together in the *chaos* or *yliados.* The *star* is the active force which gives form to matter. The *stars* are reasonable beings addicted to sodomy and adultery, like other creatures. Each of them draws at pleasure out of the *chaos,* the plant and the metal to which it has an affinity, and gives a *sideric* form to their *root.* There are two kinds of *seed ;* the *sperm* is the vehicle of the true seed. It is engendered by speculation, by imagination, by the power of the *star.* The occult, invisible, *sideric* body produces the *true seed,* and the Adamic man secretes only the visible envelope of it. Putrefaction cannot give birth to a new body: the seed must pre-exist, and it is developed during putrefaction by the power of the stars. The generation of animals is produced by the concourse of the infinite number of seeds which detach themselves from all parts of the body. Thus the seed of the nose reproduces a nose, that of the eye the eye, and so on.

With respect to the elements themselves, Paracelsus admits occasionally their influence on the functions of the body, and the theory of diseases ; but he deduces the faculties which they possess from the *stars.* It was he that first shook the doctrine of the four elements, originally contrived by Empedocles. Alchymy had introduced another set of elements, and the alchymists maintained that salt, sulphur, and mercury, were the true elements of things. Paracelsus endeavoured to reconcile these chemical elements with his cabalistic ideas, and to show more clearly their utility in the theory of medicine. He invented a *sideric salt*, which can only be perceived by the exquisite senses of a theosophist, elevated by the abnegation of all gross sensuality to a level with pure and spiritual demons. This *salt* is the cause of the consistence of bodies, and it is it which gives them the faculty of being reproduced from their ashes.

Paracelsus imagined also a *sideric sulphur*, which being vivified by the influence of the stars, gives bodies the property of growing, and of being combustible. He admits also a *sideric mercury*, the foundation of fluidity and volatilization. The concourse of these three substances forms the body. In different parts of his works, Paracelsus says, that the *elements* are composed of these three principles. In plants he calls the salt *balsam*, the sulphur *resin* and the mercury *gotaronium*. In other passages he opposes the assertion of the Galenists, that *fire* is *dry* and *hot*, *air cold* and *moist*, *earth dry* and *cold*, *water moist* and *cold*. Each of these elements, he says, is capable of admitting all qualities, so that in reality there exists a *dry water*, a *cold fire*, &c.

I must not omit another remarkable physiological doctrine of Paracelsus, namely, that there exists in the stomach a demon called *Archæus*, who presides over the chemical operations which take place in it, separating the poisonous from the nutritive part of food, and furnishing the alimentary substances with the tincture, in consequence of which they become capable of being assimilated. This *ruler of the stomach*, who changes bread into blood, is the type of the physician, who ought to keep up a good understanding with him, and lend him his assistance. To produce a change in the humours ought never to be the object of the true physician, he should endeavour to concentrate all his operations on the stomach and the ruler who reigns in it. This Archæus to whom the name of *Nature* may also be given, produces all the changes by his own power. It is he alone who cures diseases. He has a *head* and *hands*, and is nothing else than the *spirit of life*, the *sideric body* of man, and no other spirit besides exists in the body. Each part of the body has also a peculiar stomach in which the secretions are elaborated.

There are, he informs us, five different causes of diseases. The first is the *ens astrorum*. The constellations do not immediately induce diseases, but they alter and infect the air. This is what, properly speaking constitutes the *entity of the stars*. Some constellations *sulphurize* the atmosphere, others communicate to it *arsenical, saline*, or *mercurial* qualities. The arsenical astral entities injure the blood, the mercurial the head, the saline the bones and the vessels. Orpiment occasions tumours and dropsies, and the *bitter stars* induce fever.

The second morbific cause is the *ens veneni*, which proceeds from alimentary substances: when the archeus is languid putrefaction ensues, either *localiter* or *emuncturaliter*. This last takes place when those evacuations, which ought to be expelled by the nose, the intestines, or the bladder, are retained in the body. Dissolved mercury escapes through the pores of the skin, white sulphur by the nose, arsenic by the ears, sulphur diluted with water by the eyes, salt in solution by the urine, and sulphur deliquesced by the intestines.

The third morbific cause of disease is the *ens naturale;* but Paracelsus subjects to the ens astrorum the principles which the schools are in the habit of arranging among the number of natural causes. The *ens spirituale* forms the fourth species and the *ens deale* or *Christian entity* the fifth. This last class comprehends all the immediate effects of divine predestination.

It would lead us too far if I were to point out the strange methods which he takes to discover the cause

of diseases. But his doctrine concerning *tartar* is too
important, and does our fanatic too much credit to be
omitted. It is without doubt the most useful of all
the innovations which he introduced. *Tartar* according to him, is the principle of all the maladies proceeding from the thickening of the humours, the
rigidity of the solids, or the accumulation of earthy
matter. Paracelsus thought the term *stone* not suitable to indicate that matter, because it applies only to
one species of it. Frequently the principle proceeds
from mucilage, and mucilage is tartar. He calls this
principle *tartar (tartarus)* because it burns like hell-
fire, and occasions the most dreadful diseases. As
tartar (bitartrate of potash) is deposited at the bottom
of the wine-cask, in the same way *tartar* in the living
body is deposited on the surface of the teeth. It is
deposited on the internal parts of the body when the
archæus acts with too great impetuosity and in an irregular manner, and when it separates the nutritive
principle with too much impetuosity. Then the saline
spirit unites itself to it and coagulates the earthy
principle, which is always present, but often in the
state of *materia prima* without being coagulated.

In this manner tartar, in the state of *materia prima,*
may be transmitted from father to son. But it is not
hereditary and transmittable when it has already assumed the form of gout, of renal calculus, or of obstruction. The saline spirit which gives it its form,
and causes its coagulation, is seldom pure and free
from mixture; usually it contains alum, vitriol, or
common salt; and this mixture contributes also to
modify the tartarous diseases. The tartar may be
likewise distinguished according as it comes from the
blood itself, or from foreign matters accumulated in
the humours. The great number of calculi which have
been found in every part of the body, and the obstructions, confirm the generality of this morbific cause,
to which are due most of the diseases of the liver.

When the tartarous matter is increased by certain articles of food, renal calculi are engendered, a calculous paroxysm is induced, and violent pain is occasioned. It acts as an emetic, and may even give occasion to death, when the saline spirit becomes corrosive; and when the tartar coagulated by it becomes too irritating.

Tartar, then, is always an excrementitious substance, which in many cases results from the too great activity of the digestive forces. It may make its appearance in all parts of the body, from the irregularity and the activity, too energetic or too indolent, of the archeus; and then it occasions particular accidents relative to each of the functions. Paracelsus enumerates a great number of diseases of the organs, which may be explained by that one cause; and affirms, that the profession of medicine would be infinitely more useful, if medical men would endeavour to discover the tartar before they tried to explain the affections.

Paracelsus points out, also, the means by which we can distinguish the presence of tartar in urine. For this it is necessary, not merely to inspect the urine, but to subject it to a chemical analysis. He declaims violently against the ordinary ouroscopy. He divides urine into internal and external; the internal comes from the blood, and the external announces the nature of the food and drink which has been employed. To the sediment of urine he gives the new name of *alcola*, and admits three species of it, namely, *hypostasis*, *divulsio*, and *sedimen*. The first is connected with the stomach, the second with the liver, and the third with the kidneys; and tartar predominates in all the three.

The Cabala constantly directs Paracelsus in his therapeutics and materia medica. As all terrestrial things have their image in the region of the stars, and as diseases depend also on the influence of the stars, we have nothing more to do, in order to obtain a cer-

tain cure for these diseases, than to discover, by means
of the Cabala, the harmony of the constellations.
Gold is a specific against all diseases of the *heart*, because, in the mystic scale, it is in harmony with that
viscus. The *liquor of the moon* and crystal cure the
diseases of the *brain*. The liquor *alkahest* and *cheiri*
are efficacious against those of the *liver*. When we
employ vegetable substances, we must consider their
harmony with the constellations, and their magical
harmony with the parts of the body and the diseases,
each star drawing, by a sort of magical virtue, the
plant for which it has an affinity, and imparting to it
its activity. So that plants are a kind of sublunary
stars. To discover the virtues of plants, we must study
their anatomy and cheiromancy; for the leaves are
their hands, and the lines observable on them enable
us to appreciate the virtues which they possess. Thus
the anatomy of the *chelidonium* shows us that it is a
remedy for jaundice. These are the celebrated *signatures* by means of which we deduce the virtues of
vegetables, and the medicines of analogy which they
present in relation to their form. Medicines, like women, are known by the forms which they affect. He
who calls in question this principle, accuses the
Divinity of falsehood, the infinite wisdom of whom has
contrived these external characters to bring the study
of them more upon a level with the weakness of the
human understanding. On the corolla of the euphrasia
there is a black dot; from this we may conclude that
it furnishes an excellent remedy against all diseases of
the eye. The lizard has the colour of malignant ulcers,
and of the carbuncle; this points out the efficacy
which that animal possesses as a remedy.

   These signatures were exceedingly convenient for
the fanatics, since they saved them the trouble of
studying the medical virtues of plants, but enabled
them to decide the subject *à priori*. Paracelsus acted
very considerately, when he ascribed these virtues

principally to the stars, and affirmed that the observation of favourable constellations is an indispensable condition in the employment of these medicines. "The remedies are subjected to the will of the stars, and directed by them; you ought therefore to wait till heaven is favourable, before ordering a medicine."

Paracelsus considered all the effects of plants as specifics, and the use of them as secrets. The same notions explain the eulogy which he bestowed on the *elixir of long life*, and upon all the means which he employed to prolong the term of existence. He believed that these methods, which contained the *materia prima*, served to repair the constant waste of that matter in the human body. He was acquainted, he says, with four of these arcana, to which he applied the mystic terms, *mercury of life, philosopher's stone*, &c. The *polygonum persicaria* was an infallible specific against all the effects of magic. The method of using it is, to apply it to the suffering part, and then to bury it in the earth. It draws out the malignant spirits like a magnet, and it is buried to prevent these malignant spirits from making their escape.

The reformation of Paracelsus had the great advantage of representing *chemistry* as an indispensable art in the preparation of medicines. The disgusting decoctions and useless syrups gave place to *tinctures, essences*, and *extracts*. Paracelsus says, expressly, that the true use of chemistry is to prepare medicines, and not to make gold. He takes that opportunity of declaiming against cooks and innkeepers, who drown medicines in soup, and thus destroy all their properties. He blames medical men for prescribing simples, or mixtures of simples, and affirms that the object should always be to extract the quintessence of each substance; and he describes at length the method of extracting this quintessence. But he was very little scrupulous about the substances from which this quintessence was to be extracted. The heart of a hare, the bones of a hare, the bone of the heart of a stag, mother-of-pearl, coral, and various other bodies may, he says, be used indiscriminately to furnish a quintessence capable of curing some of the most grievous diseases.

Paracelsus combats with peculiar energy the method of cure employed by the disciples of Galen, directed solely against the predominating humours, and the elementary qualities. He blames them for attempting to correct the action of their medicines, by the addition of useless ingredients. Fire and chemistry, he affirmed, are the sole correctives. It was Paracelsus that first introduced *tin* as a remedy for worms, though his mode of employing it was not good.

I have been thus particular in pointing out the philosophical and medical opinions of Paracelsus, because they were productive of such important consequences, by setting medical men free from the slavish deference which they had been accustomed to pay to the dogmas of Galen and Avicenna. But it was the high rank to which he raised chemistry, by making a knowledge of it indispensable to all medical men; and by insisting that the great importance of chemistry did not consist in the formation of gold, but in the preparation of medicines, that rendered the era of Paracelsus so important in the history of chemistry; for after his time the art of chemistry was cultivated by medical men in general—it became a necessary part of their education, and began to be taught in colleges and medical schools. The object of chemistry came to be, not to discover the philosopher's stone, but to prepare medicines; and a great number of new medicines, both from the mineral and vegetable kingdom—some of more, some of less, consequence, soon issued from the laboratories of the chemical physicians.

There can be little doubt that many chemical preparations were either first introduced into medicine by Paracelsus, or at least were first openly prescribed by him : though from the nature of his writings, and the secrecy in which he endeavoured to keep his most valuable remedies, it is not easy to point out what these remedies were. Mercury is said to have been employed in medicine by Basil Valentine; but it was Paracelsus who first used it openly as a cure for the venereal disease, and who drew general attention to it by his encomiums on its medical virtues, and by the eclat of the cures which he performed by means of it, after all the Galenical prescriptions of the schools had been tried in vain.

He ascertained that alum contains, united to an acid, not a metallic oxide, but an earth.   He mentions metallic arsenic ; but there is some reason for believing that this metal was known to Geber and the Arabian physicians.   Zinc is mentioned by him, and likewise bismuth, as substances not truly metallic, but approaching to metals in their properties : for malleability and ductility were considered by him as essential to the metals. * I cannot be sure of any other chemical fact which appears in Paracelsus, and which was not known before his time.   The use of sal ammoniac in subliming several metallic calces, was familiar to him, but it had long ago been explained by Geber.   It is clear also that Geber was acquainted with aqua regia, and that he employed it to dissolve gold.   Paracelsus's reputation as a chemist, therefore, depends not upon

* Philosophiæ, tract. iv. De Mineralibus. Opera Paracelsi, ii. 282. " Quando ergo hoc modo metalla fiunt et producuntur, dum scilicet verus metallicus fluxus et ductilitas aufertur et in septem metalla distribuitur; residentia quædam manet in Ares, instar fœtûm trium primorum.   Ex hac nescitur zinetum, quod et metallum est et non est.   Sic et bisemutum et huic similia alia partim fluida, partim ductilia sunt—Zinetum maxima ex parte spuria soboles est ex cupro et bisemutum de stanno.   Ex hisce duobus omnium plurimæ fæces et remanentiæ in Ares fiunt."

any discoveries which he actually made, but upon the great importance which he attached to the knowledge of it, and to his making an acquaintance with chemistry an indispensable requisite of a medical education.

Paracelsus, as the founder of a new system of medicine, the object of which was to draw chemistry out of that state of obscurity and degradation into which it had been plunged, and to give it the charge of the preparation of medicine, and presiding over the whole healing art, deserved a particular notice; and I have even endeavoured, at some length, to lay his system of opinions, absurd as it is, before the reader. But the same attention is not due to the herd of followers who adopted his absurdities, and even carried them, if possible, still further than their master: at the same time there are one or two particulars connected with the Paracelsian sect which it would be improper to omit.

The most celebrated of his followers was Leonhard Thurneysser-zum-Thurn, who was born in 1530, at Basle, where his father was a goldsmith. His life, like that of his master, was checkered with very extraordinary vicissitudes. In 1560 he was sent to Scotland to examine the lead-mines in that country. In 1558 he commenced miner and sulphur extractor at Tarenz on the Inn, and was so successful, that he acquired a great reputation. He had turned his attention to medicine on the Paracelsian plan, and in 1568 made himself distinguished by several important cures which he performed. In 1570 he published his Quinta Essentia, with wooden cuts, in Munster; from thence he went to Frankfort on the Oder, and published his Piso, a work which treats of *waters, rivers,* and *springs*. John George, Elector of Brandenburg, was at that time in Frankfort, and was informed that the treatise of Thurneysser pointed out the existence of a great deal of riches in the March of Brandenburg, till that time unknown. His courtiers, who were anxious to establish mines in their possessions, united in re-commending the author. He was consulted about a disease under which the wife of the elector was labour-ing, and having performed a cure, he was immediately named physician to this prince.

He turned this situation to the best account. He sold Spanish white, and other cosmetics, to the ladies of the court; and instead of the disgusting decoctions of the Galenists, he administered the remedies of Paracelsus under the pompous titles of *tincture of gold, magistery of the sun, potable gold,* &c. By these methods he succeeded in amassing a prodigious fortune, but was not fortunate enough to be able to keep it. Gaspard Hoffmann, professor at Frankfort, a well-informed and enlightened man, published a treatise, the object of which was to expose the extra-vagant pretensions and ridiculous ignorance of Thur-neysser. This book drew the attention of the cour-tiers, and opened the eyes of the elector. Thur-neysser lost much of his reputation; and the methods by which he attempted to bolster himself up, served only to sink him still lower in the estimation of men of sense. Among other things, he gave out that he was the possessor of a devil, which he carried about with him in a bottle. This pretended devil was no-thing else than a scorpion, preserved in a phial of oil. The trick was discovered, and the usual consequences followed. He lost a process with his wife, from whom he was separated; this deprived him of the greatest part of his fortune. In 1584 he fled to Italy, where he occupied himself with the transmutation of metals, and he died at Cologne in 1595.

Thurneysser extols Paracelsus as the only true phy-sician that ever existed. His Quintessence is written in verse. In the first book *The Secret* is the speaker. He is represented with a padlock in his mouth, a key in his hand, and seated on a coffer in a chamber, the windows of which are shut. This personage teaches that all things are composed of salt, sulphur, and mercury, or of earth, air, and water; and consequently that *fire* is excluded from the number of the elements. We must search for the secret in the *Bible*, and then in the *stars* and the *spirits*. In the second book, *Alchymy* is the speaker. She points out the mode of performing the processes; and says that to endeavour to fix volatile substances, is the same thing as to endeavour to trace white letters on a wall with a piece of charcoal. She prohibits all long processes, because God created the world in six days.

His method of judging of the diseases from the urine of the patient deserves to be mentioned. He distilled the urine, and fixed to the receiver a tube furnished with a scale, the degrees of which consisted of all the parts of the body. The phenomena which he observed during the distillation of the urine, enabled him to draw inferences respecting the state of all these different organs.

I pass over Bodenstein, Taxites, and Dorn, who distinguished themselves as partisans of Paracelsus. Dorn derived the whole of chemistry from the first chapter of Genesis, the words of which he explained in an alchymistical sense. These words in particular, " And God made the firmament, and divided the waters which were under the firmament from the waters which were above the firmament," appeared to him to be an account of the *great work*. Severinus, physician to the King of Denmark, and canon of Roskild, was also a celebrated partisan of Paracelsus; but his writings do not show either that knowledge or stretch of thought which would enable us to account for the reputation which he acquired and enjoyed.

There were very few partisans of Paracelsus out of Germany. The most celebrated of his followers among the French, was Joseph du Chesne, better known by the name of Quercitanus, who was physician to Henry IV. He was a native of Gascony, and drew many enemies upon himself by his arrogant and over-bearing conduct. He pretended to be acquainted with the method of making gold. He was a thorough-going Paracelsian. He affirmed that diseases, like plants, spring from seeds. The word alchymy, according to him, is composed of the two Greek words ἄλς (salt) and χημεια, because the *great secret* is concealed in salt. All bodies are composed of three principles, as God is of three substances. These principles are contained in saltpetre, the salts of sulphur solid and volatile, and the volatile mercurial salt. He who possesses *sal generalis* may easily produce philosophical gold, and draw potable gold from the three kingdoms of nature. To prove the possibility of this transmutation, he cites an experiment very often repeated after him, and which some theologians have even employed as analagous to the resurrection of the dead; namely, the faculty which plants have of being produced from their ashes. His materia medica is founded on the *signatures* of plants, which he carries so far as to assert that male plants are more suitable to men, and female plants to women. Sulphuric acid, he says, has a magnetic virtue, in consequence of which it is capable of curing the epilepsy. He recommends the *magisterium cranii humani* as an excellent medicine, and boasts much of the virtues of antimony.

Du Chesne was opposed by Riolanus, who attacked chemical remedies with much bitterness. The medical faculty of Paris took up the cause of the Galenists with much zeal, and prohibited their fellows and licentiates from using any chemical medicines whatever. He had to sustain a dispute with Aubert relative to the origin and the transmutation of metals. Fenot came to the assistance of Aubert, and affirmed that gold possesses no medical properties whatever, that *crabs' eyes* are of no use when administered in intermittents, and that the laudanum of Paracelsus (being an opiate) is in reality hurtful instead of being beneficial.

The decree of the medical faculty of Paris which placed antimony among the poisons, and which occasioned that of the Parliament of Paris, was composed by Simon Pietre, the elder, a man of great erudition and the most unimpeachable probity. Had it been literally obeyed it would have occasioned very violent proceedings; because chemical remedies, as they act more promptly and with greater energy, were getting daily into more general use. In 1603 the celebrated Theodore Turquet de Mayenne was prosecuted, because, in spite of the prohibition, he had sold antimonial preparations. The decree of the faculty against him exhibits a remarkable proof of the bigotry and intolerance of the times.* However Turquet does not seem to have been molested notwithstanding this decree. He ceased indeed to be professor of chemistry, but continued to practise medicine as formerly; and two members of the faculty, Seguin and Akakia, even wrote an apology for him. At last he went to England, whither he had been invited, to accept an honourable appointment.

* It was as follows: "Collegium medicorum in Academia Parisiensi legitime congregatum, audita renunciatione sensorum, quibus demandata erat provincia examinandi apologiam sub nomine Mayerni Turqueti editam, ipsam unanimi consensu damnat, tanquam famosum libellum, mendacibus conviciis et impudentibus calumniis refertum, quæ nonnisi ab homine imperito, impudenti, temulento et furioso profiteri potuerunt. Ipsum Turquetum indignum judicat, qui usquam medicinam faciat, propter temeritatem, impudentiam et veræ medicinæ ignorantiam. Omnes vero medicos, qui ubique gentium et locorum medicinam exercent, hortatur ut ipsum Turquetum similiaque hominum et opinionum portenta, a se suisque finibus arceant et in Hippocratis ac Galeni doctrina constantes permaneant: et prohibuit ne quis ex hoc medicorum Parisiensium ordine cum Turqueto eique similibus medica consilia ineat. Qui secus fecerit, scholæ ornamentis et academiæ privilegiis privabitur, et de regentium numero expungetur.—Datum Lutetiæ in scholis superioribus, die 5 Decembris, anno salutis, 1603."

The mystical doctrines of Paracelsus are supposed to have given origin to the sect of Rosecrucians, concerning which so much has been written and so little certain is known. It is not at all unlikely that the greatest part, if not the whole that has been stated about the antiquity, and extent, and importance of this sect, is mere fiction, and that the origin of the whole was nothing else than a ludicrous performance of Valentine Andreæ, an ecclesiastic of Calwe, in the country of Wirtemburg, a man of much learning, genius, and philanthropy. From his life, written by himself, and preserved in the library of Wolfenbuttel, we learn that in the year 1603 he drew up the celebrated Noce Chimique of Christian Rosenkreuz, in order to counteract the alchymistical and the theosophistical dogmas so common at that period. He was unable to restrain his risible faculties when he saw this *ludibrium juvenilis ingenii* adopted as a true history, while he meant it merely as a satire. It is believed that the Fama Fraternitatis is a production of this ecclesiastic, and that he published it in order to correct the chemists and enthusiasts of the time. He himself was called Andreæ, Knight of the Rose-cross (*rosæ crucis*) because he had engraven on his seal a cross with four roses.

It is true that Andreæ instituted, in 1620, a *fraternitas christiana*, but with quite other views than those which are supposed to have actuated the Rosecrucians. His object was to correct the religious opinions of the times, and to separate Christian theology from scholastic controversies, with which it had been unhappily intermixed. He himself, in different parts of his writings, distinguishes carefully between the Rosecrucians and his own society, and amuses himself with the credulity of the German theosophists, who adopted so readily his fiction for a series of truths. It would appear, therefore, that this secret order of Rosecrucians, notwithstanding the brilliant origin assigned to it, really owes its birth to the pleasantry of a clergyman of Wirtemburg, who endeavoured by that means to set bounds to the chimeras of theosophy, but who unfortunately only increased still more the adherents of this absurd science.

A crowd of enthusiasts found it too advantageous to propagate the principles of the *rosa crux* not to endeavour to unite them into a sect.   Valentine Weigel, a fanatical preacher at Tschoppau, near Chemnitz, left at his death a prodigious number of followers, who were already Rosecrucians, without bearing the name. Egidius Gutmann, of Suabia, was equally a Rosecrucian, without bearing the name; he condemned all pagan medicines, and affirmed that he possessed the universal remedy which ennobles man, cures all diseases, and gives man the power of fabricating gold. " To fly in the air, to transmute metals, and to know all the sciences," says he, " nothing more is requisite than faith."

Oswald Crollius, of Hesse, must also take his station in this honourable fraternity of enthusiasts.   He was physician to the Prince of Anhalt, and afterwards a counsellor of the Emperor Rodolphus II.   The introduction to his Basilica Chymica, contains a short but exact epitome of the opinions of Paracelsus.   It is not worth while to give the reader a notion of his own opinions, which are quite as absurd and unintelligible as those of Paracelsus and his followers.   As a preparer of chemical medicines he deserves more credit; *antimonium diaphoreticum* was a favourite preparation of his, and so was sulphate of potash, which was known at the time by the name of *specificum purgans Paracelsi*: he knew chloride of silver well, and first gave it the name of *luna cornea*, or *horn silver*: fulminating gold was known to him, and called by him *aurum volatile.*

This is the place to mention Andrew Libavius, of Halle, in Saxony, where he was a physician, and a professor in the gymnasium of Coburg, who was one of the most successful opponents of the school of Paracelsus, and whose writings do him much credit. As a chemist, he deserves perhaps to occupy a higher rank than any of his contemporaries : he was, it is true, a believer in the possibility of transmuting metals, and boasted of the wonderful powers of *aurum pota bile;* but he always distinguishes between rational alchymy and the *mental* alchymy of Paracelsus. He separated, with great care, *chemistry* from the reveries of the theosophists, and stands at the head of those who opposed most successfully the progress of superstition and fanaticism, which was making such an overwhelming progress in his time. His writings are very numerous and various, and were collected and published at Frankfort, in 1615, in three folio volumes, under the title of " Opera omnia Medico-chymica." Libavius himself died in 1616. It would occupy more space than we have room for, to attempt an abstract of his very multifarious works. A few observations will be sufficient : he wrote no fewer than five different tracts to expose the quackery of George Amwald, who had boasted that he was in possession of a panacea, by means of which he was enabled to perform the most wonderful cures, and which he was in the habit of selling to his patients at an enormous price ; Li bavius showed that this boasted panacea was nothing else than *cinnabar*, which neither possessed the virtues ascribed to it by Amwald, nor deserved to be purchased at so high a price. He entered also into a controversy with Crollius, and exposed his fanatical and absurd opinions. He engaged likewise in a dispute with Henning Scheunemann, a physician in Bamberg, who was a Rosecrucian, and, like the rest of his brethren, profoundly ignorant not merely of all science, but even of philology. The expressions of Scheunemann are so obscure, that we learn more of his opinions from Libavius than from his own writings. He divides the internal nature of man into seven different degrees, from the seven changes it undergoes: these are, combustion, sublimation, dissolution, putrefaction, distillation, coagulation, and tincture. He gives us likewise an account of ten modifications which the three elements undergo; but as they are quite unintelligible, it is not worth while to state them. Libavius had the patience to analyze and expose all these gallimatias.

Libavius's system of chemistry, entitled " Alchymia è dispersis passim optimorum auctorum, veterum et recentiorum exemplis potissimum, tum etiam preceptis quibusdam operose collecta, adhibitisque ratione et experientia quanta potuit esse methodo accurate explicata et in integrum corpus redacta. Accesserunt tractati nonnulli physici chymici item methodistici." Frankfort, 1595, folio, 1597, 4to. — is really an excellent book, considering the period in which it was written, and deserves the attention of every person who is interested in the history of chemistry. I shall notice some of the most remarkable chemical facts which occur in Libavius, and which I have not observed in any preceding writer; who the actual discoverer of these facts really was, it is impossible to say, in consequence of the secrecy which at that time was affected, and the obscure terms in which chemical facts are in general stated.

He was aware that the fumes of sulphur have the property of blackening white lead. He was in the habit of purifying cinnabar by means of arsenic and oxide of lead. He knew the method of giving glass a red colour by means of gold or its oxide, and was aware of the method of making artificial gems, such as ruby, topaz, hyacinth, garnet, balass, by tinging glass by means of metallic oxides. He points out fluor spar as an excellent flux for various metals and their oxides. He knew that when metals were fused along with alkaline bodies, a certain portion of them was converted into slags, and this portion he endeavoured to recover by the addition of iron filings. He was aware of the mode of acidifying sulphur by means of nitric acid. He knew that camphor is soluble in nitric acid, and forms with it a kind of oil. Of the perchloride of tin he was undoubtedly the discoverer, as it has continued ever since his time to pass by his name; namely, *fuming liquor of Libavius.* He was aware, that alcohol or spirits could be obtained by distilling the fermented juice of a great variety of sweet fruits. He procured sulphuric acid by the distillation of alum and sulphate of iron, as Geber had done long before his time; but he determined the nature of the acid with more care than had been done, and showed, that it was the same as that obtained by the combustion of sulphur along with saltpetre. To him, therefore, in some measure, are we indebted for the process of preparing sulphuric acid which is at present practised by manufacturers.

Libavius found a successor in Angelus Sala, of Vicenza, physician to the Duke of Mecklenburg-Schwerin, worthy of his enlightened views and indefatigable exertions to oppose the torrent of fanaticism which threatened to overwhelm all Europe. Sala was still more addicted to chemical remedies than Libavius himself; but he had abjured a multitude of prejudices which had distinguished the school of Paracelsus. He discarded *aurum potabile,* and considered fulminating gold as the only remedy of that metal that deserved to be prescribed by medical men. He treated the notion of the existence of a universal remedy with contempt. He described sulphuret of gold and glass of antimony with a good deal of precision. He recommended sulphuric acid as an excellent remedy, and showed that it might be formed indifferently from sulphur, or by distilling blue vitriol or green vitriol. He affirmed, that the essential salts obtained from plants had not the same virtues as the plants from which they are obtained. He showed that sal ammoniac is a compound of muriatic acid and ammonia. To him, therefore, we are indebted for the first accurate mention of ammonia. It could not but have been noticed before by chemists, as it is procured with so much ease by the distillation of animal substances; but Sala is the first person who seems to have examined it with attention, and to have recognised its peculiar properties, and the readiness with which it saturates the different acids. He showed that iron has the property of precipitating copper from acid solutions : he pointed out also various precipitations of metals by other metals. He seems to have been acquainted with calomel, and to have been aware of at least some of its medical properties. He says, that fulminating gold loses its fulminating property when mixed with its own weight of sulphur, and the sulphur is burnt off it. Many other curious chemical facts occur in his writings, which it would be too tedious to particularize here. His works were collected and published in a quarto volume at Frankfort, in 1647, under the title of " Opera Medico-chymica, quæ extant omnia." There was another edition in the same place in 1682, and an edition was published at Rome in 1650.

CHAPTER V.

OF VAN HELMONT AND THE IATRO-CHEMISTS.

PARACELSUS first raised the dignity of chemistry, by pointing out the necessity of it for medical men, and by showing the superiority of chemical medicines over the disgusting decoctions of the Galenists. Libavius and Angelus Sala had carefully separated chemistry from the fanatical opinions of the followers of Paracelsus and the Rosecrucians. But matters were not doomed to remain in this state. Chemistry underwent a new revolution at this period, which shook the Spagirical system to its foundation; substituted other principles, and gave to medicine an aspect entirely new. This revolution was in a great measure due to the labours of Van Helmont.

John Baptist Van Helmont was a gentleman of Brabant, and Lord of Merode, of Royenboch, of Oorschot, and of Pellines. He was born in Brussels in 1577, and studied scholastic philosophy in Louvain till the age of seventeen. After having finished his *humanity* (as it was termed), he ought, according to the usage of the place, to have taken his degree of master of arts; but, having reflected on the futility of these ceremonies, he resolved never to solicit any academical honour. He next associated himself to the Jesuits, who then delivered courses of philosophy at Louvain, to the great displeasure of the professors of

N 2

that city. One of the most celebrated of the Jesuits, Martin del Rio, even taught him magic. But Van Helmont was disappointed in his expectations: instead of that true wisdom which he hoped to acquire, he met with nothing but scholastic dialectics, with all its usual subtilties. He was no better satisfied with the doctrines of the Stoics, who taught him his own weakness and misery.

At last the works of Thomas à Kempis, and John Taulerus fell into his hands. These sacred books of mysticism attracted his attention : he thought that he perceived that wisdom is the gift of the Supreme Being; that it must be obtained by prayer ; and that we must renounce our own will, if we wish to participate in the influence of the divine grace. From this moment he imitated Jesus Christ, in his humility. He abandoned all his property to his sister, renouncing the privileges of his birth, and laying aside the rank which he had hitherto occupied in society. It was not long before he reaped the fruit of these abnegations. A genius appeared to him in all the important circumstances of his life. In the year 1633 his own soul appeared to him under the figure of a resplendent crystal.

The desire which he had of imitating in every respect the conduct of Christ, suggested to him the idea of practising medicine as a work of charity and benevolence. He began, as was then the custom of the time, by studying the art of healing in the writings of the ancients. He read the works of Hippocrates and Galen with avidity; and made himself so well acquainted with their opinions, that he astonished all the medical men by the profundity of his knowledge. But as his taste for mysticism was insatiable, he soon became disgusted with the writings of the Greeks; an accident led him to abandon them for ever. Happening to take up the glove of a young girl afflicted with the *itch*, he caught that disagreeable disease. The Galenists whom he consulted, attributed it to the combustion of the bile, and the saline state of the phlegm. They prescribed a course of purgatives which weakened him considerably, without effecting a cure. This circumstance disgusted him with the system of the humorists, and led him to form the resolution of reforming medicine, as Paracelsus had done. The works of this reformer, which he read with attention, awakened in him a spirit of reformation, but did not satisfy him; because his knowledge, being much greater than that of Paracelsus, he could not avoid despising the disgusting egotism, and the ridiculous ignorance of that fanatic. Though he had already refused a canonicate, he took the degree of doctor of medicine, in 1599, and afterwards travelled through the greatest part of France and Italy; and he assures us, that during his travels, he performed a great number of cures. On his return, he married a rich Brabantine lady, by whom he had several children; among others a son, afterwards celebrated under the name of Francis Mercurius, who edited his father's works, and who went a good deal further than his father had done, in all the branches of theosophy. Van Helmont passed the rest of his life on his estate at Vilvorde, almost constantly occupied with the processes of his laboratory. He died in the year 1644, on the 13th of December, at six o'clock in the evening, after having nearly reached the age of sixty-seven years.

The system of Van Helmont has for its basis the opinions of the spiritualists. He arranged even the influence of evil genii, the efforts of sorcerers, and the power of magicians among the causes which produce diseases. The archeus of Paracelsus constituted one of the capital points of his theory; but he ascribed to it a more substantial nature than Paracelsus had done. This archeus is independent of the elements; it has no form; for form constitutes the object of generation, or of production. These ideas are obviously borrowed from the ancients. The *form* of Aristotle is not the μορφη, but the ἐνεργεια (*the power of acting*) which matter does not possess.

The archeus draws all the corpuscles of matter to the aid of *fermentation*. There are, properly speaking, only two causes of things; the cause *ex qua*, and the cause *per quam*. The first of these causes is *water*. Van Helmont considered water as the true principle of every thing which exists; and he brought forward very specious arguments in favour of his opinion, drawn both from the animal and vegetable kingdom. The reader will find his arguments on the subject, in his treatise entitled " Complexionum atque Mistionum elementalium Figmentum."* The only one of his experiments that, in the present state of our knowledge, possesses much plausibility, is the follow‧ ing : He took a large earthen vessel, and put into it 200 lbs. of earth, previously dried in an oven. This earth he moistened with rain-water, and planted in it a willow which weighed five pounds. After an interval of five years, he pulled up his willow and found that its weight amounted to 169 pounds, and about three ounces. During these five years, the earth in the pot was duly watered with rain or distilled water. To prevent the earth in which the willow grew from being mixed with new earth blown upon it by the winds, the pot was covered with tin plate, pierced with a great number of holes to admit the air freely. The leaves which fell every autumn during the vegetation of the willow in the pot, were not reckoned in the 169 lbs. 3 oz. The earth in the pot being again dried in the oven, was found to have lost about two ounces of its original weight. Thus 164 lbs. of wood, bark,

* J. B. Van Helmont, Opera Omnia, p. 100. The edition which I quote from was printed at Frankfort, in 1682, at the expense of John Justus Erythropilus, in a very thick quarto volume.

roots, &c., were produced from water alone.* This, and several other experiments which it is needless to state, satisfied him that all vegetable substances are produced from water alone. He takes it for granted that fish live (ultimately at least) on water alone; but they contain almost all the peculiar animal substances that exist in the animal kingdom. Hence he concludes that animal substances are derived also from pure water.† His reasoning with respect to sulphur, glass, stone, metals, &c., all of which he thinks may ultimately be resolved into water, is not so satisfactory.

Water produces elementary earth, or pure quartz; but this elementary earth does not enter into the composition of organic bodies. Van Helmont excludes *fire* from the number of elements, because it is not a substance, nor even the essential form of a substance. The matter of fire is compound, and differs entirely from the matter of light. Water gives origin also to the three chemical principles, salt, sulphur, and mercury, which cannot be considered as elements or active principles. I do not see clearly how he gets rid of *air;* for he says, that though water may be elevated in the form of vapour, yet that these vapours are no more air than the dust of marble is water.

According to Van Helmont, a particular disposition of matter, or a particular mixture of that matter is not necessary for the formation of a body. The archeus, by its sole power, draws all bodies from water, when the *ferment* exists. This *ferment*, in its quality of a mean which determines the action of the archeus, is not a formal being; it can neither be called a *substance*, nor an *accident*. It pre-exists in the seed which is developed by it, and which contains in itself a second ferment of the seed, the product of the first. The ferment exhales an odour, which attracts the generating spirit of the archeus. This spirit consists in an

* Van Helmont, Opera Omnia, p. 104. † Ibid., p. 105.

aura vitalis, and it creates the bodies of nature in its own image, after its own *idea*. It is the true foundation of life, and of all the functions of organized bodies; it disappears only at the instant of death to produce a new creation of the body, which enters then, for the second time, into fermentation. The seed, then, is not indispensable to enable an animal to propagate its species; it is merely necessary that the archeus should act upon a suitable ferment. Animals produced in this manner are as perfect as those which spring from eggs.

When water, as an element, ferments, it develops a vapour, to which Van Helmont gave the name of *gas*, and which he endeavours to distinguish from *air*. This gas contains the chemical principles of the body from which it escapes in an aerial form by the impulse of the archeus. It is a substance intermediate between spirit and matter, the principle of action of life, and of generation of all bodies; for its production is the first result of the action of the vital spirit on the torpid ferment, and it may be compared to the *chaos* of the ancients.

The term *gas*, now in common use among chemists, and applied by them to all elastic fluids which differ in their properties from common air, was first employed by Van Helmont: and it is evident, from different parts of his writings, that he was aware that different species of gas exist. His *gas sylvestre* was evidently our *carbonic acid gas*, for he says, that it is evolved during the fermentation of wine and beer; that it is formed when charcoal is burnt in air; and that it exists in the Grotto del Cane. He was aware that this gas extinguishes a lighted candle. But he says that the gases from dung, and those formed in the large intestines, when passed through a candle, catch fire, and exhibit a variety of colours, like the rainbow.* To

* De Flatibus, sect. 49. Opera Van Helmont, p. 405.

these combustible gases he gave the names of *gas pingue, gas siccum, gas fuliginosum,* or *endimicum.*

Sal ammoniac, he says, may be distilled alone, without danger, and so may aqua fortis *(aqua chrysulca),* but if they be mixed together so much gas sylvestre is produced, that the vessels employed, however strong, will burst asunder, unless an opening be left for the escape of this gas.* In the same way cream of tartar cannot be distilled in close vessels without breaking them in pieces, an opening must be left for the escape of the *gas sylvestre,* which is generated in such abundance.† He says, also, that when carbonate of lime is dissolved in distilled vinegar, or silver in nitric acid, abundance of gas sylvestre is extricated. From these, and many other passages which might be quoted, it is evident that Van Helmont was aware of the evolution of gas during the solution of carbonates and metals in acids, and during the distillation of various animal and vegetable substances, that he had anticipated the experiments made so many years after by Dr. Hales, and for which that philosopher got so much credit. But it would be going too far to say, as some have done, that Van Helmont knew accurately the differences which characterize the different gases which he produced, or indeed that he distinguished accurately between them. For it is evident, from the passages quoted and from many others which occur in his treatise, De Flatibus, that carbonic acid, protoxide of azote, and deutoxide of azote, and probably also muriatic acid gas were all considered by him as constituting one and the same gas. How, indeed, could he distinguish between different gases when he was not acquainted with the method of collecting them, or of determining their properties? These observations of Van Helmont, then, though they do him much credit, and

* Ibid., p. 408.   † Ibid., p. 409.

show how far his chemical knowledge was superior to that of the age in which he lived, take nothing from the merit or the credit of those illustrious chemists who, in the latter half of the eighteenth century, devoted themselves to the investigation of this part of chemistry, at that time attended with much difficulty, but intimately connected with the subsequent progress which the science has made.

Van Helmont was aware, also, that the bulk of air is diminished when bodies are burnt in it. He considered respiration to be necessary in this way : the air was drawn into the blood by the pulmonary arteries and veins, and occasioned a fermentation in it requisite for the continuance of life.

Gas, according to Van Helmont, has an affinity with the principle of the movement of the stars, to which he gave the name of *blas*. It had, he supposed, much influence on all sublunary bodies. He admitted in the ferment which gives birth to plants, a substance which, after the example of Paracelsus, he called *pessas*, and to the metallic ferment he gave the name of *bur*.*

The archeus of Van Helmont, like that of Paracelsus, has its seat in the stomach. It is the same thing as the sentient soul. This notion of the nature and seat of the archeus was founded on the following experiment : He swallowed a quantity of *aconitum (henbane)*. In two hours he experienced the most disagreeable sensation in his stomach. His feeling and understanding seemed to be concentrated in that organ, for he had no longer the free use of his mental faculties. This feeling induced him to place the seat of understanding in the stomach, of volition in the

* In his Magnum Oportet, sect. 39, p. 151, he gives an account of the origin of metals in the earth, and in that section there is a description of *bur*, which those who are anxious to understand the ideas of the author on this subject may con-- sult

heart, and of memory in the brain. The faculty of desire, to which the ancients had assigned the liver as its organ, he placed in the spleen. What confirmed him still more in the idea that the stomach is the seat of the soul, is the fact, that life sometimes continues after the destruction of the brain, but never, he alleges, after that of the stomach. The sentient soul acts constantly by means of the *vital spirits*, which are of a resplendent nature, and the nerves serve merely to moisten these spirits which constitute the mediums of sensation. By virtue of the archeus man is much nearer to the realm of spirits and the father of all the genii, than to the world. He thinks that Paracelsus's constant comparison of the human body with the world is absurd. Yet Van Helmont, at least in his youth, was a believer in magnetism, which he employed as a method of explaining the effect of sympathy.

The archeus exercises the greatest influence on digestion, and he has chiefly the stomach and spleen under his superintendence. These two organs form a duumvirate in the body; for the stomach cannot act alone and without the concurrence of the spleen. Digestion is produced by means of an acid liquor, which dissolves the food, under the superintendence of the archeus. Van Helmont assures us that he had himself tasted this acid liquor in the stomach of birds. Heat, strictly speaking, does not favour digestion; for we see no increase of the digestive powers during the most ardent fever. Nor are the powers of digestion wanting in fishes, although they want the animal heat which is requisite for mammiferous animals. Certain birds even digest fragments of glass, which, certainly, simple heat would not enable them to do. The pylorus is, in some measure, the director of digestion. It acts by a peculiar and immaterial power, in virtue of a *blas*, and not as a muscle. It opens and shuts the stomach according to the orders

of the archeus.   It is in it, therefore, that the causes
of derangement of digestion must be sought for.

The duumvirate just spoken of is the cause of
natural sleep, which does not belong to the soul,
as far as it resides in the stomach.   Sleep is a natural
action, and one of the first vital actions.   Hence the
reason why the embryo sleeps without ceasing.   At
any rate it is not true that sleep is owing to vapours
which mount to the brain.   During sleep the soul
is naturally occupied, and it is then that the deity
approaches most intimately to man.   Accordingly,
Van Helmont informs us, that he received in dreams
the revelation of several secrets, which he could not
have learnt otherwise.

The duumvirate operates the *first* digestion, of
which, Van Helmont enumerates six different species.
When the acid, which is prepared for digestion,
passes into the duodenum it is neutralized by the
bile of the gall-bladder.   This constitutes the second
digestion.   To the bile of the gall-bladder, Van Helmont gave the name of *fel*, and he carefully distinguished it from the biliary principle in the mass
of the blood.   This last he called *bile*.   The *fel* is
not an excrementitious matter, but a humour necessary to life, a true vital balsam.   Van Helmont
endeavoured to show by various experiments that it
is not *bitter*.

The *third* digestion takes place in the vessels of
the mesentery, into which the gall-bladder sends the
prepared fluid.   The *fourth* digestion is operated in
the heart, where the red blood becomes more yellow
and more volatile by the addition of the vital spirits.
This is owing to the passage of the vital spirit from
the posterior to the anterior ventricle, through the
pores of the septum.   At the same time the pulse
is produced, which of itself develops heat; but does
not regulate it in any manner, as the ancients pretended that it did.   The *fifth* digestion consists in the

conversion of the arterial blood into vital spirit. It takes place principally in the brain, but is produced also throughout all the body. The *sixth* digestion consists in the elaboration of the nutritive principle in each member, where the archeus prepares its own nourishment by means of the vital spirits. Thus, there are six digestions: the number seven has been chosen by nature for a state of repose.

From the preceding sketch of the physiology of Van Helmont, it is evident that he paid little or no regard to the structure of the parts in explaining the functions. In his pathology we find the same passion for spiritualism. He admitted, indeed, the importance of anatomy, but he regretted that the pathological part of that science had been so little cultivated. As the archeus is the foundation of life and of all the functions, it is plain that the diseases can neither be derived from the four cardinal humours, nor from the disposition or the action of opposite things; the proximate cause of diseases must be sought for in the sufferings, the anger, the fear, and the other affections of the archeus, and their remote cause may be considered as the ideal seed of the archeus. Disease, in his opinion, is not a negative state or a mere absence of health, it is a substantial and active thing as well as a state of health. Most of the diseases which attack certain parts or members of the body result from an error in the archeus, who sends his ferment from the stomach in which he resides into the other parts of the body. Van Helmont explained in this way not only the epilepsy and madness, but likewise the *gout*, which does not proceed from a flux, and has not its seat in the limb in which the pain resides, but is always owing to an error in the vital spirit. It is true that the character of the gout acts upon the semen in which the vital spirit principally manifests its action, and that in this way diseases are propagated in the act of generation; but if, during life, instead of altering the semen it is carried to the liquid of the articulations, this is a proof of the prudence of nature, which lavishes all her cares on the preservation of the species, and loves better to alter the humours of the articulations than the semen itself. The gout acidifies the liquors of the articulations, which is then coagulated by the acids. The duumvirate is the cause of apoplexy, vertigo, and particularly of a species of asthma, which Van Helmont calls *caducus pulmonalis.* Pleurisy is produced in a similar way. The archeus, in a movement of rage, sends acrid acids to the lungs, which occasion an inflammation. Dropsy is also owing to the anger of the archeus, who prevents the secretions of the kidneys from going on in the usual way.

Of all the diseases, fever appeared to him most conformable to his notions of the unlimited power of the archeus. The causes of fever are all much more proper to offend the archeus, than to alter the structure of parts and the mixture of humours. The cold fit is owing to a state of fear and consternation, into which the archeus is thrown, and the hot stage results from his disordered movements. All fevers have their peculiar seat in the duumvirate.

Van Helmont was in general much more successful in refuting the scholastic opinions by which the practice of medicine was regulated in his time, than in establishing his own. We are struck with the force of his arguments against the Galenical doctrine of fever, and against the influence of the cardinal humours on the different kinds of fever. He refuted no less vehemently the idea of the putridity of the blood, while that liquid circulates in the vessels. Perhaps he carried the opposite doctrine too far; but his opinions have had a good effect upon subsequent medical theory, and medical men learned from them to make less use of the term putridity. The phrase *mixture of humours,* not

more intelligible, however, came to be substituted
for it.

Van Helmont's theory of urinary calculi deserves
peculiar attention, because it exhibits the germ of a
more rational explanation of these concretions than
had been previously attempted by physiologists. Van
Helmont was aware that Paracelsus, who ascribed
these concretions to tartar, had formed an idea of
their nature, which a careful chemical analysis would
immediately refute. He satisfied himself that urinary
calculi differ completely from common stones, and
that they do not exist in the food or drink which the
calculous person had taken. Tartar, he says, precipitates from wine, not as an earth, but as a crystallized salt. In like manner, the natural salt of urine
precipitates from that liquid, and gives origin to calculi. We may imitate this natural process by mixing
spirit of urine with rectified alcohol. Immediately an
offa alba is precipitated.

It is needless to observe that Van Helmont was
mistaken, in supposing that this *offa* was the matter
of calculus. Spirit of urine was a strong solution of
carbonate of ammonia. The alcohol precipitated this
salt; so that his *offa* was merely *carbonate of ammonia*. Nor is there the shadow of evidence that alcohol,
as Van Helmont thought it did, ever makes its way
into the mass of humours; yet his notion of the origin
of calculi is not less accurate, though of course he
was ignorant of the chemical nature of the various
substances which constitute these calculi. From this
reasoning Van Helmont was induced to reject the
term *tartar*, employed by Paracelsus. To avoid all
false interpretations he substitutes the word *duelech*,
to denote the state in which the spirit of urine precipitates and gives origin to these calculous concretions.

As all diseases proceeded in his opinion from the
archeus, the object of his treatment was to calm the
archeus, to stimulate it, and to regulate its movements.

To accomplish these objects he relied upon dietetics, and upon acting on the imaginations of his patients. He considered *certain words* as very efficacious in curing the diseases of the archeus. He admitted the existence of the universal medicine, to which he gave the names of *liquor alkahest, ens primum salium, primus metallus.* Mercurials, antimonials, opium, and wine, are particularly agreeable to the archeus, when in a state of delirium from fever.

Among the mercurial preparations, he praises what he calls *mercurius diaphoreticus* as the best. He gives no account of the mode of preparing it; but from some circumstances I think it must have been *calomel.* He considers it as a sovereign remedy in fevers, dropsies, diseases of the liver, and ulcers of the lungs. He employed the red oxide of mercury as an external application to ulcers. The principal antimonial preparations which he employed were the hydrosulphuret, or *golden sulphur*, and the deutoxide, or *antimonium diaphoreticum.* This last medicine was used in scruple doses—a proof of its great inertness compared with the protoxide of antimony.

Opium he considered as a fortifying and calming medicine. It contains an acrid salt and a bitter oil, which give it the virtue of putting a stop to the errors of the archeus, when it was sending its acid ferment into other acid parts of the body. Van Helmont assures us that he wrought many important cures by the employment of wine.

Such is a very short statement of the opinions of a man, who, notwithstanding his attachment to the fanatical opinions which distinguished the time in which he lived, had the merit of overturning a vast number of errors, both theoretical and practical; and of laying down many principles, which, for want of erudition, have been frequently assigned to modern writers. Van Helmont has been frequently placed on the same level with Paracelsus, and treated like him with contempt.

But his claims upon the medical world are much higher, and his merits infinitely greater. His notions, it is true, were fanatical; but his erudition was great, his understanding excellent, and his industry indefatigable. His writings did not become known till rather a late period; for, with the exception of a single tract, they were not published till 1648, by his son, after his death.

The decided preference given to chemical medicines by Van Helmont, and the uses to which he applies chemical theory, had a natural tendency to raise chemistry to a higher rank in the eyes of medical men than it had yet reached. But the man to whom the credit of founding the iatro-chemical sect is due, is Francis de le Boé Sylvius, who was born in the year 1614. While a practitioner of medicine at Amsterdam, he studied with profound attention the system of Van Helmont, and the rival and much more popular theory of Descartes : upon these he founded his own theory, which, in reality, contains little entitled to the name of original, notwithstanding the tone in which he speaks of it, and his repeated declarations that he had borrowed from no one. He was appointed professor of the theory and practice of medicine in the University of Leyden, where he taught with such eclat, and drew after him so great a number of pupils, that Bóerhaave alone surpassed him in this respect. It was he that first introduced the practice of giving clinical lectures in the hospitals, on the cases treated in the presence of the pupils. This admirable innovation has been productive of much benefit to medicine. He greatly promoted anatomical studies, and inspected, himself, a vast number of dead bodies. This is the more remarkable, because his own system, like that of Van Helmont, from whom it was borrowed, was quite independent of the structure of the parts.

Every thing was explained by him according to the principles of chemistry, as they were then understood.

The celebrity of the university in which he taught, and the vast number of his pupils, contributed to spread this theory into every part of the world, and to give it an eclat which is really surprising, when we consider it with attention. But he possessed the talents just suited for securing the reception of his opinions by his pupils as infallible oracles, and of being the idol of the university. Yet it is melancholy to be obliged to add, that few persons ever more abused the favours of nature, or the advantages of situation and elocution.

To form a clear idea of the principles of this founder of iatro-chemistry, we have only to call to mind the ferments of Van Helmont, which constitute the foundation-stone of the whole system. We cannot, says he, conceive a single change in the mixture of the humours, which is not the consequence of fermentation; and yet he assigns to this fermentation conditions which are scarcely to be found united in the living body. Digestion, in his opinion, is a true fermentation produced by the application of a ferment. Like Van Helmont, he admits a *triumvirate;* but places it in the humours; the effervescence or fermentation of which enabled him to explain most of the functions of the body. Digestion is the result of the mixture of the saliva with the pancreatic juice and the bile, and the fermentation of these humours. The saliva, as well as the pancreatic juice, contains an acidulous salt easily recognised by the taste. Here Sylvius derives advantage from the experiments of Regnier de Graaf on the pancreatic juice, which he had constantly found acid.

Sylvius, who affirmed that the bile contained an alkali, united with an oil and a volatile spirit, supposes an effervescence from the union of the alkali of the bile with the acid of the pancreatic juice, and this *fermentation* he considered as the cause of digestion. By this fermentation the *chyle* is produced, which is nothing else than the *volatile spirit* of the food accompanied by an *oil* and an alkali, neutralized by a weak acid. The blood is more than completed (*plus quam perficitur*) in the spleen. It acquires its highest perfection by the addition of a certain quantity of vital spirits. The *bile* is not drawn from the blood in the liver, but pre-exists in the circulating fluid. It mixes with that fluid anew to be carried to the heart together with the *lymph*, equally mixed with the blood, and there it gives origin to a vital fermentation. In this way the blood becomes the centre of reunion of all the humours of the secretions, which mix together or separate, without the solids taking the smallest share in the operations. Indeed, so completely are the solids banished from the system of Sylvius that he attends to nothing whatever except the humours.

The formation and motion of the blood is explained by the fermentation of the oily volatile salt of the bile, and the dulcified acid of the lymph, which develops the vital heat, by which the blood is attenuated and becomes capable of circulating. This vital fire, quite different from ordinary fire is kept up in its turn by the uniform mixture of the blood. It attenuates the humours, not because it is *heat* but because it is composed of *pyramids*. This last notion is obviously borrowed from Descartes, just as the fermentation in the heart, as the cause of the motion of the blood, reminds us of the opinions of Van Helmont.

Sylvius explains the preparation of the vital spirits in the encephalos by distillation, and he finds a great resemblance between their properties and those of spirit of wine. The nerves conduct these spirits to the different parts, and they spread themselves in the substance of the organs to render them sensible. When they insinuate themselves into the glands the addition of the acid of the blood produces a liquid analogous to naphtha, which constitutes the *lymph*. Lymph, then, is a compound of the vital spirit and the acid of the blood. *Milk* is formed in the mammæ by the afflux of a very mild acid, which gives a white colour to the red humour of the blood.

The theory of the natural functions was no less chemical. Even the diseases themselves were explained upon chemical principles. Sylvius first introduced the word *acridity* to denote a predominance of the chemical elements of the humours, and he looked upon these *acridities* as the proximate cause of all diseases. But as every thing acrid may be referred to one or other of two classes, acids and alkalies, there are only two great classes of diseases; namely, those proceeding from an *acid acridity*, and those proceeding from an *alkaline*.

Sylvius was not altogether ignorant of the constituent parts of the animal humours; but it is obvious, from the account of his opinions just given, that this knowledge was very incomplete; indeed the whole of his chemical science resolves itself into a comparison of the humours of the living body with chemical liquids. Perhaps his notions respecting such of the *gases*, as he had occasion to observe, were somewhat clearer than those of Van Helmont. He called them *halitus*, and takes some notice of their different chemical properties, and states the influence which he supposes them to exert in certain diseases.

In the human body he saw nothing but a magna of humours continually in fermentation, distillation, effervescence, or precipitation; and the physician was degraded by him to the rank of a distiller or a brewer.

Bile acquires different acridities, when bad food, altered air, or other similar causes act apon the body. It becomes *acid* or *alkaline*. In the former case it thickens and occasions obstructions; in the latter it excites febrile heat; and the viscid vapours elevated from it are the cause of the cold fit with which fever commences. All acute and continued fevers have their origin in this acridity of the bile. The vicious mixture of the bile with the blood, or its specific acridity, produces *jaundice*, which is far from being always owing to obstructions in the liver. The vicious effervescence of the bile with the pancreatic juice produces almost all other diseases. But all these assertions of Sylvius are unsupported by evidence.

The acid acridity of the pancreatic juice, and the obstruction of the pancreatic ducts, which are produced by it, are considered by him as the cause of intermittent fevers. When the acid of the pancreatic juice acquires still more acridity, hypochondriasis and hysteria are the consequences of it. If, during the morbid effervescence of the pancreatic juice with the bile an acid and viscid humour arise, the vital spirits of the heart are overwhelmed during a certain time. This occasions syncope, palpitation of the heart, and other nervous affections.

When the acid acridity of the pancreatic juice or of the lymph (for both are similar) is deposited on the nerves, the consequence is spasms or convulsions; epilepsy in particular depends upon the acrid vapours produced by the morbid effervescence of the pancreatic juice with acrid bile. Gout has the same origin as intermittent fevers, for we must look for it in the obstruction of the pancreas and the lymphatic glands, accompanied with an acid acridity of the lymph. Rheumatism is owing to the acrid acid, deprived of the oil which dulcifies it. The smallpox is occasioned by an acid acridity in the lymph, which gives origin to the pustules. Indeed all suppuration in general is owing to a coagulating acid in the lymph. Syphilis results from a caustic acid in the lymph. The itch is produced by an acid acridity of the lymph. Dropsies are produced by the same acid acridity of the lymph. Urinary calculi are the consequences of a coagulating acid existing in the lymph and the pancreatic juice. Corrosive acids, and the loss of volatile spirits, occasion leucorrhœa.

From the preceding statement it would appear that almost all diseases proceed from acids. However, Sylvius informs us that malignant fevers are owing to a superabundance of volatile salts and to a too great tenuity of the blood. The vital spirits themselves give occasion to diseases. They are sometimes too aqueous, sometimes they effervesce too violently, and sometimes not at all. Hence all the nervous diseases, which Sylvius never considers as existing by themselves; but as always derived from the acid, acrid, or alkaline vapours which trouble the vital spirits.

The method of cure which Sylvius deduced from these absurd and contemptible hypotheses, was worthy of the hypotheses themselves; and certainly constitute the most detestable mode of treatment that ever has disgraced medical science. To diseases produced by the effervescence of the bile he opposed purgatives; because in his opinion emetics produced injurious effects. The reason was, that the emetics which he employed were too violent, consisting of antimonial preparations, particularly *powder of Algcrotti*, or an impure protoxide of antimony. For though *emetic tartar* had been discovered in 1630, it does not seem to have come into use till a much later period. We do not find any notice of it in the *praxis chymiatrica* of Hartmann published in 1647, at Geneva.

He endeavoured to moderate the acridity of the bile by opiates and other narcotics. It will scarcely be believed, though it was a natural consequence of his opinions, when we state that he recommended ammoniacal preparations, particularly his oleaginous volatile salt, and spirit of hartshorn, &c., as cures for almost all diseases. Sometimes they were employed to correct the acidity of the lymph, sometimes to destroy the acid acridity of the pancreatic juice, sometimes to correct the inertness of the vital spirits, sometimes to promote the secretions, and to induce a flow of the menses. Volatile spirit of amber and opium were prescribed by him in intermittent fevers; and volatile salts in almost all acute diseases. He united them with antivenomous potions, angelica, contrayerva, bezoard, crabs' eyes, and other similar substances. These absorbents seemed to him very necessary to correct the acidity of the pancreatic juice, and the acridity of the bile. In administering them he paid no attention to the regular course which acute diseases usually run; he neither inquired into the remote nor proximate causes of disease, nor to the symptoms: every thing was neglected connected with induction, and his whole proceedings regulated by wild speculations and absurd theories, quite inconsistent with the phenomena of nature.

To attempt to refute these wild notions of Sylvius would be loss of time. It is extraordinary, and almost incredible, that he could have regulated his practice by them: and it is a still more incredible thing, and exhibits a very humiliating view of human nature, that these crudities and absurdities were swallowed with avidity by crowds of students, who placed a blind reliance on the dogmas of their master, and were initiated by him into a method of treating their patients, better calculated than any other that could easily have been devised, to aggravate all their diseases, and put an end to their lives. If any of the patients of the iatro-chemists ever recovered their health, well might it be said that their recovery was not the consequence of the prescriptions of their physicians, but that it took place in spite of them.*

---

* As an example of the prescriptions of Sylvius, we give the following for malignant fever:

        R. Theriac. veter. ʒij
           Antim. diaphor. ʒj
           Syrup. Card. Benedic. ʒij
           Aq. prophylact. ʒj
           — Cinnam. ʒss
           — Scabios. ʒij
                   M. D.

It is a very remarkable circumstance, and shows clearly that mankind in general had become disgusted with the dogmas of the Galenists, that iatro-chemistry was adopted more or less completely by almost all physicians. There were, indeed, a few individuals who raised their voices against it; but, what is curious and inexplicable, they never attempted to start objections against the principles of the iatro-chemists, or to point out the futility of their hypothesis, and their inconsistency with fact. They combated them by arguments not more solid than those of their antagonists.

During the presidency of Riolan over the Medical College of Paris, that learned body set itself against all innovations. Guy Patin, who was a medical professor in the University of Paris, and a man of great celebrity, opposed the chemical system of medicine with much zeal. In his Martyrologium Antimonii he collects all the cases in which the use of antimony, as a medicine, had proved injurious to the patient. But in the year 1666, the dispute relative to antimony, and particularly relative to tartar emetic, became so violent, that all the doctors of the faculty of Paris were assembled by an order of the parliament, under the presidency of Dean Vignon, and after a long deliberation, it was concluded by a majority of ninety-two votes, that tartar emetic, and other antimonials, should not only be permitted, but even recommended. Patin after this decision pretended no longer to combat chemical medicine; but he did not remain inactive. One of his friends, Francis Blondel, demanded the resolution to be cancelled; but his exertions were unsuccessful; nor were the writings of Guillemeau and Menjot, who were also keen partisans of the views of Patin, attended with better success.

In England iatro-chemistry assumed a direction quite peculiar. It was embraced by a set of men who had cultivated anatomy with the most marked success, and who were quite familiar with the experimental method of investigating nature. The most eminent of all the English supporters of iatro-chemistry was Thomas Willis, who was a contemporary of Sylvius.

Dr. Willis was born at Great Bodmin, in Wiltshire, in 1621. He was a student at Christchurch College, in Oxford, when that city was garrisoned for King Charles I. Like the other students, he bore arms for his Majesty, and devoted his leisure hours to the study of physic. After the surrender of Oxford to the parliament, he devoted himself to the practice of medicine, and soon acquired reputation. He appropriated a room as an oratory for divine service, according to the forms of the church of England, to which most of the loyalists of Oxford daily resorted. In 1660, he became Sedleian professor of natural philosophy, and the same year he took the degree of doctor of physic. He settled ultimately in London, and soon acquired a higher reputation, and a more extensive practice, than any of his contemporaries. He died in 1675, and was buried in Westminster Abbey. He was a first-rate anatomist. To him we are indebted for the first accurate description of the brain and nerves.

But it is as an iatro-chemist that he claims a place in this work. His notions approach nearer to those of Paracelsus than to the hypotheses of Van Helmont and Sylvius. He admits the three chemical elements of Paracelsus, salt, sulphur, and mercury, in all the bodies in nature, and employs them to explain their properties and changes; but he gives the name of *spirit* to the *mercury* of Paracelsus. He ascribes to it the virtue of volatilizing all the constituent parts of bodies : salt, on the other hand, is the cause of fixity in bodies; *sulphur* produces colour and heat, and unites the *spirit* to the *salt*. In the stomach there occurs an acid ferment, which forms the chyle with the sulphur of the aliments : this chyle enters into effervescence in the heart, because the salt and sul-

phur take fire together. From this results the vital
flame, which penetrates every thing. The vital spirits
are secreted in the brain by a real distillation. The
vessels of the testes draw an elixir from the constituent
parts of the blood; but the spleen retains the earthy
part, and communicates a new igneous ferment to the
circulating fluid. On this account the blood must be
considered as a humour, constantly disposed to fermentation, and in this respect it may be compared to
wine. Every humour in which salt, sulphur, and
spirit predominates in a certain manner, may be converted into a *ferment*. All diseases proceed from a
morbid state or action of this ferment; and a physician may be compared to a wine-merchant; for, like
him, he has nothing to do but to watch that the necessary fermentations take place with regularity, and
that no foreign substance come to derange the operation.

At this period the mania of explaining every thing
had proceeded to such a length, that no distinction
was made between dead and living bodies. The chemical facts which were at that time known, were applied without hesitation to explain all the functions
and all the diseases of the living body. According to
Willis, fever is the simple result of a violent and pre-
ternatural effervescence of the blood and the other
humours of the body, either produced by external
causes, or by internal ferments, into which the chyle
is converted when it mixes with the blood. The effervescence of the vital spirits is the source of quotidians;
that of salt and sulphur produces continued fever;
and external ferments of a malignant nature produce
malignant fevers. Thus the smallpox is owing to the
seeds of fermentation set in activity by an external
principle of contagion. Spasms and convulsions are
produced by an explosion of the salt and sulphur
with the animal spirits. Hypochondriacal affections
and hysteria depend originally on the morbid putrifac-

tion of the blood in the spleen, or on a bad fermentes-cible principle, loaded with salt and sulphur, which unites with the vital spirits and deranges them. Scurvy is owing to an alteration of the blood, which may then be compared to vapid or stale wine. The gout is merely the coagulation of the nutritive juices altered by the acidified animal spirits; just as sulphuric acid forms a coagulum with carbonate of potash.

The action of medicines is easily explained by the effects which they produce on the nourishing principles. Sudorifics are considered as cordials, because they augment the sulphur of the blood, which is the true food of the vital flame. Cordials purify the animal spirits, and fix the too volatile blood. Willis dis-agrees with the other iatro-chemists of his time in one thing: he recommends bleeding in the greater num-ber of diseases, as an excellent method of diminishing unnatural fermentation.

Dr. Croone, a celebrated Fellow of the Royal So-ciety, was another English iatro-chemist, who attempt-ed to explain muscular motion by the effervescence of the nervous fluid, or animal spirits.

It is not worth while to notice the host of writers—English, French, Italian, Dutch, and German, who exerted themselves to maintain, improve, and defend, the chemical doctrines of medicine. The first person who attempted to overturn these absurd doctrines, and to introduce something more satisfactory in their place, was Mr. Boyle, at that time in the height of his celebrity.

Robert Boyle was born at Youghall, in the pro-vince of Munster, on the 25th of January, 1627. He was the seventh son, and the fourteenth child of Richard, Earl of Cork. He was partly educated at home, and partly at Eton, where he was under the tuition of Sir Henry Wotton. At the age of eleven, he travelled with his brother and a French tutor through France to Geneva, where he pursued his

studies for twenty-one months, and then went to
Italy. During this period, he acquired the French
and Italian languages; and, indeed, talked in the former with so much fluency and correctness, that he
passed, when he thought proper, for a Frenchman. In
1642, his father's finances were deranged, by the
breaking out of the great Irish rebellion. His tutor,
who was a Genevese, was obliged to borrow, on his
own credit, a sum of money sufficient to carry him
home. On his arrival, he found his father dead; and,
though two estates had been left to him, such was the
state of the times, that several years elapsed before he
could command the requisite sum of money to supply
his exigencies. He retired to an estate at Stalbridge,
in Dorsetshire.

In 1654 he went to Oxford, where he associated
himself with a number of eminent men (Dr. Willis
among others), who had constituted themselves into a
combination for experimental investigations, distinguished by the name of the *Philosophical College.*
This society was transferred to London; and, in 1663,
was incorporated by Charles II. under the name of the
Royal Society. In 1668 Mr. Boyle took up his re
sidence in London, where he continued till the last day
of December, 1691, assiduously occupied in experimental investigations, on which day he died, in the
sixty-fifth year of his age.

We are indebted to Mr. Boyle for the first introduction of the air-pump and the thermometer into
Britain, and for contributing so much, by means of
Dr. Hooke, to the improvement of both. His hydro-
statical and pneumatical investigations and experiments constitute the foundation of these two sciences.
The thermometer was first made an accurate instrument of investigation by Sir Isaac Newton, in 1701.
This he did by selecting as two fixed points the temperatures at which water freezes and boils; marking
these upon the stem of the thermometer, and dividing

the interval between them into a certain number of de-grees. All thermometers made in this way will stand at the same point when plunged into bodies of the same temperature. The number of divisions between the freez-ing and boiling points constitute the cause of the differ-ences between different thermometers. In Fahrenheit's thermometer, which is used in Great Britain, the num-ber of degrees, between the freezing and boiling points of water, is 180; in Reaumur's it is 80; in Celsius's, or the centigrade, it is 100 ; and in De Lisle's it is 150.

But my reason for mentioning Mr. Boyle here was, the attempt which he made in 1661, by the publica-tion of his Sceptical Chemist, to overturn the absurd opinions of the iatro-chemists. He raises doubts, not only respecting the existence of the elements of the Peripatetics, but even of those of the chemists. The first elements of bodies, in his opinion, are *atoms*, of different shapes and sizes ; the union of which gives origin to what we vulgarly call *elements*. We cannot restrain the number of these to four, as the Peripatetics do ; nor to three, with the chemists : neither are they immutable, but convertible into each other. Fire is not the means that ought to be employed to obtain them ; for the *salt* and *sulphur* are formed during its action by the union of different simple bodies.

Boyle shows, besides, that the chemical theory of qualities is exceedingly inaccurate and uncertain ; be-cause it takes for granted things which are very doubt-ful, and in many cases directly contrary to the pheno-mena of nature. He endeavours to prove the truth of these ideas, and particularly the production of the chemical principles, by a great number of convincing and conclusive experiments.

In another treatise, entitled " The Imperfections of the Chemical Doctrine of Qualities,"* he points out, in the second section, the insufficiency of the hypotheses of

* Shaw's Boyle, iii. 424.

Sylvius relative to the generality of acids and alkalies. He shows that the offices ascribed to them are arbitrary, and the notions respecting them unsettled; that the hypotheses respecting them are needless, and insufficient, and afford but an unsatisfactory solution of the phenomena.

These arguments of Boyle did not immediately shake the credit of the chemical system. In the year 1691, a chemical academy was founded at Paris by Nicolas de Blegny, the express object of which was to examine these objections of Boyle, which by this time had attracted great attention. Boyle's experiments were repeated and confirmed; but the academicians, notwithstanding, came to the conclusion, that it is unnecessary to have recourse to the true elements of bodies; and that the phenomena which occur in the animal economy may be explained by the predominance of acids or alkalies. Various other publications appeared, all on the same side.

In Germany, Hermann Conringius, the most skilful physician of his time, opposed the chemical theory; and his opinions were impugned by Olaus Borrichius, who defended not only alchymy, but the chemical theory of medicine, with equal erudition and zeal.*

Towards the end of the sixteenth century, the chemists thought of examining the liquids of the living body, to ascertain whether they really contained the acids and alkalies which had been assigned them, and considered as the cause of all diseases. But at that time chemistry had made so little progress, and such was the want of skill of those who undertook these investigations, that they readily obtained every thing that was wanted to confirm their previous notions. John Viridet, a physician of Geneva, announced that he had found an acid in the saliva and the pancreatic juice, and an alkali in the gastric juice and the bile.

* De Ortu et Progressu Chemiæ. *Hafniæ*, 1674.

But the most celebrated experiments of that period were those of Raimond Vieussens, undertaken in 1698, in order to discover the presence of an acid spirit in the blood. His method was, to mix blood with a species of clay, called *bole*, and to subject the mixture to distillation, He found that the liquid distilled over was acid. Charmed with this discovery, which he considered as of first-rate importance, he announced it by letter to the different academies, and colleges in Europe. Some doubts being raised about the accuracy of his experiment, it having been alleged that the acid came from the clay which he had mixed with the blood, and not from the blood itself, Vieussens purified the *bole* from all the acid which it could contain, and repeated his experiment again. The result was the same—the acrid salt of the fluid yielded an acid spirit.

It would be needless in the present state of our knowledge to point out the inaccuracy of such an experiment, or how little it contributed to prove that blood contains a free acid. It is now well known to chemists, that blood is remarkably free from acids; and, that if we except a little common salt, which exists in all the liquids of the human body, there is neither any acid nor salt whatever in that liquid.

Michael Ettmuller, at Leipsic, who was a chemist of some eminence in his day, and published a small treatise on the science, which was much sought after, was also a zealous iatro-chemist; but his opinions were obviously regulated by the researches of Boyle. He denies the existence of acids and alkalies in certain bodies, and distinguishes carefully between acid and putrid fermentation.

One of the most formidable antagonists to the iatrochemical doctrines was Dr. Archibald Pitcairne, first a professor of medicine in the University of Leyden, and afterwards of Edinburgh, and one of the most eminent physicians of his time. He was born in Edinburgh, on the 25th of December, 1652. After finishing his school education in Dalkeith, he went to the University of Edinburgh, where he improved himself in classical learning, and completed a regular course of philosophy. He turned his attention to the law, and prosecuted his studies with so much ardour and intensity that his health began to suffer. He was advised to travel, and set out accordingly for the South of France : by the time he reached Paris he was so far recovered that he determined to renew his studies; but as there was no eminent professor of law in that city, and as several gentlemen of his acquaintance were engaged in the study of medicine, he went with them to the lectures and hospitals, and employed himself in this way for several months, till his affairs called him home.

On his return he applied himself chiefly to mathematics, in which, under the auspices of his friend, the celebrated Dr. David Gregory, he made uncommon progress. Struck with the charms of this science, and hoping by the application of it to medicine to reduce the healing art under the rigid rules of mathematical demonstration, he formed the resolution of devoting himself to the study of medicine. There was at that time no medical school in Edinburgh, and no hospital at which he could improve himself; he therefore repaired to Paris, and devoted himself to his studies with a degree of ardour that ensured an almost unparalleled success. In 1680 he received from the faculty of Rheims the degree of doctor of medicine, a degree also conferred on him in 1699 by the University of Aberdeen.

In the year 1691 his reputation was so high that the University of Leyden solicited him to fill the medical chair, at that time vacant; he accepted the invitation, and delivered a course of lectures at Leyden, which was greatly admired by all his auditors, among whom were Boerhaave and Mead. At the close of the session he set out for Scotland, to marry the daughter of Sir Archibald Stevenson: his friends in his own country would not consent to part with him, and thus he was reluctantly obliged to resign his chair in the University of Leyden.

He settled as a physician in Edinburgh, where he was appointed titular professor of medicine. His practice extended beyond example, and he was more consulted by foreigners than any Edinburgh physician either before or after his time. He died in October, 1713, admired and regretted by the whole country. He was a zealous supporter of iatro-mathematics, and as such a professed antagonist of the iatro-chemists. He refuted their opinions with much strength of reasoning, while his high reputation gave his opinions an uncommon effect; so that he contributed perhaps as much as any one, to put a period to the most disgraceful, as well as dangerous, set of opinions that ever overspread the medical horizon.

Into the merits of the iatro-mathematicians it is not the business of this work to enter; they at least display science, and labour, and erudition, and in all these respects are far before the iatro-chemists. Perhaps their own opinions were not more agreeable to the real structure of the human body, nor their practice more conformable to reason, or more successful than those of the chemists. Probably the most valuable of all Dr. Pitcairne's writings, is his vindication of the claims of Hervey to the great discovery of the circulation.

Boerhaave, the pupil of Pitcairne, and afterwards a professor in Leyden, was a no less zealous or successful opponent of the iatro-chemists.

Herman Boerhaave, perhaps the most celebrated physician that ever existed, if we except Hippocrates, was born at Voorhout, a village near Leyden, in 1668, where his father was the parish clergyman. At the age of sixteen he was left without parents, protection, advice, or fortune. He had already studied theology, and the other branches of knowledge that are considered as requisite for a clergyman, to which situation he aspired; and while occupied with these studies he supported himself at Leyden by teaching mathematics to the students—a branch of knowledge to which he had devoted himself with considerable ardour while living in his father's house. But, a report being raised that he was attached to the doctrines of Spinoza, the clamour against him was so loud that he thought it requisite to renounce his intention of going into *orders*.* He turned his studies to medicine, and the branches of science connected with that pursuit, and these delightful subjects soon engrossed the whole of his attention. In 1693 he was created doctor of medicine, and began to practise. He continued to teach mathematics for some time, till his practice increased sufficiently to enable him to live by his fees. His spare money was chiefly laid out upon books; he also erected a chemical laboratory, and though he had no garden he paid great attention to the study of plants. His reputation increased with considerable rapidity; but his fortune rather slowly. He was invited to the Hague by a nobleman, who stood high in the favour of William III., King of Great Britain; but he declined the invitation. His three great friends, to whom he was in some measure indebted for his success, were James Trigland, professor of theology,

* While travelling in a tract-boat, one of his fellow-travellers more orthodox than well informed, attacked the system of Spinoza with so little spirit, that Boerhaave was tempted to ask him if he had ever read Spinoza. The polemic was obliged to confess that he had not; but he was so much provoked at this public exposure of his ignorance, that he propagated the report of Boerhaave's attachment to Spinozism, and thus blasted his intention of becoming a clergyman.

Daniel Alphen, and John Van den Berg, both of them successively chief magistrates of Leyden, and men of great influence.

Van den Berg recommended him to the situation of professor of medicine in the University of Leyden, to which chair he was raised, fortunately for the reputation of the university, on the death of Drelincourt, in 1702. He not only gave public lectures on medicine, but was in the habit also of giving private instructions to his pupils. His success as a teacher was so great, that a report having been spread of his intention to quit Leyden, the curators of the university added considerably to his salary on condition that he would not leave them.

This first step towards fortune and eminence having been made, others followed with great rapidity. He was appointed successively professor of botany and of chemistry, while rectorships and deanships were showered upon him with an unsparing hand. And such was the activity, the zeal, and the ability with which he filled all these chairs, that he raised the University of Leyden to the very highest rank of all the universities of Europe. Students flocked to him from all quarters—every country of Europe furnished him with pupils; Leyden was filled and enriched by an unusual crowd of strangers. Though his class-rooms were large, yet so great was the number of students, that it was customary for them to keep places, just as is done in a theatre when a first-rate actor is expected to perform. He died in the year 1738, while still filling the three different chairs with undiminished reputation.

It is not our object here to speak of Boerhaave as a physician, or as a teacher of medicine, or of botany; though in all these capacities he is entitled to the very highest eulogium; his practice was as unexampled as his success as a teacher. It is solely as a chemist that he claims our attention here. His system of chemistry, published in two quarto volumes in 1732, and of which we have an excellent English translation by Dr. Shaw, printed in 1741, was undoubtedly the most learned and most luminous treatise on chemistry that the world had yet seen; it is nothing less than a complete collection of all the chemical facts and processes which were known in Boerhaave's time, collected from a thousand different sources, and from writings equally disgusting from their obscurity and their mysticism. Every thing is stated in the plainest way, stripped of all mystery, and chemistry is shown as a science and an art of the first importance, not merely to medicine, but to mankind in general. The processes given by him are too numerous and too tedious to have been all repeated by one man, how laborious soever he may have been: many of them have been taken upon trust, and, as no distinction is made in the book, between those which are stated upon his own authority and those which are merely copied from others, this treatise has been accused, and with some justice, as not always to be depended on. But the real information which it communicates is prodigious, and when we compare it with any other system of chemistry that preceded it, the superiority of Boerhaave's information will appear in a very conspicuous point of view.

After a short but valuable historical introduction he divides his work into two parts; the first treats of the *theory of chemistry*, the second of the *practical processes*.

He defines chemistry as follows: " Chemistry is an art which teaches the manner of performing certain physical operations, whereby bodies cognizable to the senses, or capable of being rendered cognizable, and of being contained in vessels, are so changed by means of proper instruments, as to produce certain determinate effects; and at the same time discover the causes thereof; for the service of various arts."

This definition is not calculated to throw much light on chemistry to those who are unacquainted with its nature and object. Neither is it conformable to the modern notions entertained of chemistry; but it is requisite to keep in mind Boerhaave's definition of chemistry, when we examine his system, that we may not accuse him of omissions and imperfections, which are owing merely to the state of the science when he gave his system to the world.

In his theory of chemistry he begins with the metals, which he treats of in the following order: Gold, mercury, lead, silver, copper, iron, tin. The account of them, though imperfect, is much fuller and more satisfactory than any that preceded it. He then treats of the salts, which are, common salt, saltpetre, borax, sal ammoniac and alum. This it will be admitted is but a meagre list. However other salts occur in different parts of the book which are not described here. He next gives an account of sulphur. Here he introduces *white arsenic*, obtained, he says, from cobalt, and not known for more than two hundred years. He considers it as a real sulphur, and takes no notice of metallic arsenic, though it had been already alluded to by Paracelsus. He then treats of bitumens, including under the name not merely bitumens liquid and solid, but likewise pit-coal, amber, and ambergris. An account of stones and earths comes next, and constitutes the most defective part of the book. It is very surprising that in this part of his work he takes no notice of *lime*. The semi-metals come next: they are, antimony, bismuth, zinc. Here he gives an account of the three vitriols or sulphates of iron, copper, and zinc. He knew the composition of sulphate of iron; but was ignorant of that of sulphate of copper and sulphate of zinc. He considers semi-metals as compounds of a true metal and sulphur, and therefore enumerates cinnabar among the semi-metals. Lastly he treats of vegetables and animals; and it is needless to say that his account is very imperfect.

He next treats of the utility of chemistry, and shows its importance in natural philosophy, medicine, and the arts. Afterwards he describes the instruments of chemistry. This constitutes the longest and the most important part of the whole work. He first treats of fire at great length. Here we have an account of the thermometer, of the expansion produced by heat, of steam, and in fact the germ of many of the most important parts of the science of heat, which have since been expanded and applied to the improvement, not merely of chemistry, but of the arts and resources of human industry. The experiments of Fahrenheit related by him, on the change of temperature induced by agitating water and mercury together at different degrees of heat, gave origin to the whole doctrine of specific heats. Though Boerhaave himself seemed not aware of the importance of these experiments, or indeed even to have considered them with any attention. But when afterwards analyzed by Dr. Black, these experiments gave origin to one of the most important parts of the whole science of heat.

He next treats at great length on *fuel*. Here his opinions are often very erroneous, from his ignorance of a vast number of facts which have since come to light. It is curious that during the whole of his very long account of combustion he makes no allusion to the peculiar opinions of Stahl on the subject; though they were known to the public, and had been admitted by chemists in general, before his work was published. To what are we to ascribe this omission? It could scarcely have been owing to ignorance, Stahl's reputation being too high to allow his opinions to be treated with neglect. We must suppose, I think, that Boerhaave did not adopt Stahl's doctrine of combustion; but at the same time did not think it proper to enter into any controversy on the subject.

He next treats of the heat produced when different liquids are mixed, as alcohol and water, &c. He gives many examples of such increase of temperature, and describes the phenomena very correctly. But he was unable to assign the cause of the evolution of this heat. The subject was elucidated many years after by Dr. Irvine, who showed that it was owing to a diminution of the specific heat which takes place when liquids combine chemically together. It is in this part of his work that he gives an account of phosphorus, of the action of nitric acid on volatile oils, and he concludes, from all the facts which he states, that elementary fire is a corporeal body. His explanation of the combustion of Homberg's pyrophorus and of common phosphorus, shows clearly that he had no correct notion of the reason why air is necessary to maintain combustion, nor of the way in which that elastic fluid performs its part in the great phenomena of nature.

He next treats of the mode of regulating fire for chemical purposes : then he treats of *air*, his account being chiefly taken from Boyle. He ascribes the discovery of the law of the elasticity of air both to Boyle and Mariotte. Boyle, I believe, was the first discoverer of it. The French are in the habit of calling it the law of Mariotte. He then treats of *water*, and lastly of *earth ;* but even here no mention whatever is made of lime. In the last part of the theory of chemistry he treats at great length of menstruums. These are water, oils, alcohol, alkalies, acids, and neutral salts. He mentions potash and ammonia, but takes no notice of soda; the difference between potash and soda not being accurately known. Nor can we expect any particular account of the difference between the properties of mild and caustic potash; as this subject was not understood till the time of Dr. Black. The only acids which he mentions are the *acetic, sulphuric, nitric, muriatic,* and *aqua regia.* He subjoins a disquisition on the alcahest or universal solvent, which it is obvious enough, however, from the way in which he speaks of it, that he was not a believer in. The object of his practical part is to teach the method of making all the different chemical substances known when he wrote. This he does in two hundred and twenty-seven processes, in which all the manipulations are described with considerable minuteness. This part of the work must have been long considered as of great utility, and must have been long resorted to by the student as a mine of practical information upon almost every subject that could arrest his attention. So immense is the progress that chemistry has made since the days of Boerhaave, and so different are the researches that at present occupy chemists, and so much greater the degree of precision requisite to be attained, that his processes and directions are now of little or no use to a practical student of chemistry, as they convey little or none of the knowledge which it is requisite for him to possess.

Boerhaave made a set of most elaborate experiments, to refute the ideas of the alchymists respecting the possibility of fixing mercury. He put a quantity of pure mercury into a glass vessel, and kept it for fifteen years at a temperature rather higher than 100°. It underwent no alteration whatever, excepting that a small portion of it was converted into a black powder. But this black powder was restored to the state of running mercury by trituration in a mortar. In this experiment the air had free access to the mercury. It was repeated in a close vessel with the same result, excepting that the mercury was kept hot for only six months instead of fifteen years.

To show that mercury cannot be obtained from metals by the processes recommended by the alchymists, he dissolved pure nitrate of lead in water, and, mixing the solution with sal ammoniac, chloride of lead precipitated. Of this chloride he put a quantity into a retort, and poured over it a strong lixivium of caustic potash. The whole was digested at the temperature of 96° for six months and six days. It was then distilled in a glass retort, by a temperature gradually raised to redness, but not a particle of mercury was evaporated, as it had been alleged by the alchymists would be the case.

Isaac Hollandus had stated that mercury could be easily obtained from the salt of lead made by means of distilled vinegar. To prove this he calcined a quantity of acetate of lead, ground the residue to powder, and triturated it with a very strong alkaline lixivium, and kept the lixivium over it covered with paper for months, taking care to add water in proportion as it evaporated. The calx was then distilled in a heat gradually raised to redness ; but not a particle of mercury was obtained.*

These were not the only laborious experiments which he made with this metal. He distilled it above five hundred times, and found that it underwent no alteration. When long agitated in a glass bottle it is convertible into a black acrid powder, obviously protoxide of mercury. This black powder, when distilled, is converted into running mercury. Exposure of mercury for some months in a heat of 180°, converts it also into protoxide ; and if the heat be higher than this, the mercury is converted into a red acrid substance, obviously peroxide of mercury. But this peroxide, by simple distillation, is again reduced into the state of running mercury.†

Boerhaave combated the opinions of the iatro-chemists with great eloquence, and with a weight derived from his high reputation, and the extraordinary veneration in which his opinions were held by his disciples. His efforts were assisted by those of Bohn, who combated the medical opinions by arguments drawn both from experience and observation, and perfectly irresistible ;

* Mem. Paris, 1734, p. 539.
† Phil. Trans. 1733. No. 430, p. 145.

and the ruin of the chemical sect was consummated by the exertions of the celebrated Frederick Hoffmann, the founder of the most perfect and satisfactory system of medicine that has ever appeared. His efforts were probably roused into action by a visit which he paid to England in 1683, during which he got acquainted with Boyle and with Sydenham; the former the greatest experimentalist, and the latter the greatest physician of the time; and both of whom were declared enemies to iatro-chemistry.

CHAPTER VI.

OF AGRICOLA AND METALLURGY.

I HAVE been induced by a wish to prosecute the history of the opinions first supported by Paracelsus, and carried so much further by Van Helmont and Sylvius, to give a connected view of their effects upon medical practice and medical theory; and I have come to the commencement of the eighteenth century, without taking notice of one of the most extraordinary men, and one of the greatest promoters of chemistry that ever existed : I mean George Agricola. I shall consecrate the whole of this chapter to his labours, and those of his immediate successors.

George Agricola was born at Glaucha, in Misnia, in the year 1494. When a young man he acquired such a passion for mining and minerals, by frequenting the mountains of Bohemia, that he could not be persuaded to relinquish the study. He settled, indeed, as a physician, at Joachimstal; but his favourite study engrossed so much of his attention, that he succeeded but ill in his medical capacity. This induced him to withdraw to Chemnitz, where he devoted himself to his favourite pursuits. He studied the mineralogical writings of the ancients with the most minute accuracy; but not satisfied with this, he visited the mines in person, examined the processes followed by the miners in extracting the different ores, and in washing and sorting them. He made collections of all the different ores, and studied their nature and properties attentively: he likewise collected information about the methods of smelting them, and extracting from them the metals in a state of purity. The information which he collected, respecting the mines wrought in the different countries of Europe, is quite wonderful, if we consider the period in which he lived, the little intercourse which existed between nations, and the total want of all those newspapers and journals which now carry every new scientific fact with such rapidity to every part of the world.

Agricola died at Chemnitz in the year 1555, after he had reached the sixty-first year of his age. Maurice, the celebrated Elector of Saxony, settled on him a pension, the whole of which he devoted to his metallurgic pursuits. To him we find him dedicating the edition of his works which he published in the year of his death, and which is dated the fourteenth before the calends of April, 1555. He even spent a considerable proportion of his own estate in following out his favourite investigations. In the earlier part of his life he had expressed himself rather favourable to the protestant opinions ; but in his latter days he had attacked the reformed religion. This rendered him so odious to the Lutherans, at that time predominant in Chemnitz, that they suffered his body to remain unburied for five days together ; so that it was necessary to remove it from Chemnitz to Zeitz, where it was interred in the principal church.

His great work is his treatise De Re Metallica, in twelve books. In this work he gives an account of the instruments and machines, and every thing connected with mining and metallurgy ; and even gives figures of all the different pieces of apparatus employed in his time. He has also exhibited the Latin and German names for all these different utensils. This work may be considered as a very complete treatise on metallurgy, as it existed in the sixteenth century. The first six books are occupied with an account of mining and smelting. In the seventh book he treats of *docimasy*, or the method of determining the quantity of metal which can be extracted from every particular ore. This he does so completely, that most of his processes are still followed by miners and smelters. He gives a minute and accurate account of the furnaces, mufflles, crucibles, &c., almost such as are still employed, with minute directions for preparing the ores which are to be subjected to examination, the fluxes with which they must be mixed, and the precautions necessary in order to obtain a satisfactory result. In short, this book may be considered as a complete manual of docimasy. How much of the methods given originated with Agricola it is impossible to say. He probably did little more than collect the scattered processes employed by the smelters of metals, in different parts of the world, and reduce the whole to a regular system. But this was a great deal. Perhaps it is not saying too much, that the great progress made in the chemical investigation of the metals, was owing in a great measure to the labours of Agricola. Certainly the progress made by the moderns, in the difficult arts of mining and metallurgy, must in a great measure be ascribed to the labours of Agricola.

In the eighth book he describes the mechanical preparation of the ores, and the mode of roasting them, either in the open air or in furnaces. The ninth book is occupied with an account of smelting-furnaces. It contains also a description of the processes for obtaining mercury, antimony, and bismuth, from their ores. The tenth book treats of the separation of silver and gold from each other, by means of nitric acid and aqua regia : minute directions for the preparation of which are given. The modes of purifying the precious metals by means of sulphur, antimony, and cementations,

are also described. In the eleventh book he treats of
the method of purifying silver from copper and iron,
by means of lead. He gives an account also of the
processes employed for smelting and purifying copper.
In the twelfth book he treats of the methods of preparing common salt, saltpetre, alum, and green vitriol,
or sulphate of iron : of the preparation and purification
of sulphur, and of the mode of manufacturing glass.
In short, Agricola's work De Re Metallica is beyond
comparison the most valuable. chemical work which
the sixteenth century produced, and places the author
very high indeed among the list of the improvers of
chemistry.

The other works of Agricola are his treatise De
Natura Fossilium, in ten books ; De Ortu et Causis
Subterraneorum, in five books ; De Natura eorum quæ
effluunt ex Terra, in four books ; De veteribus et novis
Metallis, in two books ; and his Bermannus sive de
re metallica Dialogus. The treatise De veteribus et
novis Metallis is amusing. He not only collects together all the historical facts on record, respecting the
first discoverers of the different metals and the first
workers of mines, but he gives many amusing anecdotes nowhere else to be found, respecting the way in
which some of the most celebrated German mines
were discovered. In the second book he takes a geographical view of every part of the known world, and
states the mines wrought and the metals found in each.
We must not suppose that all his statements in this
historical sketch are accurate : to admit it would be
to allow him a greater share of information than could
possibly belong to any one màn. He frequently gives
us the authority upon which his statements are founded;
but he often makes statements without any authority
whatever. Thus he says, that a mine of quicksilver
had been recently discovered in Scotland : the fact
however, is, that no quicksilver-mine ever existed in
any part of Britain. There was, indeed, a foolish

story circulated about thirty years ago, about a vein of quicksilver found under the town of Berwick-upon-Tweed; but it was an assertion unsupported by any authentic evidence.

Many years elapsed before much addition was made to the processes described by Agricola. In the year 1566, Pedro Fernandes de Velasco introduced a method of extracting gold and silver from their ores in Mexico and Peru by means of quicksilver. But I have never seen a description of his process. Alonzo Barba claims for himself, and seemingly with justice, the method of amalgamating the ores of gold and silver by boiling. Barba was a Spanish priest, who lived about the year 1609, at Tarabuco, a market-town in the province of Charcso, eight miles from Plata, in South America. In the year 1615 he was curate at Tiaguacano, in the Province of Pacayes, and in 1617, he lived at Lepas in Peru. He is said to have been a native of Lepe, a small township in Andalusia, and had for many years the living of the church of St. Bernard at Potosi. His work on the amalgamation of gold and silver ores appeared at Madrid in the year 1640, in quarto.* In the year 1629 a new edition of it appeared with an appendix, under the title of "Trattado de las Antiquas Minas de España de Alonzo Carillo Lasso." The English minister at the Court of Madrid, the Earl of Sandwich, published the first part of it in an English translation at London, in 1674, under the title of " The First Book of the Art of Metals, in which is declared the manner of their generation, and the concomitants of them, written in Spanish by Albaro Alonzo Barba. By E. Earl of Sandwich."

The next improver of metallurgic processes was Lazarus Erckern, who was upper bar-master at Kut-

---

* It is entitled, " El Arte de los Metales, en que se ensena el verdadero beneficio de los de oro y plata por azoque," &c.

tenberg, in the year 1588, and was superintendent of
the mines in Germany, Hungary, Transylvania, the
Tyrol, &c., to three successive emperors. His work has
been translated into English under the title of " Heta
Minor; or the laws of art and nature in knowing,
judging, assaying, fining, refining, and enlarging the
bodies of confined metals. To which are added essays
on metallic words, illustrated with sculptures. By Sir
J. Pettus. London, 1683, folio." But this translation is a very bad one. Erckern gives a plain account
of all the processes employed in his time without a
word of theory or reasoning. It is an excellent practical book; though it is obvious enough that the
author was inferior in point of abilities to Agricola.
His treatment of Don Juan de Corduba, who offered,
in 1588, to put the Court of Vienna in possession of
the Spanish method of extracting gold and silver from
the ores by amalgamation, as related by Baron Born in
his work on amalgamation, shows very clearly that
Erckern was a very illiberal-minded man, and puffed
up with an undue conceit of his own superior knowledge.* Had he condescended to assist the Spaniard,
and to furnish him with proper materials to work upon,
the Austrians might have been in possession of the process of amalgamation with all its advantages a couple
of centuries before its actual introduction.

I need not take any notice of the docimastic treatises
of Schindlers and Schlutter, which are of a much
later date, and both of which have been translated into
French, the former by Geoffroy, junior; the latter by
Hellot. This last translation, in two large quartos,
published in 1764, constitutes a very valuable book,
and exhibits all the docimastic and metallurgic processes known at that period with much fidelity and minuteness. Very great improvements have taken place

* Born's New Process of Amalgamation, translated by
Raspe, p. 11.

since that period, but I am not aware of any work published in any of the European languages, that is calculated to give us an exact idea of the present state of the various mining and metallurgic processes—important as they are to civilized society.

Gellert's Metallurgic Chemistry, so far as it goes, is an excellent book.

CHAPTER VII.

HITHERTO I have treated of the alchymists, or
iatro-chemists, and have brought the history of chemistry down to the beginning of the eighteenth century.   But during the seventeenth century there
existed several laborious chemists, who contributed
very materially by their exertions, either to extend the
bounds of the science, or to increase its popularity and
respectability in the eyes of the world.   Of some of
the most eminent of these it is my intention to give an
account in this chapter.

Of John Rudolf Glauber, the first of these meritorious men in point of time, I know very few particulars.   He was a German and a medical man, and
spent most of his time at Salzburg, Ritzingen, Frankfort on the Maine, and at Cologne.   Towards the end
of his life he went to Holland, but during the greatest
part of his residence in that country he was confined
to a sick-bed.   He died at Amsterdam in 1668, after
having reached a very advanced age. Like Paracelsus,
whom he held in high estimation, he was in open hos-
tility with the Galenical physicians of his time.   This
led him into various controversies, and induced him
to publish various apologies; most of which still remain among his writings.   One of the most curious of
these apologies is the one against Farrner.   To this
man Glauber had communicated certain secrets of his

own, which were at that time considered as of great value; Farrner binding himself not to communicate them to any person. This obligation he not only broke, but publicly deprecated the skill and integrity of Glauber, and offered to communicate to others, for stipulated sums, a set of secrets of his own, which he vaunted of as particularly valuable. Glauber examines these secrets, and shows that every one of them possessed of any value, had been communicated by himself to Farrner, and to put an end to Farrner's unfair attempt to make money by selling Glauber's secrets, he in this apology communicates the whole processes to the public

Glauber's works were published in Amsterdam, partly in Latin, and partly in the German language. In the year 1689 an English translation of them was published in London by Mr. Christopher Packe, in one large folio volume. Glauber was an alchymist and a believer in the universal medicine. But he did not confine his researches to these two particulars, but endeavoured to improve medicine and the arts by the application of chemical processes to them. In his treatise of *philosophical furnaces* he does not confine himself to a description of the method of constructing furnaces, and explaining the use of them, but gives an account of a vast many processes, and medicinal and chemical preparations, which he made by means of these furnaces. One of the most important of these preparations was muriatic acid, which he obtained by distilling a mixture of common salt, sulphate of iron, and alum, in one of the furnaces which he describes.

He makes known the method of dissolving most of the metals in muriatic acid, and the resulting chlorides, which he denominates oils of the respective metals, constitute in his opinion valuable medicines. He mentions particularly the chloride of gold, and from the mode of preparing it, the solution must have been

strong. Yet he recommends it as an internal medicine, which he says may be taken with safety, and is
a sovereign remedy in old ulcers of the mouth, tongue,
and throat, arising from the French pox, leprosy,
scorbute, &c. Thus we see the use of gold as a remedy
for the venereal disease did not originate with M.
Chretiens, of Montpelier. This chloride of gold is so
violent a poison that it is remarkable that Glauber does
not specify the dose that patients labouring under the
diseases for which he recommends it ought to take.—
The sesqui-chloride of iron he recommends as a most
excellent application to ill-conditioned ulcers and cancers. We see from this that the use of iron in cancers,
lately recommended, is not so new a remedy as has
been supposed.

He mentions the violent action of chloride of mercury (obviously corrosive sublimate), and says that
he saw a woman suddenly killed by it, being administered internally by a surgeon. Butter of antimony he
first recognised as nothing else than a combination of
chlorine and antimony; before his time it had been
always supposed to contain mercury.

He describes the method of obtaining sulphuric
acid by distilling sulphate of iron; gives an account of
the mode of obtaining sulphate of iron and sulphate
of copper, in crystals: the method of obtaining nitric acid from nitre by means of alum, was much improved by him. He gives a particular detail of the
way of obtaining fulminating gold. This fulminating
gold he says is of little use in medicine; but he gives
a method of preparing from it a red tincture of gold,
which he considers as one of the most useful and efficacious of all medicines: this tincture is nothing else
than chloride of gold. It would take up too much
space to attempt an analysis of all the curious facts
and preparations described in this treatise on philosophical furnaces; but it will repay the perusal of any
person who will take the trouble to look into it. All

the different pharmacopœias of the seventeenth century borrowed from it largely. The third part of this
treatise is peculiarly interesting. It will be seen that
Glauber had already thought of the peculiar efficacy
of applying solutions of sulphur, &c. to the skin, and
had anticipated the various vapour and gaseous baths
which have been introduced in Vienna and other
places, during the course of the present century, and
considered as new, and as constituting an important
era in the healing art. In the fourth part he not only
treats of the docimastic processes, so well described
by Agricola and Erckern, but gives us the method of
making glass, and of imitating the precious stones by
means of coloured glasses. The fifth part is peculiarly
valuable; in it he treats of the methods of preparing
lutes for glass vessels, of the construction and qualities
of crucibles, and of the vitrification of earthen vessels.

Another of his tracts is called "The Mineral Work;"
the object of which is to show the method of separating gold from flints, sand, clay, and other minerals,
by the spirit of salt (*muriatic acid*), which otherwise
cannot be purged; also a panacea, or universal antimonial medicine. This panacea was a solution of
deutoxide of antimony in pyrotartaric acid; Glauber gives a most flattering account of its efficacy in
removing the most virulent diseases, particularly all
kinds of cutaneous eruptions. The second and third
parts of The Mineral Work are entirely alchymistical.
In the treatise called "Miraculum Mundi," his chief
object is to write a panegyric on *sulphate of soda*, of
which he was the discoverer, and to which he gave the
name of *sal mirabile*. The high terms in which he
speaks of this innocent salt are highly amusing, and
serve well to show the spirit of the age, and the dreams
which still continued to haunt the most laborious
and sober-minded chemists. The *sal mirabile* was
not merely a purgative, a virtue which it certainly
possesses in a high degree, being as mild a pur-

gative, perhaps the very best, of all the saline preparations yet tried; but it was a universal medicine, a panacea, a cure for all diseases: nor was Glauber contented with this, but pointed out many uses in the various arts and manufactures for which in his opinion it was admirably fitted. But by far the fullest account of this *sal mirabile* is given by him in his treatise on the nature of salts.

I shall satisfy myself with giving the titles of his other tracts. Every one of them contains facts of considerable importance, not to be found in any chemical writings that preceded him; but to attempt to connect these facts into one point of view would be needless, because they are not such as would be likely to interest the general reader.

1. The Consolation of Navigators. This gives an account of a method by which sailors may carry with them a great deal of nourishment in very small bulk. The method consists in evaporating the wort of malt to dryness, and carrying the dry extract to sea. This method has been had recourse to in modern times, and has been found to furnish an effectual remedy against the scurvy. He recommends also the use of muriatic acid as a remedy for thirst, and a cure for the scurvy.

2. A true and perfect Description of the extracting good Tartar from the Lees of Wine.

3. The first part of the Prosperity of Germany; in which is treated of the concentration of wine, corn, and wood, and the more profitable use of them than has hitherto been.

4. The second part of the Prosperity of Germany; wherein is shown by what means minerals may be concentrated by nitre, and turned into metallic and better bodies.

5. The third part of the Prosperity of Germany; in which is delivered the way of most easily and plentifully extracting saltpetre out of various subjects, every where obvious and at hand. Together with a

succinct explanation of Paracelsus's prophecy; that is to say, in what manner it is to be understood the northern lion will institute or plant his political or civil monarchy; and that Paracelsus himself will not abide in his grave; and that a vast quantity of riches will offer itself. Likewise who the artist Elias is, of whose coming in the last days, and his disclosing abundance of secrets, Paracelsus and others have predicted.

6. The fourth part of the Prosperity of Germany; in which are revealed many excellent, useful secrets, and such as are serviceable to the country; and withal several preparations of efficacious cates extracted out of the metals and appointed to physical uses; as also various confections of golden potions. To which is also adjoined a small treatise which maketh mention of my laboratory; in which there shall be taught and demonstrated (for the public good and benefit of mankind) wonderful secrets, and unto every body most profitable but hitherto unknown.

7. The fifth part of the Prosperity of Germany; clearly and solidly demonstrating and as it were showing with the fingers, what alchymy is, and what benefit may, by the help thereof, be gotten every where and in most places of Germany. Written and published to the honour of God, the giver of all good things, primarily; and to the honour of all the great ones of the country; and for the health, profit, and assistance against foreign invasions, of all their inhabitants that are by due right and obedience subject unto them.

8. The sixth and last part of the Prosperity of Germany; in which the arcanas already revealed in the fifth part, are not only illustrated and with a clear elucidation, but also such are manifested as are most highly necessary to be known for the defence of the country against the Turks. Together with an evident demonstration adjoined, showing, that both a particular and universal transmutation of the imperfect metals into more perfect ones by salt and fire, is

most true; and withal, by what means any one, that is
endued with but a mean knowledge in managing the
fire, may experimentally try the truth hereof in twenty-four hours' space.

9. The first century of Glauber's wealthy Storehouse
of Treasures.—Many of the processes given in this
treatise are mystically stated, or even concealed.

10. The second, third, fourth, and fifth century of
Glauber's wealthy Storehouse of Treasures.

11. New chemical Light; being a revelation of a
certain new invented secret, never before manifested
to the world.—This was a method of extracting gold
from stones.   Probably the gold found by Glauber in
his processes existed in some of the reagents employed; this, at least, is the most natural way of accounting for the result of Glauber's trials.

15. The spagyrical Pharmacopœia, or Dispensatory.
—In this book he treats chiefly of medicines peculiarly
his own; one of those, on which he bestows the greatest
praise, is *secret sa. ammoniac*, or sulphate of ammonia.   He describes the method of preparing this salt,
by saturating sulphuric acid with ammonia.   He informs us that it was much employed by Paracelsus
and Van Helmont, who distinguished it by the name
of *alkahest*.

13. Book of Fires.—Full of enigmas.

14. Treatise of the three Principles of Metals; viz.
sulphur, mercury, and salt of philosophers; how they
may be profitably used in medicine, alchymy, and
other arts.

15. A short Book of Dialogues.   Chiefly relating
to alchymy.

16. Proserpine, or the Goddess of Riches.

17. Of Elias the Artist.

18. Of the three most noble Stones generated by
three Fires.

19. Of the Purgatory of Philosophers.

20. Of the secret Fire of Philosophers.

21. A Treatise concerning the Animal Stone.

John Kunkel, who acquired a high reputation as a chemist, was born in the Duchy of Sleswick, in the year 1630 : his father was a trading chemist, or apothecary ; and Kunkel himself had, in his younger years, paid great attention to the business of an apothecary : he had also diligently studied the different processes of glass-making ; and had paid particular attention to the assaying of metals. In the year 1659, he was chamberlain, chemist, and superintendent of apothecaries to the dukes Francis Charles and Julius Henry, of Lauenburg. While in this situation, he examined many pretended transmutations of metals, and undertook other researches of importance. From this situation he was invited, by John George II., Elector of Saxony, on the recommendation of Dr. Langelott and Counsellor Vogt, as chamberlain and superintendent of the elector's laboratory, with a considerable salary. From this situation he went to Berlin, where he was chemist to the elector Frederick William ; after whose death, his laboratory and glass-house were accidentally burnt. From Berlin he was invited to Stockholm by Charles XI., King of Sweden, who gave him the title of counsellor of metals, and raised him to the rank of a nobleman : here he died, in 1702, in the seventy-second year of his age. Kunkel's greatest discovery was, the method of extracting phosphorus from urine. This curious substance had been originally discovered by Brandt, a chemist, of Hamburg, in the year 1669, as he was attempting to extract from human urine a liquid capable of converting silver into gold. He showed a specimen of it to Kunkel, with whom he was acquainted : Kunkel mentioned the fact as a piece of news to one Kraft, a friend of his in Dresden, where he then resided : Kraft immediately repaired to Hamburg, and purchased the secret from Brandt for 200 rix-dollars, doubtless exacting from him, at the same time, a promise not to reveal it to any other person. Soon after, he exhibited the phosphorus publicly in Britain and in France; whether for money, or not, does not appear. Kunkel, who had mentioned to his friend his intention of getting possession of the process, being vexed at the treacherous conduct of Kraft, attempted to discover it himself, and, after three or four years labour, he succeeded, though all that he knew from Brandt was, that urine was the substance from which the phosphorus was procured. In consequence of this success, phosphorus was at first distinguished by the epithet of *Kunkel* added to the name.

Kunkel published, in 1678, a treatise on phosphorus, in which he describes the properties of this substance, at that time a subject of great wonder and curiosity. In this treatise, he proposes phosphorus as a remedy of some efficacy, and gives a formula for preparing pills of it, to be taken internally. It is therefore erroneous to suppose, as has been done, that the introduction of this dangerous remedy into medicine is a modern discovery. Kunkel appears to have been acquainted with nitric ether. One of the most valuable of his books, is his treatise on glass-making, which was translated into French; and which, till nearly the end of the eighteenth century, constituted by far the best account of glass-making in existence. The following is a list of the most important of his works:

1. Observations on fixed and volatile Salts, potable Gold and Silver, Spiritus Mundi, &c.; also of the colour and smell of metals, minerals, and bitumens.— This tract was published at Hamburg, in 1678, and has been several times reprinted since.

2. Chemical Remarks on the chemical Principles, acid, fixed and volatile alkaline Salts, in the three kingdoms of nature, the mineral, vegetable, and animal; likewise concerning their colour and smell, &c.; with a chemical appendix against non-entia chymica.

3. Treatise of the Phosphorus mirabilis, and its wonderful shining Pills; together with a discourse on

what was formerly rightly named nitre, but is now called the *blood of nature*.

4 An Epistle against Spirit of Wine without an acid.

5. Touchstone de Acido et Urinoso, Sale calido et frigido.

6. Ars Vitraria experimentalis.

7. Collegium Physico-chymicum experimentale, *or* Laboratorium chymicum.*

Nicolas Lemery, the first Frenchman who completely stripped chemistry of its mysticism, and presented it to the world in all its native simplicity, deserves our particular attention, in consequence of the celebrity which he acquired, and the benefits which he conferred on the science. He was born at Rouen on the 17th of November, 1645. His father, Julian Lemery, was *procureur* of the Parliament of Normandy, and a protestant. His son, when very young, showed a decided partiality for chemistry, and repaired to an apothecary in Rouen, a relation of his own, in hopes of being initiated into the science; but finding that little information could be procured from him, young Lemery left him in 1666, and went to Paris, where he boarded himself with M. Glaser, at that time demonstrator of chemistry at the Jardin du Roi.

Glaser was a *true chemist*, according to the meaning at that time affixed to the term—full of obscure notions—unwilling to communicate what knowledge he possessed—and not at all sociable. In two months Lemery quitted his house in disgust, and set out with a resolution to travel through France, and pick up chemical information as he best could, from those who were capable of giving him information on the subject. He first went to Montpelier, where he boarded in the house of M. Vershant, an apothecary in that town.

---

* I have never seen a copy of this last work; it must have been valuable, as it was the book from which Scheele derived the first rudiments of his knowledge.

With his situation there he was so much pleased, that he continued in it for three years : he employed himself assiduously in the laboratory, and in teaching chemistry to a number of young students who boarded with his host. Here his reputation gradually increased so much, that he drew round him the professors of the faculty of medicine of Montpelier, and all the curious of the place, to witness his experiments. Here, too, he practised medicine with considerable success.

After travelling through all France, he returned to Paris in 1672. Here he frequented the different scientific meetings at that time held in that capital, and soon distinguished himself by his chemical knowledge. In a few years he got a laboratory of his own, commenced apothecary, and began to give public lectures on chemistry, which were speedily attended by great crowds of students from foreign countries. For example, we are told that on one occasion forty Scotchmen repaired to Paris on purpose to hear his lectures, and those of M. Du Verney on anatomy. The medicines which he prepared in his laboratory became fashionable, and brought him a great deal of money. The magistery of bismuth (or pearl-white), which he prepared as a cosmetic, was sufficient, we are told, to support the whole expense of his house. In the year 1675 he published his Cours de Chimie, certainly one of the most successful chemical books that ever appeared ; it ran through a vast number of editions in a few years, and was translated into Latin, German, Spanish, and English.

In 1681 he began to be troubled in consequence of his religious opinions. Louis XIV. was at that time in the height of his glory, entirely under the control of his priests, and zealously bent upon putting an end to the reformed religion in his dominions. Indeed, from the infamous conduct of Charles II. of England, and the bigotry of his successor, a prospect was opened to him, and of which he was anxious to avail himself, of annihilating the reformed religion altogether, and of plunging Europe a second time into the darkness of Roman catholicism.

Lemery found it expedient, in 1683, to pass over into England. Here he was well received by Charles II. : but England was at that time convulsed with those religious and political struggles, which terminated five years afterwards in the revolution. Lemery, in consequence of this state of things, found it expedient to leave England, and return to France. He took a doctor's degree at Caen, in Normandy; and, returning to Paris, he commenced all at once practitioner in medicine and surgery, apothecary, and lecturer on chemistry. The edict of Nantes was revoked in 1685, when James II. had assured Louis of his intention to overturn the established religion, and bring Great Britain again under the dominion of the pope. Lemery was obliged to give up practice and conceal himself, in order to avoid persecution. Finding his success hopeless, as long as he continued a protestant, he changed his religion in 1686, and declared himself a Roman catholic. This step secured his fortune : he was now as much caressed and protected by the court and the clergy, as he had been formerly persecuted by them. In 1699 when the Academy of Sciences was new modelled, he was appointed associated chemist, and, on the death of Bourdelin, before the end of that year, he became a pensioner. He died on the 19th of June, 1715, at the age of seventy, in consequence of an attack of palsy, which terminated in apoplexy.

Besides his System of Chemistry, which has been already mentioned, he published the following works :

1. Pharmacopée universelle, contenant toutes les Operations de Pharmacie qui sont en usage dans la Médicine.

2. Traité universelle des Drogues simples mis en ordre alphabétique.

3. Traité de l'Antimoine, contenant l'analyse chimique de ce mineral.

Besides these works, five different papers by Lemery were printed in the Memoirs of the French Academy, between 1700 and 1709 inclusive. These are as follow:

1. Explication physique et chimique des Feux souterrains, des tremblemens de Terre, des Ouragans, des Eclairs et du Tonnere.—This explanation is founded on the heat and combustion produced by the mutual action of iron filings and sulphur on each other, when mixed in large quantities.

2. Du Camphre.

3. Du Miel et de son analyse chimique.

4. De l'Urine de Vache, de ses effets en médicine et de son analyse chimique.

5. Reflexions et Experiences sur le Sublimé Corrosive.—It appears from this paper, that in 1709, when Lemery wrote, corrosive sublimate was considered as a compound of mercury with the sulphuric and muriatic acids. Lemery's statement, that he made corrosive sublimate simply by heating a mixture of mercury and decrepitated salt, is not easily explained. Probably the salt which he had employed was impure. This is the more likely, because, from his account of the matter which remained at the bottom of the matrass after sublimation, it must have either contained peroxide of iron or peroxide of mercury, for its colour he says was red.

M. Lemery left a son, who was also a member of the French Academy; an active chemist, and author of various papers, in which he endeavours to give a mechanical explanation of chemical phenomena.

Another very active member of the French Academy, at the same time with Lemery, was M. William Homberg, who was born on the 8th of January, 1652, at Batavia, in the island of Java. His father, John

Homberg, was a Saxon gentleman, who had been stripped of all his property during the thirty years war. After receiving some education by the care of a relation, he went into the service of the Dutch East India Company, and got the command of the arsenal at Batavia. There he married the widow of an officer, by whom he had four children, of whom William was the second.

His father quitted the service of the India Company and repaired to Amsterdam with his family. Young Homberg studied with avidity: he devoted himself to the law, and in 1674 was admitted advocate of Magdeburg; but his taste for natural history and science was great. He collected plants in the neighbourhood, and made himself acquainted with their names and uses. At night he studied the stars, and learned the names and positions of the different constellations. Thus he became a self-taught botanist and astronomer. He constructed a hollow transparent celestial globe, on which, by means of a light placed within, the principal fixed stars were seen in the same relative positions as in the heavens.

Otto Guericke was at that time burgomaster of Magdeburg. His experiments on a vacuum, and his invention of the air-pump, are universally known. Homberg attached himself to Otto Guericke, and this philosopher, though fond of mystery, either explained to him his secrets, in consequence of his admiration of his genius, or was unable to conceal them from his penetration. At last Homberg, quite tired of his profession of advocate, left Magdeburg and went to Italy. He sojourned for some time at Padua, where he devoted himself to the study of medicine, anatomy, and botany. At Bologna he examined the famous Bologna stone, the nature of which had been almost forgotten, and succeeded in making a pyrophorus out of it. At Rome he associated particularly with Marc-Antony Celio, famous for the large glasses

for telescopes which he was able to grind.   Nor
did he neglect painting, sculpture, and music; pursuits in which, at that time, the Italians excelled all
other nations.

From Italy he went to France, and thence passed
into England, where he wrought for some time in the
laboratory of Mr. Boyle, at that time one of the most
eminent schools of science in Europe.   He then
passed into Holland, studied anatomy under De
Graaf, and after visiting his family, went to Wittem-
berg, where he took the degree of doctor of medicine.

After this he visited Baldwin and Kunkel, to get
more accurate information respecting the phosphorus
which each had respectively discovered.   He purchased a knowledge of Kunkel's phosphorus, by
giving in exchange a meteorological toy of Otto
Guericke, now familiarly known, by which the moisture or dryness of the air was indicated—a little man
came out of his house and stood at the door in dry
weather, but retired under cover in moist weather. He
next visited the mines of Saxony, Bohemia, and
Hungary : he even went to Sweden, to visit the copper-mines of that country. At Stockholm he wrought
in the chemical laboratory, lately established by the
king, along with Hjerna, and contributed considerably to the success of that new establishment.

He repaired a second time to France, where he
spent some time, actively engaged with the men of
science in Paris.   His father strongly pressed him to
return to Holland and settle as a physician : he at
last consented, and the day of his departure was
come, when, just as he was going into his carriage, he
was stopped by a message from M. Colbert on the
part of the king.   Offers of so advantageous a nature
were made him if he would consent to remain in
France, that, after some consideration, he was induced to embrace them.

In 1682 he changed his religion and became Roman catholic: this induced his father to disinherit him. In 1688 he went to Rome, where he practised medicine with considerable success. A few years after he returned to Paris, where his knowledge and discoveries gave him a very high reputation. In 1691 he became a member of the Academy of Sciences, and got the direction of the laboratory belonging to the academy: this enabled him to devote his undivided attention to chemical investigations. In 1702 he was taken into the service of the Duke of Orleans, who gave him a pension, and put him in possession of the most splendid and complete laboratory that had ever been seen. He was presented with the celebrated burning-glass of M. Tchirnhaus, by the Duke of Orleans, and was enabled by means of it to determine many points that had hitherto been only conjectural.

In 1704 he was made first physician to the Duke of Orleans, who honoured him with his particular esteem. This appointment obliging him to reside out of Paris, would have made it necessary for him to resign his seat in the academy, had not the king made a special exemption in his favour. In 1708 he married a daughter of the famous M. Dodart, to whom he had been long attached. Some years after he was attacked by a dysentery, which was cured, but returned from time to time. In 1715 it returned with great violence, and Homberg died on the 24th of September.

His knowledge was uncommonly great in almost every department of science. His chemical papers were very numerous; though there are few of them, in this advanced period of the science that are likely to claim much attention from the chemical world. His pyrophorus, of which he has given a description in the Mémoires de l'Académie,* was made by mixing

---

* For 1711, p. 238.

together human fæces and alum, and roasting the mixture till it was reduced to a dry powder. It was then exposed in a matrass to a red heat, till every thing combustible was driven off. Any combustible will do as a substitute for human fæces—gum, flour, sugar, charcoal, may be used. When a little of this phosphorus is poured upon paper, it speedily catches fire and kindles the paper. Davy first explained the nature of this phosphorus. The potash of the alum is converted into potassium, which, by its absorption of oxygen from the atmosphere, generates heat, and sets fire to the charcoal contained in the powder.

Homberg's papers printed in the Memoirs of the French Academy amount to thirty-one. They are to be found in the volumes for 1699 to 1714 inclusive.

M. Geoffroy, who was a member of the academy about the same time with Lemery and Homberg, though he outlived them both, and who was an active chemist for a considerable number of years, deserves also to be mentioned here.

Stephen Francis Geoffroy was born in Paris on the 13th of February, 1672, where his father was an apothecary. While a young man, regular meetings of the most eminent scientific men of Paris were held in his father's house, at which he was always present. This contributed very much to increase his taste for scientific pursuits. After this he studied botany, chemistry, and anatomy in Paris. In 1692 his father sent him to Montpelier, to study pharmacy in the house of a skilful apothecary, who at the same time sent his son to Paris, to acquire the same art in the house of M. Geoffroy, senior. Here he attended the different classes in the university, and his name began to be known as a chemist. After spending some time in Montpelier, he travelled round the coast to see the principal seaports, and was at St. Malo's in 1693, when it was bombarded by the British fleet.

In 1698 Count Tallard being appointed ambassador

extraordinary to London, made choice of M. Geoffroy as his physician, though he had not taken a medical degree. Here he made many valuable acquaintances, and was elected a fellow of the Royal Society. From London he went to Holland, and thence into Italy, in 1700, where he went in the capacity of physician to M. de Louvois. The great object of M. Geoffroy was always natural history, and materia medica. In 1693 he had subjected himself to an examination, and he had been declared qualified to act as an apothecary; but his own object was to be a physician, while that of his father was that he should succeed himself as an apothecary: this in some measure regulated his education. At last he declared his intentions, and his father agreed to them; he became bachelor of medicine in 1702, and doctor of medicine in 1704.

In 1709 he was made professor of medicine in the Royal College. In 1707 he began to lecture on chemistry, at the Jardin du Roi, in place of M. Fagan, and continued to teach this important class during the remainder of his life. In 1726 he was chosen dean of the faculty of medicine; and, after the two years for which he was elected was finished, he was again chosen to fill the same situation. There existed at that time a lawsuit between the physicians and surgeons in Paris; a kind of civil war very injurious to both; and the mildness and suavity of his manners fitted him particularly for being at the head of the body of physicians during its continuance. He became a member of the academy in 1699, and died on the 6th of January, 1731.

The most important of all his chemical labours, and for which he will always be remembered in the annals of the science, was the contrivance which he fell upon, in 1718, of exhibiting the order of chemical decompositions under the form of a table.* This method

---

* Mem. Paris, 1718, p. 202; and 1720, p. 20.

was afterwards much enlarged and improved. Such tables are now usually known by the name of *tables of affinity ;* and, though they have been of late years somewhat neglected, there can be but one opinion of their importance when properly constructed.

M. Geoffroy first communicated to the French chemists the mode of making Prussian blue, as Dr. Woodward did to the English.

Claude Joseph Geoffroy, the younger brother of the preceding, was also a member of the Academy of Sciences, and a zealous cultivator of chemistry. Many of his chemical papers are to be found in the memoirs of the French Academy. He demonstrated the composition of sal ammoniac, which however was known to Glauber. He made many experiments upon the combustion of the volatile oils, by pouring nitric acid on them. He explained the pretended property which certain waters have of converting iron into copper, by showing that in such cases copper was held in solution in the water by an acid, and that the iron merely precipitated the copper, and was dissolved and combined with the acid in its place. He pointed out the constituents of the three vitriols, the green, the blue, and the white ; showing that the two former were combinations of sulphuric acid with oxides of iron and copper, and the latter a solution of lapis calaminaris (*carbonate of zinc*) in the same acid. He has also a memoir on the emeticity of antimony, tartar emetic, and kermes mineral ; but it is rather medical than chemical. He determined experimentally the nature of the salt of Seignette, or Rochelle salt, and showed that it was obtained by saturating cream of tartar with carbonate of soda, and crystallizing. It is curious that this discovery was made about the same time by M. Boulduc. I have noticed only a few of the papers of M. Geoffroy, junior; because, though they all do him credit, and contributed to the improvement of chemistry, yet none of them contain any of those great discoveries, which stand as landmarks in the progress of science, and constitute an era in the history of mankind. For the same reason I omit several other names that, in a more minute history of chemistry, would deserve to be particularized.

CHAPTER VIII.

OF THE ATTEMPTS TO ESTABLISH A THEORY IN CHEMISTRY.

BACON, Lord Verulam, as early as the commencement of the 17th century, had pointed out the importance of chemical investigations, and had predicted the immense advantages which would result from the science, when it came to be properly cultivated and extended; but he did not himself attempt either to construct a theory of chemistry, or even to extend it beyond the bounds which it had reached before he began to write. Neither did Boyle, notwithstanding the importance of his investigations, and his comparative freedom from the prejudices of the alchymists, attempt any thing like a theory of chemistry; though the observations which he made in his Sceptical Chemist, had considerable effect in overturning, or at least in hastening the downfal of the absurd chemical opinions which at that time prevailed, and the puerile hypotheses respecting the animal functions, and the pathology and treatment of diseases founded on these opinions. The first person who can with propriety be said to have attempted to construct a theory of chemistry, was Beccher.

John Joachim Beccher, one of the most extraordinary men of the age in which he lived, was born at Spires, in Germany, in the year 1635. His father, as

Beccher himself informs us, was a very learned Lutheran preacher. As he lost his father when he was very young, and as that part of Germany where he lived had been ruined by the thirty years' war, his family was reduced to great poverty. However, his passion for information was so great, that he contrived to educate himself by studying what books he could procure, and in this way acquired a great deal of knowledge. Afterwards he travelled through the greatest part of Germany, Italy, Sweden, and Holland.

In the year 1666 he was appointed public professor of medicine in the University of Mentz, and soon after chief physician to the elector. In that capacity he took up his residence in Munich, where he was furnished by the elector with an excellent laboratory: but he soon fell into difficulties, the nature of which does not appear, and was obliged to leave the place. He took refuge in Vienna, where, from his knowledge of finance, he was appointed chamberlain to Count Zinzendorf, and through him acquired so much importance in the eyes of the court, that he was named a member of the newly-erected College of Commerce, and obtained the title of imperial commercial counsellor and chamberlain. But here also he speedily raised up so many enemies against himself, that he found it necessary to leave Vienna, and to carry with him his wife and children. He repaired to Holland, and settled at Haerlem in 1678. Here he was likely to have been successful; but his enemies from Vienna followed him, and obliged him to leave Holland. In 1680 we find him in Great Britain, where he examined the Scottish lead-mines, and smelting-works; and in 1681, and 1682, he traversed Cornwall, and studied the mines and smelting-works of that great mining county; here he suggested several improvements and ameliorations. Soon after this an advantageous proposal was made to him by the Duke of Mecklenburg Gustrow, by means of Count Zinzendorf; but all his projects were arrested by his death, which took place in the year 1682. It is said that he died in London, but I have not been able to find any evidence of this.

It would be a difficult task to particularize his various discoveries, which are scattered through a multiplicity of writings. He was undoubtedly the first discoverer of boracic acid, though the credit of the discovery has usually been given to Homberg.[*] But then he gives no account of boracic acid, nor does he seem to have attended to its qualities. The following is a list of Beccher's writings:

1. Metallurgia, or the Natural Science of Metals.
2. Institutiones Chymicæ.
3. Parnassus Medicinalis illustrata.
4. Œdipus Chymicus seu Institutiones Chymicæ.
5. Acta laboratorii Chymici Monacensis seu Physica Subterranea.—This, which is the most important of all his works, is usually known by the name of " Physica Subterranea." This is the sole title affixed to it in the edition published at Leipsic, in 1703, to which Stahl has prefixed a long introduction. It is divided into seven sections. In the first he treats of the creation of the world; in the second he gives a chemical account of the motions and changes which are constantly going on in the earth; in the third he treats of the three principles of all bodies, which he calls *earths*. The first of these principles of metals and stones is the *fusible* or *stony earth*; the second principle of minerals is the *fat earth*, improperly called *sulphur*; the third principle is the *fluid earth*, improperly called *mercury*; in the fourth section he treats of the action

---

[*] In the sixth chemical thesis, in the second supplement to the Physica Subterranea (page 791, Stahl's Edition. Lipsiæ, 1703), he says, " ubi etiam, continuato igne, ipsum sal volatile acquires, quod eadem methodo cum vitriolo seu spiritu aut oleo vitrioli, et oleo tartari, vel *borace* succedit."

of subterraneous principles, or the formation of *mixts;* in the fifth he treats of the solution of the three classes of mixts, animals, vegetables, and metals; in the sixth he treats of *mixts,* in which he gives their chemical constituents. This section is very curious, because it gives Beccher's views of the constitution of compound bodies. It will be seen from it that he had much more correct notions of the real objects of chemistry, than any of his contemporaries. In the seventh and last section he treats of the accidents and physical affections of subterraneous bodies.

6. Experimentum Chymicum novum quo artificialis et instantanea metallorum generatio et transmutatio, ad oculum demonstratur.—This constitutes the first supplement to the Physica Subterranea.

7. Supplementum secundum in Physicam subterraneam, demonstratio philosophica seu Theses Chymicæ, veritatem et possibilitatem transmutationis metallorum in aurum evincentes.

8. Trifolium Beccherianum Hollandicum.

9. Experimentum novum et curiosum de Minera arenaria perpetua, sive prodromus historiæ seu propositionis Præp. D.D. Hollandiæ ordinibus ab authore factæ, circa auri extractionem mediante arena littorali per modum mineræ perpetuæ seu operationis magnæ fusoriæ cum emolumento. Loco supplementi tertii in Physicam suam subterraneam.

10. Chemical Luckpot, or great chemical agreement; in a collection of one thousand five hundred chemical processes.

11. Foolish Wisdom and wise Folly.

12. Magnalia Naturæ.

13. Tripus Hermeticus fatidicus pandens oracula chemica; seu I. Laboratorium portatile, cum methodo vere spagyricæ seu juxta exigentiam naturæ laborandi. Accessit pro praxi et exemplo; II. Centrum mundi concatenatum seu Duumviratus hermeticus s. magnorum duorum productorum nitri et salis textura et ana-

tomia atque in omnium præcedentium confirmationem adjunctum est; III. Alphabetum Minerale seu viginti quatuor theses de subterraneorum mineralium genesi, textura et analysi; his accessit concordantia mercurii lunæ et menstruorum.

14. Chemical Rose-garden.

15. Pantaleon delarvatus.

16. Beccheri, Lancelotti, etc. Epistolæ quatuor Chemicæ.

Beccher's great merit was the contrivance of a chemical theory, by which all the known facts were connected together and deduced from one general principle. But as this theory was adopted and considerably modified by Stahl, it will be better to lay a sketch of it before the reader, after mentioning a few particulars of the life and labours of one of the most extraordinary men whom Germany has produced; a man who, in spite of the moroseness and haughtiness of his character, and in spite of the barbarity of his style, raised himself to the very first rank as a man of science; and had the rare or almost unique fortune of giving laws at the same time to two different and important sciences, which he cultivated together, without letting his opinions respecting the one influence him with regard to the other. These sciences were chemistry and medicine.

George Ernest Stahl was born at Anspach, in the year 1660. He studied medicine at Jena under George Wolfgang Wedel; and got his doctor's degree at the age of twenty-three. Immediately after this he began his career as a public lecturer. In 1687 the Duke of Weimar gave him the title of physician to the court. In 1694 he was named, at the solicitation of Frederick Hoffmann, second professor of medicine in the University of Halle, which had just been established. Hoffmann and he were at that time great friends, though they afterwards quarrelled. Both of them were men of the very highest talents and both were the founders of medical systems which, of course, each was anxious to support. Hoffmann had greatly the superiority in elegance and clearness of style, and in all the amenities of polite manners. But perhaps the moroseness of Stahl, and the obscurity, or rather mysticism of his style, contributed equally with the more amiable qualities of Hoffmann to excite the attention and produce the veneration with which he was viewed by his pupils, and, indeed, by the world at large.

At Halle he continued as a teacher of medicine for twenty-two years. In 1716 he was appointed physician to the King of Prussia. In consequence of this appointment he left Halle, and resided in Berlin, where he died in the year 1734, in the seventy-fifth year of his age. Notwithstanding the great figure that Stahl made as a chemist, there is no evidence that he ever taught that science in any public school. The Berlin Academy had been founded under the superintendence of Leibnitz, who was its first president; and therefore existed when Stahl was in Berlin : but, till it was renovated in 1745 by Frederick the Great, this academy possessed but little activity, and could scarcely, therefore, have stimulated Stahl to attend to chemical science. However, his Chymia rationalis et experimentalis was published in 1720, while he resided in Berlin. The same date is appended to the preface of his Fundamenta Chymiæ ; but, from some expressions in that preface, it must, I should think, have been written, not by Stahl, but by some other person.* I suspect that the book had been written by some of his pupils, from the lectures of the author while at Halle. If this was really the case, it is obvious

---

* "Primus in his facem prætulit Beccherus ; eumque magno cum artis progressu sequentem videmus in ostendenda corporum analysi et synthesi chymica versatissimum et acutissimum—Stahlium."

that Stahl must have taught chemistry as well as medicine in the University of Halle.

Stahl's medical theory is not less deserving of notice than his chemical. But it is not the object of this work to enter into medical speculations. Like Van Helmont, he resolved all diseases into the actions of the *soul*, which was not merely the former of the body, but its ruler and regulator. When any of the functions are deranged, the soul exerts itself to restore them again to their healthy state; and she accomplishes this by what in common language is called disease. The business of a medical man, then, is not to prevent diseases, or to stop them short when they appear; because they are the efforts of the soul, the *vis medicatrix naturæ*, to restore the deranged state of the functions : but he must watch these diseases, and prevent the symptoms from becoming too violent. He must assist nature to produce the intended effect, and check her exertions when they become abnormal. It was a kind of modification of this theory, or rather a mixture of the Stahlian and Hoffmannian theories, that Dr. Cullen afterwards taught in Edinburgh with so much eclat. And these opinions, so far as medical theories have any influence on practice, still continue in some measure prevalent. Indeed, much of the vulgar practice followed by medical men, chiefly in consequence of the education which they have received, is deduced from these two theories. But it would be too great a digression from the object of this work to enter into any details : suffice it to say, that the rival theories of Hoffmann and Stahl for many years divided the medical world in Germany, if not in the greater part of Europe. It was no small matter of exultation to so young a medical school as Halle, to have at once within its walls two such eminent teachers as Hoffmann and Stahl.

Let us turn our attention to the chemical writings of Stahl. Of these the most important is his Fundamenta

Chymiæ dogmaticæ et experimentalis. It is divided, like the chemistry of Boerhaave, into a theoretical and practical part. The perusal of it is very disagreeable, as it is full of German words and phrases, and symbols are almost constantly substituted for words, as was at that time the custom.

His definition of chemistry is much more exact than Boerhaave's. It is, according to him, the art of resolving compound bodies into their constituents, and of again forming them by uniting these constituents together.

He is inclined to believe with Beccher, that the simple principles are four in number. The *mixts* are compounds of these principles; and he shows by the doctrine of permutations that if we suppose the simple principles four, then the number of mixts will be 40,340. He treats in the first place of *mixts, compounds*, and *aggregates*.

The first object of chemistry is *corruption*, the second *generation*. Of these he treats at considerable length, giving an account of the different chemical processes, and of the apparatus employed.

He next treats of *salts*, which he defines mixts composed of water and earth, both simple and pure, and intimately united. The salts are vitriol, alum, nitre, common salt, and sal ammoniac. He next treats of more compound salts. These are sugar, tartar, salts from the animal and salts from the mineral kingdom, and quicklime.

After this comes sulphur, cinnabar, antimony, the sulphur of vitriol, the sulphur of nitre, resins, and distilled oils. Then he treats of water, which he divides into aqua *humida* or common water, and aqua *sicca* or mercury. Next he treats of earths, which are of two kinds, viz., *friable earths*, such as *clay*, *loam*, sand, &c., and metallic earths constituting the bases of the metals.

He next treats of the metals; and, as a preliminary,

we have a description of the method of smelting, and operating upon the different metals. The metals are then described successively in the following order: Gold, silver, copper, iron, tin, lead, bismuth, zinc, antimony.

To this part of the system are added three sections. The first treats of mercuries, the second of the philosopher's stone, and the third of the universal medicine. We must not suppose that Stahl was a believer in these ideal compositions; his object is merely to give a history of the different processes which had been recommended by the alchymists.

The second part of his work is divided into two *tracts.* The first tract contains three sections. The first of these treats of the nature of solids and fluids, of solutions and menstrua, of the effects of heat and fire, of effervescence and boiling, of volatilization, of fusion and liquefaction, of distillation, of precipitation, of calcination and incineration, of detonation, of amalgamation, of crystallization and inspissation, and of the fixity and firmness of bodies. In the second section we have an account of salts, and of their generation and transmutation, of sulphur and inflammability, of phosphorus, of colours, and of the nature of metals and minerals. In this article he gives short definitions of these bodies, and shows how they may be known. The bodies thus defined are gold, silver, iron, copper, lead, tin, mercury, antimony, sulphur, arsenic, vitriol, common salt, nitre, alum, sal ammoniac, alkalies, and salts ; viz., muriatic acid, sulphuric, nitric, and sulphurous.

In the third section he treats of the method of reducing metallic calces, of the mode of separating metals from their scoriæ, of the mode of making artificial gems, and finally of the mode of giving copper a golden colour.

The second tract is divided into two parts. The first part is subdivided into four sections. In the first section he treats of the instruments of chemical motion, of fire, of air, of water, of the most subtile earth or salt. In the second section he treats *de subjectis*, under the several heads of dissolving aggregates, of triturations and solutions, and of calcinations and combustions. In the third section he treats of the object of chemistry under the following heads: Of chemical corruption, consisting of compounds from liquids, of the separation of solids and fluids, of mixts, of the solution of compounds from solids. In the fourth section he treats of fermentation.

The second part of this second tract treats of chemical generation, and is divided into two sections. In the first section he treats of the aggregate collection of bodies into fluids and solids. The section treats of compositions under the heads of volatile and solid bodies. He gives in the last article an account of the combination of mixts.

The third and last part of this elaborate work discusses three subjects; viz. *zymotechnia* or *fermentation*, *halotechnia*, or the production and properties of salts, and *pyrotechnia*, in which the whole of the Stahlian doctrine of *phlogiston* is developed. This third part has all the appearance of having been notes written down by some person during the lectures of Stahl : for it consists of alternate sentences of Latin and German. It is not at all likely that Stahl himself would have produced such a piebald work; but if he lectured in Latin, as was at that time the universal custom, it was natural for a person occupied in taking down the lectures, to write as far as was possible in Latin, but when any of the Latin phrases were lost, or did not immediately occur to memory, it were equally natural to write down the meaning of what the professor stated in the language most familiar to the writer, which was undoubtedly the German.

Another of Stahl's works is entitled "Opusculum Chymico-physico-medicum," published at Halle in a

thick quarto volume, in the year 1715. It contains a
great number of tracts, partly chemical and partly
medical, which it is needless to specify. Perhaps the
most curious of them all is his dissertation to show the
way in which Moses ground the golden calf to powder,
dissolved it in water, and obliged the children of Israel
to drink it. He shows that a solution of hepar sul-
phuris (*sulphuret of potassium*), has the property of
dissolving gold, and he draws as a conclusion from his
experiments that this was the artifice employed by
Moses. We have in the same volume a pretty detailed
treatise on metallurgic pyrotechny and docimasy. This
is the more curious, because Stahl never appears to
have frequented the mines and smelting-houses of
Germany. He must, therefore, have drawn his information from books and from experiment.

Another of his books is entitled " Experimenta, Ob--
servationes, Animadversiones, CCC. Numero." An
octavo volume, printed at Berlin in 1731. Another of
his books is entitled "Specimen Beccherianum." There
are also two chemical books of Stahl, which I have
seen only in a French translation, viz., *Traité de
Soufre* and *Traité de Sels*. These are the only chemical writings of Stahl that I have seen. There are
probably others; indeed I have seen the titles of several other chemical works ascribed to him. But as it
is doubtful whether he really wrote them or not, I
think it unnecessary to specify them here.

Stahl's writings evince the great progress which
chemistry had made even since the time of Beccher.
But it is difficult to say what particular new facts,
which appear first in his writings were discovered by
himself, and what by others. I shall not, therefore,
attempt any enumeration of them. His reasoning is
more subtile, and his views much more extensive and
profound than those of his predecessors. The great
improvement which he introduced into chemistry was
the employment of *phlogiston*, to explain the phe-

nomena of combustion and calcination. This theory
had been originally broached by Beccher, from whom
Stahl evidently borrowed it, but he improved and simplified it so much that the whole credit of it was given
to him. It was called the Stahlian theory, and raised
him to the highest rank among chemists. The sole
objects of chemists for thirty or forty years after his
time was to illucidate and extend his theory. It applied
so happily to all the known facts, and was supported
by experiments, which appeared so decisive that nobody thought of calling it in question, or of interrogating nature in any other way than he had pointed
out. It will be requisite, therefore, before proceeding
further with this historical sketch, to lay the outlines
of the phlogistic theory before the reader.

It was conceived by Beccher and Stahl that all
combustible bodies are compounds. One of the constituents they supposed to be dissipated during the
combustion, while the other constituent remained behind. Now when combustible bodies are subjected to
combustion, some of them leave an acid behind them ;
while others leave a fixed powdery matter, possessing
the properties of an *earth*, and called usually the
calx of the combustible body. The metals are the
substances which leave a calx behind them when
burnt, and sulphur and phosphorus leave an acid.
With respect to those bodies that would not burn,
chemists did not speculate much at first; but afterwards they came to think that they consisted of the
fixed substance that remained after combustion.
Hence the conclusion was natural, that they had
already undergone combustion. Thus quicklime
possessed properties very similar to the calces of metals.
It was natural, therefore, to consider it as a calx, and
to believe that if the matter dissipated during combustion could be again restored, lime would be converted into a substance similar to the metals.

Combustibility then, according to this view of the

subject, depends upon a principle or material substance, existing in every combustible body, and dissipated during the combustion. This substance was considered to be absolutely the same in all combustible bodies whatever; hence the difference between combustible bodies proceeded from the other principle or number of principles with which this common substance is combined. In consequence of this identity Stahl invented the term *phlogiston*, by which he denoted this common principle of combustible bodies. Inflammation, with the several phenomena that attend it, depended on the gradual separation of this principle, which being once separated, what remained of the body could no longer be an inflammable substance, but must be similar to the other kinds of matter. It was this opinion that combustibility is owing to the presence of phlogiston, and inflammation to its escape, that constituted the peculiar theory of Beccher, and which was afterwards illustrated by Stahl with so much clearness, and experiments to prove its truth were advanced by him of so much force, that it came to be distinguished by the name of the Stahlian theory.

The identity of phlogiston in all combustible bodies was founded upon observations and experiments of so decisive a nature, that after the existence of the principle itself was admitted, they could not fail to be satisfactory. When phosphorus is made to burn it gives out a strong flame, much heat is evolved, and the phosphorus is dissipated in a white smoke: but if the combustion be conducted within a glass vessel of a proper shape, this white smoke will be deposited on the inside of the glass; it quickly absorbs moisture from the atmosphere, and runs into an acid liquid, known by the name of phosphoric acid. If this liquid be put into a platinum crucible, and gradually heated to redness, the water is dissipated, and a substance remains which, on cooling, congeals into a transparent colourless body like glass: this is dry *phosphoric* acid. If now we mix phosphoric acid with a quantity of charcoal powder, and heat it sufficiently in a glass retort, taking care to exclude the external air, a *portion* or the *whole* of the charcoal will disappear, and phosphorus will be formed possessed of the same properties that it had before it was subjected to combustion. The conclusion deduced from this process appeared irresistible; the charcoal, or a portion of it, had combined with the phosphoric acid, and both together had constituted phosphorus.

Now, in changing phosphoric acid into phosphorus, we may employ almost any kind of combustible substance that we please, provided it be capable of bearing the requisite heat; they will all equally answer, and will all convert the acid into phosphorus. Instead of charcoal we may take lamp-black, or sugar, or resin, or even several of the metals. Hence it was concluded that all of these bodies contain a common principle which they communicate to the phosphoric acid; and since the new body formed is in all cases identical, the principle communicated must also be identical. Hence combustible bodies contain an identical principle, and this principle is phlogiston.

Sulphur by burning is converted into sulphuric acid; and if sulphuric acid be heated with charcoal, or phosphorus, or even sulphur, it is again converted into sulphur. Several of the metals produce the same effect. The reasoning here was the same as with regard to phosphoric acid, and the conclusion was similar.

When lead is kept nearly at a red heat in the open air for some time, being constantly stirred to expose new surfaces to the air, it is converted into the beautiful pigment called *red lead;* this is a calx of lead. To restore this calx again to the state of metallic lead, we have only to heat it in contact with almost any combustible matter whatever. Pit-coal, peat, charcoal, sugar, flour, iron, zinc, &c., all these bodies then must contain one common principle, which they communicate to red lead, and by so doing convert it into lead. This common principle is phlogiston.

These examples are sufficient to show the reader the way in which Stahl proved the identity of phlogiston in all combustible bodies. And the demonstration was considered as so complete that the opinion was adopted by every chemist without exception.

When we inquire further, and endeavour to learn what qualities phlogiston was supposed to have in its separate state, we find this part of the subject very unsatisfactory, and the opinions very unsettled. Beccher and Stahl represented phlogiston as a dry substance, or of an earthy nature, the particles of which are exquisitely subtile, and very much disposed to be agitated and set in motion with inconceivable velocity. This was called by Stahl *motus verticillaris*. When the particles of any body are agitated with this kind of motion, the body exhibits the phenomena of heat or ignition, or inflammation, according to the violence and rapidity of the motion.

This very crude opinion of the earthy nature of phlogiston, appears to have been deduced from the insolubility of most combustible substances in water. If we except alcohol, and ether, and gums, very few of them are capable of being dissolved in that liquid. Thus the metals, sulphur, phosphorus, oils, resins, bitumens, charcoal, &c., are well known to be insoluble. Now, at the time that Beccher and Stahl lived, insolubility in water was considered as a character peculiar to earthy bodies; and as those bodies which contain a great deal of phlogiston are insoluble in water, though the other constituents be very soluble in that liquid, it was natural enough to conclude that phlogiston itself was of an earthy nature.

But though the opinions of chemists about the nature and properties of phlogiston in a separate state were unsettled, no doubts were entertained respecting

its existence, and respecting its identity in all combustible bodies. Its presence or its absence produced
almost all the changes which bodies undergo. Hence
chemistry and combustion came to be in some measure
identified, and a theory of combustion was considered
as the same thing with a theory of chemistry.

Metals were compounds of *calces* and phlogiston.
The different species of metals depend upon the different species of calx which each contains; for there
are as many *calces* (each simple and peculiar) as there
are metals. These calces are capable of uniting with
phlogiston in indefinite proportions. The calx united
to a little phlogiston still retains its earthy appearance
—a certain additional portion restores the calx to the
state of a metal. An enormous quantity of phlogiston
with which some calces, as calx of manganese, are
capable of combining, destroys the metallic appearance of the body, and renders it incapable of dissolving in acids.

The affinity between a metallic calx and phlogiston
is strong; but the facility of union is greatly promoted
when the calx still retains a little phlogiston. If we
drive off the whole phlogiston we can scarcely unite
the calx with phlogiston again, or bring it back to the
state of a metal: hence the extreme difficulty of reducing the calx of zinc, and even the red calx of iron.

The various colours of bodies are owing to phlogiston, and these colours vary with every alteration in the
proportion of phlogiston present.

It was observed very early that when a metal was
converted into a calx its weight was increased. But
this, though known to Beccher and Stahl, does not
seem to have had any effect on their opinions. Boyle,
who does not seem to have been aware of the phlogistic theory, though it had been broached before his
death, relates an experiment on tin which he made.
He put a given weight of it into an open glass vessel,
and kept it melted on the fire till a certain portion of

it was converted into a calx : it was now found to have increased considerably in weight. This experiment he relates in order to prove the materiality of heat : in his opinion a certain quantity of heat had united to the tin and occasioned the increase of weight. This opinion of Boyle was incompatible with the Stahlian theory : for the tin had not only increased in weight, but had been converted into a calx. It was therefore the opinion of Boyle that calx of tin was a combination of *tin* and *heat*. It could not consequently be true that calx of tin was tin deprived of phlogiston.

When this difficulty struck the phlogistians, which was not till long after the time of Stahl, they endeavoured to evade it by assigning new properties to phlogiston. According to them it is not only destitute of weight, but endowed with a principle of levity. In consequence of this property, a body containing phlogiston is always lighter than it would otherwise be, and it becomes heavier when the phlogiston makes its escape : hence the reason why calx of tin is heavier than the same tin in the metallic state. The increase of weight is not owing, as Boyle believed, to the fixation of heat in the tin, but to the escape of phlogiston from it.

Those philosophic chemists, who thus refined upon the properties of phlogiston, did not perceive that by endowing it with a principle of levity, they destroyed all the other characters which they had assigned to it. What is gravity ? Is it not an attraction by means of which bodies are drawn towards each other, and remain united ? And is there any reason for supposing that chemical attraction differs in its nature from the other kinds of attraction which matter possesses ? If, then, phlogiston be destitute of gravity, it cannot possess any attraction for other bodies ; if it be endowed with a principle of levity, it must have the property of repelling other bodies, for that is the only meaning that can be attached to the term. But if phlogiston has the property of repelling all other substances, how comes it to be fixed in combustible bodies ? It must be united to the calces or the acids, which constitute the other principle of these bodies ; and it could not be united, and remain united, unless a principle of attraction existed between it and these bases ; that is to say, unless it possessed a principle the very opposite of levity.

Thus the fact, that calces are heavier than the metals from which they are formed, in reality overturned the whole doctrine of phlogiston; and the only reason why the doctrine continued to be admitted after the fact was known is, that in these early days of chemistry, the balance was scarcely ever employed in experimenting : hence alterations in weight were little attended to or entirely overlooked. We shall see afterwards, that when Lavoisier introduced a more accurate mode of experimenting, and rendered it necessary to compare the original weights of the substances employed, with the weights of the products, he made use of this very experiment of Boyle, and a similar one made with mercury, to overturn the whole doctrine of phlogiston.

The phlogistic school being thus founded by Stahl, in Berlin, a race of chemists succeeded him in that capital, who contributed in no ordinary degree to the improvement of the science. The most deservedly celebrated of these were Neumann, Pott, Margraaf, and Eller.

Caspar Neumann was born at Zullichau, in Germany, in 1682. He was early received into favour by the King of Prussia, and travelled at the expense of that monarch into Holland, England, France, and Italy. During these travels he had an opportunity of making a personal acquaintance with the most eminent men of science in all the different countries which he visited. On his return home, in 1724, he was appointed professor of chemistry in the Royal College of Physic and Surgery at Berlin, where he delivered a course of lectures annually. During the remainder of his life he enjoyed the situation of superintendent of the Royal Laboratory, and apothecary to the King of Prussia. He died in 1737. He was a Fellow of the Royal Society, and several papers of his appeared in the Transactions of that learned body. The following is a list of these papers, all of which were written in Latin :

1. Disquisitio de camphora.
2. De experimento probandi spiritum vini Gallici, per quam usitato, sed revera falso et fallaci.

Some merchants in Holland, England, Hamburg, and Dantzic, were in possession of what they considered an infallible test to distinguish French brandy from every other kind of spirit. It was a dusky yellowish liquid. When one or two drops of it were let fall into a glass of French brandy, a beautiful blue colour appeared at the bottom of the glass, and when the brandy is stirred, the whole liquid becomes azure. But if the spirit tried be malt spirit, no such colour appears in the glass. Neumann ascertained that the test liquid was merely a solution of sulphate of iron in water, and that the blue colour was the consequence of the brandy having been kept in oak casks, and thus having dissolved a portion of tannin. Every spirit will exhibit the same colour, if it has been kept in oak casks.

3. De salibus alkalino-fixis.
4. De camphora thymi.
5. De ambragrysea.

His other papers, published in Germany, are the following :

In the Ephemerides.

1. De oleo distillato formicorum æthereo.
2. De albumine ovi succino simili.

In the Miscellania Berolinensia.

1. Meditationes in binas observationes de aqua per

putrefactionem rubra, vulgo pro tali in sanguinem versa habita.

2. Succincta relatio exactis Pomeraniis de prodigio sanguinis in palude viso.

3 De prodigio sanguinis ex Pomeranio nunciato.

4. Disquisitio de camphora.

5. De experimento probandi spiritum vini Gallicum.

6. De spiritu urinoso caustico.

7. Demonstratio syrupum violarum ad probanda liquida non sufficere.

8. Examen correctionis olei raparum.

9. De vi caustica et conversione salium alkalino-fixorum aëri expositorum in salia neutra.

He published separately,

1. De salibus alkalino-fixis et camphora.

2. De succino, opio, caryophyllis aromaticis et castoreo.

3. On saltpetre, sulphur, antimony, and iron.

4. On tea, coffee, beer, and wine.

5. Disquisitio de ambragrysea.

6. On common salt, tartar, sal ammoniac and ants.

After Neumann's death, two copies of his chemical lectures were published. The first consisting of notes taken by one of his pupils, intermixed with incoherent compilations from other authors, was printed at Berlin in 1740. The other was printed by the booksellers of the Orphan Hospital of Zullichau (the place of Neumann's birth), and is said to have been taken from the original papers in the author's handwriting. Of this last an excellent translation, with many additions and corrections, was published by Dr. Lewis, in London, in the year 1759; it was entitled, " The Chemical Works of Caspar Neumann, M. D., Professor of Chemistry at Berlin, F. R. S., &c. Abridged and methodized; with large additions, containing the later discoveries and improvements made in Chemistry, and the arts depending thereon. By William Lewis, M.B.,

F.R.S. London, 1759." This is an excellent book, and contains many things that still retain their value, notwithstanding the improvements which have been made since in every department of chemistry.

I have reason to believe that the laborious part of this translation and compilation was made by Mr. Chicholm, whom Dr. Lewis employed as his assistant. Mr. Chicholm, when a young man, went to London from Aberdeen, where he had studied at the university, and acquired a competent knowledge of Greek and Latin, but no means of supporting himself. On his arrival in London, one of the first things that struck his attention was a Greek book, placed open against the pane of a bookseller's window. Chicholm went up to the window, at which he continued standing till he had perused the whole Greek page thus exposed to his view. Dr. Lewis happened to be in the shop: he had been looking out for a young man whom he could employ to take charge of his laboratory, and manage his processes, and who should possess sufficient intelligence to read chemical works for him, and collect out of each whatever deserved to be known, either from its novelty or ingenuity. The appearance and manners of Chicholm struck him, and made him think of him as a man likely to answer the purposes which he had in view. He called him into the shop, and after some conversation with him, took him home, and kept him all his life as his assistant and operator. Chicholm was a laborious and painstaking man, and by continually working in Lewis's laboratory, soon acquired a competent knowledge of chemistry. He compiled several manuscript volumes, partly consisting of his own experiments, and partly of collections from other authors. At Dr. Lewis's death, all his books were sold by auction, and these manuscript volumes among the rest. They were purchased by Mr. Wedgewood, senior, who at the same time took Mr. Chicholm into his service, and gave him the charge of his own laboratory. It was Mr. Chicholm that was the constructor of the well-known piece of apparatus known by the name of Wedgewood's pyrometer. After his death the instrument continued still to be constructed for some time; but so many complaints were made of the unequal contraction of the pieces, that Mr. Wedgewood, junior, who had succeeded to the pottery in consequence of the death of his father, put an end to the manufacture of them altogether.

John Henry Pott was born at Halberstadt, in the year 1692. He was a scholar of Hoffmann and Stahl, and from this last he seems to have imbibed his taste for chemistry. He settled at Berlin, where he became assessor of the Royal College of Medicine and Surgery, inspector of medicines, superintendent of the Royal Laboratory, and dean of the Academy of Sciences of Berlin. He was chosen professor of theoretical chemistry at Berlin; and on the death of Neumann, in 1737, he succeeded him as professor of practical chemistry. He was beyond question the most learned and laborious chemist of his day. His erudition, indeed, was very great; and his historical introductions to his dissertation displays the extent of his reading on every subject of which he had occasion to treat. It has often struck me that the historical introductions which Bergmann has prefixed to his papers, are several of them borrowed from Pott. The Lithogeognosia of Pott is one of the most extraordinary productions of the age in which he lived. It was the result of a request of the King of Prussia, to discover the ingredients of which Saxon porcelain was made. Mr. Pott, not being able to procure any satisfactory information relative to the nature of the substances employed at Dresden, resolved to undertake a chemical examination of all the substances that were likely to be employed in such a manufacture. He tried the effect of fire upon all the stones, earths, and minerals, that he could procure, both separately and mixed together in various proportions. He made at least thirty thousand experiments in six years, and laid the foundation for a chemical knowledge of these bodies.* It is to this work of Pott that we are indebted for our knowledge of the effects of heat upon various earthy bodies, and upon mixtures of them. Thus he found that pure white clay, or mixtures of pure clay and quartz-sand, would not fuse at any temperature which he could produce ; but clay, mixed with lime or with oxide of iron, enters speedily into fusion. Clay also fuses with its own weight of borax ; it forms a compact mass with half its weight, and does not concrete into a hard body when mixed with a third of its weight of that salt. Clay fuses easily with fluor spar ; it fuses, also, with twice its weight of protoxide of lead, and with its own weight of sulphate of lime, but with no other proportion tried. It was a knowledge of these mutual actions of bodies on each other, when exposed to heat, that gradually led to the methods of examining minerals by the blowpipe. These methods were brought to the present state of perfection by Assessor Gahn, of Fahlun, the result of whose labours has been published by Berzelius, in his treatise on the blowpipe. Pott died in 1777, in the eighty-fifth year of his age.

His different chemical works (his Lithogeognosia excepted) were collected and translated into French by M. Demachy, in the year 1759, and published in four small octavo volumes. The chemical papers contained in these volumes are thirty-two in number. Some of these papers cannot but appear somewhat extraordinary to a modern chemist: for example, M. Duhamel had

* There is a French translation of this work, entitled " Litheognosie, ou Examen Chymique des Pierres et des Terres en général, et du Talc de la Topaz, et de la Steatite en particulier ; avec une Dissertation sur le Feu et sur la Lumière." Paris, 1753. With a continuation, constituting a second volume, in which all the experiments in the first volume are exhibited in the form of tables.

published in the memoirs of the French Academy, in the year 1737, a set of experiments on common salt, from which he deduced that its basis was a fixed alkali, which possessed properties different from those of potash, and which of course required to be distinguished by a peculiar name. It is sufficiently known that the term *soda* was afterwards applied to this alkali; by which name it is known at present. Pott, in a very elaborate and long dissertation on the base of common salt, endeavours to refute these opinions of Duhamel. The subject was afterwards taken up by Margraaf, who demonstrated, by decisive experiments, that the base of common salt is *soda;* and that soda differs essentially in its properties from potash.

Pott's dissertation on *bismuth* is of considerable value. He collects in it the statements and opinions of all preceding writers on this metal, and describes its properties with considerable accuracy and minuteness. The same observations apply to his dissertation on zinc.

John Theodore Eller, of Brockuser, was born on the 29th of November, 1689, at Pletzkau, in the principality of Anhalt Bernburg. He was the fourth son of Jobst Hermann Eller, a man of a respectable family, whose ancestors were proprietors of considerable estates in Westphalia and the Netherlands. Young Eller received the rudiments of his education in his father's house, from which he went to the University of Quedlinburg; and from thence to the University of Jena, in 1709. He was sent thither to study law; but his passion was for natural philosophy, which led him to devote himself to the study of medicine. From Jena he went to Halle, and finally to Leyden, attracted by the reputation of the older Albinus, of Professor Sengerd and the celebrated Boerhaave, at that time in the height of his reputation. The only practical anatomist then in Leyden, was M. Bidloo, an old man of eighty, and of course

unfit for teaching. This induced Eller to repair to
Amsterdam, to study under Rau, and to inspect the
anatomical museum of Ruysch. Bidloo soon dying,
Rau was appointed his successor at Leyden, whither
Eller followed him, and dissected under him till the
year 1716. After taking his degree at Leyden, Eller
returned to Germany, and devoted a considerable
time to the study and examination of the mines of
Saxony and the Hartz, and of the metallurgic processes connected with these mines. From these mines
he repaired to France, and resumed his anatomical
studies under Du Verney and Winslow. Chemistry
also attracted a good deal of his attention, and he frequented the laboratories of Grosse, Lemery, Bolduc,
and Homberg, at that time the most eminent chemists
in Paris.

From Paris he repaired to London, where he formed
an acquaintance with the numerous medical men of
eminence who at that time adorned this capital. On
returning to Germany in 1721, he was appointed physician to Prince Victor Frederick of Anhalt Bernburg.
From Bernburg he went to Magdeburg; and the
King of Prussia called him to Berlin in 1724, to teach
anatomy in the great anatomic theatre which had been
just erected. Soon after he was appointed physician
to the king, a counsellor and professor in the Royal
Medico-Chirurgical College, which had been just
founded in Berlin. He was also appointed dean of
the Superior College of Medicine, and physician to
the army and to the great Hospital of Frederick. In
the year 1755 Frederick the Great made him a privy-
counsellor, which is the highest rank that a medical
man can attain in Prussia. The same year he was
made director of the Royal Academy of Sciences of
Berlin. He died in the year 1760, in the seventy-first
year of his age. He was twice married, and his second
wife survived him.

Many chemical papers of Eller are to be found in

the memoirs of the Berlin Academy. They were of sufficient importance, at the time when he published them, to add considerably to his reputation, though not sufficiently so to induce me to give a catalogue of them here. I am not aware of any chemical discovery for which we are indebted to him ; but have been induced to give this brief notice of him, because he is usually associated with Pott and Margraaf, making with them the three celebrated chemists who adorned Berlin, during the splendid reign of Frederick the Great.

Andrew Sigismund Margraaf was born in Berlin, in the year 1709, and acquired the first principles of chemistry from his father, who was an apothecary in that city. He afterwards studied under Neumann, and travelling in quest of information to Frankfort, Strasburg, Halle, and Freyburg, he returned to Berlin enriched with all the knowledge of his favourite science which at that time existed. In 1760, on the death of Eller, he was made director of the physical class of the Berlin Academy of Sciences. He died in the year 1782, in the seventy-third year of his age. He gradually acquired a brilliant reputation in consequence of the numerous chemical papers which he successively published, each of which usually contained a new chemical fact, of more or less importance, deduced from a set of experiments generally satisfactory and convincing. His papers have a greater resemblance to those of Scheele than of any other chemist to whom we can compare them. He may be considered as in some measure the beginner of chemical analysis; for, before his time, the chemical analysis of bodies had hardly been attempted. His methods, as might have been expected, were not very perfect; nor did he attempt numerical results. His experiments on phosphorus and on the method of extracting it from urine are valuable; they communicated the first accurate notions relative to this

substance and to phosphoric acid. He first determined the properties of the earth of alum, now known
by the name of *alumina*; showed that it differed from
every other, and that it existed in clay, and gave
to that substance its peculiar properties. He demonstrated the peculiar nature of soda, the base of
common salt, which Pott had called in question, and
thus verified the conclusions of Duhamel. He gives
an easy process for obtaining pure silver from the
chloride of that metal: his method is to dissolve
the pure chloride of silver in a solution of caustic
ammonia, and to put into the liquid a sufficient
quantity of pure mercury; the silver is speedily reduced
and converted into an amalgam, and when this
amalgam is exposed to a red heat the mercury is
driven off and pure silver remains. The usual method
of reducing the chloride of silver is to heat it in a
crucible with a sufficient quantity of carbonate of
potash, a process which was first recommended by
Kunkel. But it is scarcely possible to prevent the
loss of a portion of the silver when the chloride is
reduced in this way. The modern process is undoubtedly the simplest and the best, to reduce it
by means of hydrogen. If a few pieces of zinc be
put into the bottom of a beer-glass and some dilute
sulphuric acid be poured over it an effervescence
takes place, and hydrogen gas is disengaged. Chloride of silver, placed above the zinc in the same
glass, is speedily reduced by this hydrogen and converted into metallic silver.

Margraaf's chemical papers, down to the time of
publication, were collected together, translated into
French and published at Paris in the year 1762,
in two very small octavo volumes, they consist of
twenty-six different papers: some of the most curious
and important of which are those that have been
just particularized. Several other papers written by
him appeared in the memoirs of the Berlin Academy,

after this collection of his works was published, particularly " A demonstration of the possibility of drawing fixed alkaline salts from tartar by means of acids, without employing the action of a violent fire." It was this paper, probably, that led Scheele, a few years after, to his well-known method of obtaining tartaric acid, a modification of which is still followed by manufacturers.

" Observations concerning a remarkable volatilization of a portion of a kind of stone known by the names of flosse, flusse, fluor spar, and likewise by that of hesperos: which volatilization was effectuated by means of acids." Pott had already shown the value of fluor spar as a flux. Three years after the appearance of Margraaf's paper, Scheele discovered the nature of fluor spar, and first drew the attention of chemists to the peculiar properties of fluoric acid.

In France, in consequence chiefly of the regulations established in the Academy of Sciences, in the year 1699, a race of chemists always existed, whose specific object was to cultivate chemistry, and extend and improve it. The most eminent of these chemical labourers, after the Stahlian theory was fully admitted in France till its credit began to be shaken, were Reaumur, Hellot, Duhamel, Rouelle, and Macquer. Besides these, who were the chief chemists in the academy, there were a few others to whom we are indebted for chemical discoveries that deserve to be recorded.

René Antoine Ferchault, Esq., Seigneur de Reaumur, certainly one of the most extraordinary men of his age, was born at Rochelle, in 1683. He went to the school of Rochelle, and afterwards studied philosophy under the Jesuits at Poitiers. Hence he went to Bourges, to which one of his uncles, canon of the holy chapel in that city, had invited him. At this time he was only seventeen years of age, yet his parents ventured to intrust a younger brother to his care, and this care he discharged with all the fidelity and sagacity of a much older man. Here he devoted himself to mathematics and physics, and he soon after went to Paris to improve the happy talents which he had received from nature. He was fortunate enough to meet with a friend and relation in the president, Henault, equally devoted to study with himself, equally eager for information, and possessed of equal honour and integrity, and equally promising talents.

He came to Paris in 1703. In 1708 he was admitted into the Academy of Sciences, in the situation of *élève* of M. Varignon, vacant by the promotion of M. Saurin to the rank of associate.

The first papers of his which were inserted in the Memoirs of the Academy were geometrical : he gave a general method of finding an infinity of curves, described by the extremity of a straight line, the other extremity of which, passing along the surface of a given curve, is always obliged to pass through the same point. Next year he gave a geometrical work on Developes ; but this was the last of his mathematical tracts. He was charged by the academy with the task of giving a description of the arts, and his taste for natural history began to draw to that study the greatest part of his attention. His first work as a naturalist was his observations on the formation of shells. It was unknown whether shells increase by intussusception, like animal bodies, or by the exterior and successive addition of new parts. By a set of delicate observations he showed that shells are formed by the addition of new parts, and that this was the cause of the variety of colour, shape, and size which they usually affect. His observations on snails, with a view to the way in which their shells are formed, led him to the discovery of a singular insect, which not only lives on snails, but in the inside of their bodies, from which it never stirs till driven out by the snail.

During the same year, he wrote his curious paper on the silk of spiders. The experiments of M. Bohn had shown that spiders could spin a silk that might be usefully employed. But it remained to be seen whether these creatures could be fed with profit, and in sufficiently great numbers to produce a sufficient quantity of silk to be of use. Reaumur undertook this disagreeable task, and showed that spiders could not be fed together without attacking and destroying one another.

The next research which he undertook, was to discover in what way certain sea-animals are capable of attaching themselves to fixed bodies, and again disengaging themselves at pleasure. He discovered the various threads and pinnæ which some of them possess for this purpose, and the prodigious number of limbs by which the sea-star is enabled to attach itself to solid bodies. Other animals employ a kind of cement to glue themselves to those substances to which they are attached, while some fix themselves by forming a vacuum in the interval between themselves and the solid substances to which they are attached.

It was at this period that he found great quantities of the buccinum, which yielded the purple dye of the ancients, upon the coast of Poitou. He observed, also, that the stones and little sandy ridges round which the shellfish had collected were covered with a kind of oval grains, some of which were white, and others of a yellowish colour, and having collected and squeezed some of these upon the sleeve of his shirt, so as to wet it with the liquid which they contained, he was agreeably surprised in about half an hour to find the wetted spot assume a beautiful purple colour, which was not discharged by washing. He collected a number of these grains, and carrying them to his apartment, bruised and squeezed different parcels of them upon bits of linen; but to his great surprise, after two or three hours, no colour appeared on the wetted part; but, at the same time, two or three spots of the plaster at the window, on which drops of the liquid had fallen, had become purple; though the day was cloudy. On carrying the pieces of linen to the window, and leaving them there, they also acquired a purple colour. It was the action of light, then, on the liquor, that caused it to tinge the linen. He found, likewise, that when the colouring matter was put into a phial, which filled it completely, it remained unchanged; but when the phial was not full, and was badly corked, it acquired colour. From these facts it is evident, that the purple colour is owing to the joint action of the light and the oxygen of the atmosphere upon the liquor of the shellfish.

About this time, likewise, he made experiments upon a subject which attracted the attention of mechanicians—to determine whether the strength of a cord was greater, or less, or equal to the joint strength of all the fibres which compose it. The result of Reaumur's experiments was, that the strength of the cord is less than that of all the fibres of which it is composed. Hence it follows, that the less that a cord differs from an assemblage of straight fibres, the stronger it is. This, at that time considered as a singular mechanical paradox, was afterwards elucidated by M. Duhamel.

It was a popular opinion of all the inhabitants of the sea-shore, that when the claws of crabs, lobsters, &c., are lost by any means, they are gradually replaced by others, and the animal in a short time becomes as perfect as at first. This opinion was ridiculed by men of science as inconsistent with all our notions of true philosophy. Reaumur subjected it to the test of experiment, by removing the claws of these animals, and keeping them alone for the requisite time in sea-water: new claws soon sprang out, and perfectly replaced those that had been removed. Thus the common opinion was verified, and the contemptuous smile of the half-learned man of science was shown to be the result of ignorance, not of knowledge.

Reaumur was not so fortunate in his attempts to explain the nature of the shock given by the torpedo; which we now know to be an electric shock produced by a peculiar apparatus within the animal. Reaumur endeavoured to prove, from dissection, that the shock was owing to the prodigious rapidity of the blow given by the animal in consequence of a peculiar structure of its muscles.

The turquoise was at that time, as it still is, considerably admired in consequence of the beauty of its colour. Persia was the country from which this precious stone came, and it was at that time considered as the only country in the universe where it occurred. Reaumur made a set of experiments on the subject and showed that the fossil bones found in Languedoc, when exposed to a certain heat, assume the same beautiful green colour, and become turquoises equally beautiful with the Persian. It is now known, that the true Persian turquoise, the *calamite* of mineralogists, is quite different from fossil bones coloured with copper. So far, therefore, Reaumur deceived himself by these experiments; but at that time chemical knowledge was too imperfect to enable him to subject Persian turquoise to an analysis, and determine its constitution.

About the same period, he undertook an investigation of the nature of imitation pearls, which resemble the true pearls so closely, that it is very difficult, from appearances, to distinguish the true from the false. He showed that the substance which gave the false pearls their colour and lustre, was taken from a small fish called by the French *able*, or *ablette*. He likewise undertook an investigation of the origin of true pearls, and showed that they were indebted for their production to a disease of the animal. It is now known, that the introduction of any solid body, as a grain of sand, within the shell of the living pearl-shellfish, gives occasion to the formation of pearl. Linnæus boasted that he knew a method of forming artificial pearls; and doubtless his process was merely introducing some solid particle of matter into the living shell. Pearls consist of alternate layers of carbonate of lime and animal membrane; and the colour and lustre to which they owe their value depends upon the thinness of the alternate coats.

The next paper of Reaumur was an account of the rivers in France whose sand yielded gold-dust, and the method employed to extract the gold. This paper will well repay the labour of a perusal; it owes its interest in a great measure to the way in which the facts are laid before the reader.

His paper on the prodigious bank of fossil shells at Touraine, from which the inhabitants draw manure in such quantities for their fields, deserves attention in a geological point of view. But his paper on flints and stones is not so valuable; it consists in speculations, which, from the infant state of chemical analysis when he wrote, could not be expected to lead to correct conclusions.

I pass over many of the papers of this most indefatigable man, because they are not connected with chemistry; but his history of insects constitutes a charming book, and contains a prodigious number of facts of the most curious and important nature. This book alone, supposing Reaumur had done nothing else, would have been sufficient to have immortalized the author.

In the year 1722 he published his work on the *art of converting iron into steel, and of softening cast-iron*. At that time no steel whatever was made in France; the nation was supplied with that indispensable article from foreign countries, chiefly from Germany. The object of Reaumur's book was to teach his countrymen the art of making steel, and, if possible, to explain the nature of the process by which iron is changed into steel. Reaumur concluded from his experiments, that steel is iron impregnated with *sulphureous* and *saline* matters. The word *sulphureous*, as at that time used, was nearly synonymous with our present term *combustible*. The process which he found to answer, and which he recommends to be followed, was to mix together

        4   parts of soot
        2   parts of charcoal-powder
        2   parts of wood-ashes
        1½ parts of common salt.

The iron bars to be converted into steel were surrounded with this mixture, and kept red-hot till converted into steel. Reaumur's notion of the difference between iron and steel was an approximation to the truth. The saline matters which he added do not enter into the composition of steel; and if they did, so far from improving, they would injure its qualities. But the charcoal and soot, which consist chiefly of carbon, really produce the desired effect; for steel is a combination of *iron* and *carbon*.

In consequence of these experiments of Reaumur, it came to be an opinion entertained by chemists, that steel differed from iron merely by containing a greater proportion of phlogiston; for the charcoal and soot with which the iron bars were surrounded was considered as consisting almost entirely of phlogiston; and the only useful purpose which they could serve, was supposed to be to furnish phlogiston. This opinion continued prevalent till it was overturned towards the end of the last century, first by the experiments of Bergmann, and afterwards by those of Berthollet, Vandermond, and Monge, published in the Memoirs of the French Academy for 1786 (page 132). In this elaborate memoir the authors take a view of all the different processes followed in bringing iron from the ore to the state of steel: they then give an account of the researches of Reaumur and of Bergmann; and lastly relate their own experiments, from which they finally draw, as a conclusion, that steel is a compound of iron and carbon.

The regent Orleans, who at that time administered the affairs of France, thought that this work of Reaumur was deserving a reward, and accordingly offered him a pension of 12,000 livres. Reaumur requested of the regent that this pension should be given in the name of the academy, and that after his death it should continue, and be devoted to defray the necessary expenses towards bringing the arts into a state of perfection. The request was granted, and the letters patent made out on the 22d of December, 1722.

At that time tin-plate, as well as steel, was not made in France; but all the tin-plates wanted were brought from Germany, where the processes followed were kept profoundly secret. Reaumur undertook to discover a method of tinning iron sufficiently cheap to admit the article to be manufactured in France—and he succeeded. The difficulty consisted in removing the scales with which the iron plates, as prepared, were always covered. These scales consist of a vitrified oxide of iron, to which the tin will not unite. Reaumur found, that when these plates are steeped in water acidulated by means of bran, and then allowed to rust in stoves, the scales become loose, and are easily detached by rubbing the plates with sand. If after being thus cleansed they are plunged into melted tin, covered with a little tallow to prevent oxidizement, they are easily tinned. In consequence of this explanation of the process by Reaumur, tin-plate manufactories were speedily established in different parts of France. It was about the same time, or only a little before it, that tin-plate manufactories were first started in England. The English tin-plate was much more beautiful than the German, and therefore immediately preferred to it; because in Germany the iron was converted into plates by hammering, whereas in England it was rolled out. This made it much smoother, and consequently more beautiful.

Another art, at that time unknown in France, and indeed in every part of Europe except Saxony, was the art of making porcelain, a name given to the beautiful translucent stoneware which is brought from China and Japan. Reaumur undertook to discover the process employed in making it. He procured specimens of porcelain from China and Japan, and also of the imitations of those vessels at that time made in various parts of France and other European countries. The true porcelain remained unaltered, though exposed to the most violent heat which he was capable of producing; but the imitations, in a furnace heated by no means violently, melted into a perfect glass. Hence he concluded, that the imitation-porcelains were merely glass, not heated sufficiently to be brought into fusion; but true porcelain he conceived to be composed of two different ingredients, one of which is capable of resisting the most violent heat which can be raised, but the other, when heated sufficiently, melts into a glass. It is this last ingredient that gives porcelain its translucency, while the other makes it refractory in the fire. This opinion of Reaumur was soon after confirmed by Father d'Entrecolles, a French missionary in China, who sent some time after a memoir to the academy, describing the mode followed by the Chinese in the manufactory of their porcelain. Two substances are employed by them, the one called *kaolin* and the other *petunse*. It is now known that *kaolin* is what we call porcelain-clay, and that *petunse* is a fine white felspar. Felspar is fusible in a violent heat, but porcelain-clay is refractory in the highest temperatures that we have it in our power to produce in furnaces.

Reaumur made another curious observation on glass, which has been, since his time, employed very successfully to explain the appearances of many of our trap-rocks. If a glass vessel, properly secured in sand, be raised to a red heat, and then allowed to cool very slowly, it puts off the appearance of glass and assumes that of stoneware, or porcelain. Vessels thus altered have received the name of *Reaumur's porcelain*. They are much more refractory than glass, and therefore may be exposed to a pretty strong red heat without any danger of softening or losing their shape. This change is occasioned by the glass being kept long in a soft state: the various substances of which it is composed are at liberty to exercise their affinities and to crystallize. This makes the vessel lose its glassy structure altogether. In like manner it was found by Sir James Hall and Mr. Gregory Watt, that when common greenstone was heated sufficiently, and then rapidly cooled, it melted and concreted into a glass; but if after having been melted it was allowed to cool exceedingly slowly, the constituents again crystallized and arranged themselves as at first—so that a true greenstone was again formed. In the same way lavas from a volcano either assume the appearance of slag or of stone, according as they have cooled rapidly or slowly. Many of the lavas from Vesuvius cannot be distinguished from our *greenstones*.

Reaumur's labours upon the thermometer must not be omitted here ; because he gave his name to a thermometer, which was long used in France and in other parts of Europe. The first person that brought thermometers into a state capable of being compared with each other was Sir Isaac Newton, in a paper published in the Philosophical Transactions for 1701. Fahrenheit, of Amsterdam, was the first person that put Newton's method in practice, by fixing two points on his scale, the freezing-water point and the boiling-water point, and dividing the interval between them into one hundred and eighty degrees.

But no fixed point existed in the thermometers employed in France, every one graduating them according to his fancy ; so that no two thermometers could be compared together. Reaumur graduated his thermometers by plunging them into freezing water or a mixture of snow and water. This point was marked zero, and was called the freezing-water point. The liquid used in his thermometers was spirit of wine : he took care that it should be always of the same strength, and the interval between the point of freezing and boiling water was divided into eighty degrees. Deluc afterwards rectified this thermometer, by substituting mercury for spirit of wine. This not only enabled the thermometer to be used to measure higher temperatures, but corrected an obvious error which existed in all the thermometers constructed upon Reaumur's principle : for spirit of wine cannot bear a temperature of eighty degrees Reaumur without being dissipated into vapour—absolute alcohol boiling at a hundred and sixty-two degrees two-thirds. It is obvious from this, that the boiling point in Reaumur's thermometer could not be accurate, and that it would vary, according to the quantity of empty space left above the alcohol.

Finally, he contrived a method of hatching chickens by means of artificial heat, as is practised in Egypt.

We are indebted to him also for a set of important observations on the organs of digestion in birds. He showed, that in birds of prey, which live wholly upon animal food, digestion is performed by solvents in the stomach, as is the case with digestion in man : while those birds that live upon vegetable food have a very powerful stomach or gizzard, capable of triturating the seeds which they swallow. To facilitate this triturating process, these fowls are in the habit of swallowing small pebbles.

The moral qualities of M. Reaumur seem not to have been inferior to the extent and variety of his acquirements. He was kind and benevolent, and remarkably disinterested. He performed the duties of intendant of the order of St. Louis from the year 1735 till his death, without accepting any of the emoluments of the office, all of which were most religiously given to the person to whom they belonged, had she been capable of performing the duties of the place. M. Reaumur died on the 17th of October, 1756, after having lived very nearly seventy-five years.

John Hellot was born in Paris in the year 1685, on the 20th of November. His father, Michael Hellot, was of a respectable family, and the early part of his son's education was at home: it seems to have been excellent, as young Hellot acquired the difficult art of writing on all manner of subjects in a precise, clear, and elegant style. His father intended him for the church; but his own taste led him decidedly to the study of chemistry. He had an uncle a physician, some of whose papers on chemical subjects fell into his hands. This circumstance kindled his natural taste into a flame : he formed an acquaintance with M. Geoffroy, whose reputation as a chemist was at that time high, and this friendship was afterwards cemented by Geoffroy marrying the niece of M. Hellot.

His circumstances being easy, he went over to England, to form a personal acquaintance with the many eminent philosophers who at that time adorned that country. His fortune was considerably deranged by Law's celebrated scheme during the regency of the Duke of Orleans. This obliged him to look out for some resource : he became editor of the Gazette de France, and continued in this employment from 1718 to 1732. During these fourteen years, however, he did not neglect chemistry, though his progress was not so rapid as it would have been, could he have devoted to that science his undivided attention. In 1732 he was put forward by his friends as a candidate for a place in the Academy of Sciences ; and in the year 1735 he was chosen adjunct chemist, vacant by the promotion of M. de la Condamine to the place of associate. Three years after he was declared a supernumerary pensioner, without passing through the step of associate. His reputation as a chemist was already considerable, and after he became a member of the academy, he devoted himself to the investigations connected with his favourite science.

His first labours were on zinc; in two successive papers he endeavoured to decompose this metal, and to ascertain the nature of its constituents. Though his labour was unsuccessful, yet he pointed out many new properties of this metal, and various new compounds into which it enters. Neither was he more successful in his attempt to account for the origin of the red vapours which are exhaled from nitre in certain circumstances. He ascribed them to the presence of ferruginous matters in the nitre; whereas they are owing to the expulsion and partial decomposition of the nitric acid of the nitre, in consequence of the action of some more powerful acid.

His paper on sympathetic ink is of more importance. A German chemist had shown him a saline solution of a red colour which became blue when heated: this led him to form a sympathetic ink, which was pale red, while the paper was moist, but became blue upon drying it by holding it to the fire. This sympathetic ink was a solution of cobalt in muriatic acid. It does not appear from Hellot's paper that he was exactly aware of the chemical constitution of the liquid which constituted his sympathetic ink; though it is clear he knew that cobalt constitutes an essential part of it.

Kunkel's phosphorus, though it had been originally discovered in Germany, could not be prepared by any of the processes which had been given to the public. Boyle had taught his operator, Godfrey Hankwitz, the method of making it. This man had, after Boyle's death, opened a chemist's shop in London, and it was he that supplied all Europe with this curious article:

on that account it was usually distinguished by the name of *English phosphorus*. But in the year 1737 a stranger appeared in Paris, who offered for a stipulated reward to communicate the method of manufacturing this substance to the Academy of Sciences. The offer was accepted by the French government, and a committee of the academy, at the head of which was Hellot, was appointed to witness the process, and ascertain all its steps. The process was repeated with success ; and Hellot drew up a minute detail of the whole, which was inserted in the Memoirs of the Academy, for the year 1737. The publication of this paper constitutes an era in the preparation of phosphorus : it was henceforward in the power of every chemist to prepare it for himself. A few years after the process was much improved by Margraaf ; and, within little more than twenty years after, the very convenient process still in use was suggested by Scheele. Hellot's experiments on the comparative merits of the salts of Peyrac, and of Pecais were of importance, because they decided a dispute—they may also perhaps be considered as curiosities in an historical point of view ; because we see from them the methods which Hellot had recourse to at that early period in order to determine the purity of common salt. They are not entitled, however, to a more particular notice here.

In the year 1740 M. Hellot was charged with the general inspection of dyeing; a situation which M. du Foy had held till the time of his death in 1739. It was this appointment, doubtless, which turned his attention to the theory of dyeing, which he tried to explain in two memoirs read to the academy in 1740 and 1741. The subject was afterwards prosecuted by him in subsequent memoirs which were published by the academy.

In 1745 he was named to go to Lyons in order to examine with care the processes followed for refining gold and silver. Before his return he took care to give to these processes the requisite precision and exactness. Immediately after his return to Paris he was appointed to examine the different mines and assay the different ores in France; this appointment led him to turn his thoughts to the subject. The result of this was the publication of an excellent work on assaying and metallurgy, entitled " De la Fonte des Mines, des Fonderies, &c. Traduit de l'Allemand de Christophe-André Schlutter." The first volume of this book appeared in 1750, and the second in 1753. Though this book is called by Hellot a translation, it contains in fact a great deal of original matter; the arrangement is quite altered; many processes not noticed by Schlutter are given, and many essential articles are introduced, which had been totally omitted in the original work. He begins with an introduction, in which he gives a short sketch of all the mines existing in every part of France, together with some notice of the present state of each. The first volume treats entirely of docimasy, or the art of assaying the different metallic ores. Though this art has been much improved since Hellot's time, yet the processes given in this volume are not without their value. The second volume treats of the various metallurgic processes followed in order to extract metals from their ores. This volume is furnished with no fewer than fifty-five plates, in which all the various furnaces, &c. used in these processes are exhibited to the eye.

While occupied in preparing this work for the press he was chosen to endeavour to bring the porcelain manufactory at Sevre to a greater state of perfection than it had yet reached. In this he was successful. He even discovered various new colours proper for painting upon porcelain; which contributed to give to this manufactory the celebrity which it acquired.

In the year 1763 a phenomenon at that time quite new to France took place in the coal-mine of Briançon. A quantity of carburetted hydrogen gas had collected in the bottom of the mine, and being kindled by the lights employed by the miners, it exploded with great violence, and killed or wounded every person in the mine. This destructive gas, distinguished in this country by the name of *fire-damp*, had been long known in Great Britain and in the Low Countries, though it had not before been known in France. The Duke de Choiseul, informed of this event, had recourse to the academy for assistance, who appointed Messrs. de Montigny, Duhamel, and Hellot, a committee to endeavour to discover the remedies proper to prevent any such accident from happening for the future. The report of these gentlemen was published in the Memoirs of the Academy ;* they give an account both of the fire-damp, and *choke-damp*, or *carbonic acid gas*, which sometimes also makes its appearance in coalmines. They very justly observe that the proper way to obviate the inconveniency of these gases is to ventilate the mine properly ; and they give various methods by which this ventilation may be promoted by means of fires lighted at the bottom of the shaft, &c.

In 1763 M. Hellot was appointed, conjointly with M. Tillet, to examine the process followed for assaying gold and silver. They showed that the cupels always retained a small portion of the silver assayed, and that this loss, ascribed to the presence of a foreign metal, made the purity of the silver be always reckoned under the truth, which occasioned a loss to the proprietor.

His health continued tolerably good till he reached his eightieth year : he was then struck with palsy, but partially recovered from the first attack ; but a second attack, on the 13th of February, 1765, refused to yield to every medical treatment, and he died on the 15th of that month, at an age a little beyond eighty.

* 1763, p. 235.

Henry Louis Duhamel du Monceau was born at Paris in the year 1700. He was descended from Loth Duhamel, a Dutch gentleman, who came to France in the suite of the infamous Duke of Burgundy, about the year 1400. Young Duhamel was educated in the College of Harcourt; but the course of study did not suit his taste. He left it with only one fact engraven on his memory—that men, by observing nature, had created a science called *physics;* and he resolved to profit by his freedom from restraint and turn the whole of his attention to that subject. He lodged near the Jardin du Roi, where alone, at that time, physics were attended to in Paris. Dufoy, Geoffroy, Lemery, Jussieu, and Vaillant, were the friends with whom he associated on coming to Paris. His industry was stimulated solely by a love of study, and by the pleasure which he derived from the increase of knowledge; love of fame does not appear to have entered into his account.

In the year 1718 saffron, which is much cultivated in that part of France formerly distinguished by the name of Gâtinois, where Duhamel's property lay, was attacked by a malady which appeared contagious. Healthy bulbs, when placed in the neighbourhood of those that were diseased, soon became affected with the same malady. Government consulted the academy on the subject; and this learned body thought they could not do better than request M. Duhamel to investigate the cause of the disease; though he was only eighteen years of age, and not even a member of the academy. He ascertained that the malady was owing to a parasitical plant, which attached itself to the bulb of the saffron, and drew nourishment from it. This plant extended under the earth, from one bulb to another, and thus infected the whole saffron plantations.

M. Duhamel formed the resolution at the commencement of his scientific career to devote himself to public utility, and to prosecute those subjects which were likely to contribute most effectually to the comfort of the lower ranks of men. Much of his time was spent in endeavouring to promote the culture of vegetables, and in rendering that culture more useful to society. This naturally led to a careful study of the physiology of trees. The fruit of this study he gave to the world in the year 1758, when his Physique des Arbres was published. This constitutes one of the most important works on the subject which has ever appeared. It contains a great number of new and original facts; and contributed very much indeed to advance this difficult, but most important branch of science : nor is it less remarkable for modesty than for value. The facts gathered from other sources, even those which make against his own opinions, are most carefully and accurately stated : the experiments that preceded his are repeated and verified with much care; and the reader is left to discover the new facts and new views of the author, without any attempt on his part to claim them as his own.

M. Duhamel had been attached to the department of the marine by M. de Maurepas, who had given him the title of *inspector-general*. This led him to turn his attention to naval science in general. The construction of vessels, the weaving of sailcloths, the construction of ropes and cables, the method of preserving the wood, occupied his attention successively, and gave birth to several treatises, which, like all his works, contain immense collections of facts and experiments. He endeavours always to discover which is the best practice, to reduce it to fixed rules, and to support it by philosophical principles; but abstains from all theory when it can be supported only by hypothesis.

From the year 1740, when he became an academician, till his death in 1781, he made a regular set of meteorological observations at Pithiviers, with details relative to the direction of the needle, to agriculture, to the medical constitution of the year, and to the time of nest-building, and of the passage of birds.

Above sixty memoirs of his were published in the Transactions of the French Academy of Sciences. They are so multifarious in their nature, and embrace such a variety of subjects, that I shall not attempt even to give their titles, but satisfy myself with stating such only as bear more immediately upon the science of chemistry.

It will be proper in conducting this review to notice the result of his labours connected with the ossification of bones; because, though not strictly chemical, they throw light upon some branches of the animal economy, more closely connected with chemistry than with any other of the sciences. He examined, in the first place, whether the ossification of bones, and their formation and reparation, did not follow the same law that he had assigned to the increments of trees, and he established, by a set of experiments, that bones increase by the ossification of layers of the periosteum, as trees do by the hardening of their cortical layers. Bones in a soft state increase in every direction, like the young branches of plants; but after their induration they increase only like trees, by successive additions of successive layers. This organization was incompatible with the opinion of those who thought that bones increased by the addition of an earthy matter deposited in the meshes of the organized network which forms the texture of bones. M. Duhamel combated this opinion by an ingenious experiment. He had been informed by Sir Hans Sloane that the bones of young animals fed upon madder were tinged red. He conceived the plan of feeding them alternately with food mingled with madder, and with ordinary food. The bones of animals thus treated were found to present alternate concentric layers of red and white, corresponding to the different periods in which the animal had been fed with food containing or not containing madder. When these bones are sawn longitudinally we see the thickness of the coloured layers, greater or less, according to the number of plates of the periosteum that have ossified. As for the portions still soft, or susceptible of extending themselves in every direction, such as the plates in the neighbourhood of the mar-row, the reservoir of which increases during a part of the time that the animal continues to grow, the red colour marks equally the progress of their ossification by coloured points more or less extended.

This opinion was attacked by Haller, and defended by M. Fougeroux, nephew of M. Duhamel; but it is not our business here to inquire how far correct.

One of the most important of M. Duhamel's papers, which will secure his name a proud station in the annals of chemistry, is that which was inserted in the Memoirs of the Academy for 1737, in which he shows that the base of common salt is a true fixed alkali, different in some respects from the alkali ex-tracted from land plants, and known by the name of *potash*, but similar to that obtained by the incinera-tion of marine plants. We are surprised that a fact so simple and elementary was disputed by the French chemists, and rather indicated than proved by Stahl and his followers. The conclusions of Duhamel were disputed by Pott; but finally confirmed by Margraaf. M. Duhamel carried his researches further, he wished to know if the difference between potash and soda depends on the plants that produce them, or on the nature of the soil in which they grow. He sowed kali at Denain-villiers, and continued his experiments during a great number of years. M. Cadet, at his request, examined the salts contained in the ashes of the kali of Denain-villiers. He found that during the first year soda pre-dominated in these ashes. During the successive years the potash increased rapidly, and at last the soda almost entirely disappeared. It was obvious from this, that the alkalies in plants are drawn at least chiefly from the soil in which they vegetate.

The memoirs of M. Duhamel on ether, at that time almost unknown, on soluble tartars, and on lime, contain many facts both curious and accurately stated; though our present knowledge of these bodies is so much greater than his—the new facts ascertained respecting them are so numerous and important, that the contributions of this early experimenter, which probably had a considerable share in the success of subsequent investigations, are now almost forgotten. Nor would many readers bear patiently with an attempt to enumerate them.

There is a curious paper of his in the Memoirs of the Academy for 1757. In this he gives the details of a spontaneous combustion of large pieces of cloth soaked in oil and strongly pressed. Cloth thus prepared had often produced similar accidents. Those who were fortunate enough to prevent them, took care to conceal the facts, partly from ignorance of the real cause of the combustion, and partly from a fear that if they were to state what they saw, their testimony would not gain credit. If the combustion had not been prevented, then the public voice would have charged those who had the care of the cloths with culpable negligence, or even with criminal conduct. The observation of M. Duhamel, therefore, was useful, in order to prevent such unjust suspicions from hindering those concerned from taking the requisite precautions. Yet, twenty years after the publication of his paper, two accidental spontaneous combustions, in Russia, were ascribed to treason. The empress Catharine II. alone suspected that the combustion was spontaneous, and experiments made by her orders fully confirmed the evidence previously advanced by the French philosopher.

One man alone would have been insufficient for all the labours undertaken by M. Duhamel; but he had a brother who lived upon his estate at Denainvilliers

(the name of which he bore), and divided his time between the performance of benevolent actions and
studying the operations of nature. M. Denainvilliers
prosecuted in his retreat the observations and experiments intrusted by his brother to his charge. Thus
in fact the memoirs of Duhamel exhibit the assiduous labours of two individuals, one of whom con-
tentedly remained unknown to the world, satisfied
with the good which he did, and the favours which he
conferred upon his country and the human race.

The works of M. Duhamel are very voluminous,
and are all written with the utmost plainness. Every
thing is elementary, no previous knowledge is taken
for granted. His writings are not addressed to philosophers, but to all those who are in quest of practical
knowledge. He has been accused of diffuseness of
style, and of want of correctness; but his style is
simple and clear; and as his object was to inform, not
philosophers, but the common people, greater conciseness would have been highly injudicious.

Neither-he nor his brother ever married, but thought
it better to devote their undivided attention to study.
Both were assiduous in no ordinary degree, but the
ardour of Duhamel himself continued nearly undiminished till within a year of his death; when, though
he still attended the meetings of the academy, he no
longer took the same interest in its proceedings. On
the 22d of July, 1781, just after leaving the academy,
he was struck with apoplexy, and died after lingering
twenty-two days in a state of coma.

He was without doubt one of the most eminent men
of the age in which he lived; but his merits as a chemist will chiefly be remembered in consequence of his
being the first person who demonstrated by satisfactory evidence the peculiar nature of soda, which had
been previously confounded with potash. His merits
as a vegetable physiologist and agriculturist were of a
very high order.

Peter Joseph Macquer was born at Paris, in 1718. His father, Joseph Macquer, was descended from a noble Scottish family, which had sacrificed its property and its country, out of attachment to the family of the Stuarts.* Young Macquer made choice of medicine as a profession, and devoted himself chiefly to chemistry, for which he showed early a decided taste. He was admitted a member of the Academy of Sciences in the year 1745, when he was twenty-seven years of age. Original researches in chemistry, the composition of chemical elementary works, and the study of the arts connected with chemistry, occupied the whole remainder of his life.

His first paper treated of the effect produced by heating a mixture of saltpetre and white arsenic. It was previously known, that when such a mixture is distilled nitric acid comes over tinged with a blue colour; but nobody had thought of examining the residue of this distillation. Macquer found it soluble in water and capable of crystallizing into a neutral salt composed of potash (the base of saltpetre), and an acid into which the arsenic was changed by the nitric acid communicating oxygen to it.

Macquer found that a similar salt might be obtained with soda or ammonia for its base. Thus he was the first person who pointed out the existence of arsenic acid, and ascertained the properties of some of the salts which it forms. But he made no attempt to obtain arsenic acid in a separate state, or to determine its properties. That very important step was reserved for Scheele, for Macquer seems to have had no suspicion of the true nature of the salt which he had formed.

---

* I do not know what the true name was of which Macquer is a corruption. Ker is a Scottish name belonging to two noble families, the Duke of Roxburgh and the Marquis of Lothian; but I am not aware of M'Ker being a Scottish name : besides, neither of these families was attached to the house of Stuart.

His next set of experiments was on Prussian blue. He made the first step towards the discovery of the nature of the principle to which that pigment owes its colour. Prussian blue had been accidentally discovered by Diesbach, an operative chemist of Berlin, in 1710, but the mode of producing it was kept secret till it was published in 1724, by Dr. Woodward in the Philosophical Transactions. It consisted in mixing potash and blood together, and heating the mixture in a covered crucible, having a small hole in the lid, till it ceased to give out smoke. The solution of this mixture in water, when mixed with a solution of sulphate of iron, threw down a green powder, which became blue when treated with muriatic acid : this blue matter was *Prussian blue*. Macquer ascertained that when Prussian blue is exposed to a red heat its blue colour disappears, and it is converted into common peroxide of iron. Hence he concluded that Prussian blue is a compound of oxide of iron, and of something which is destroyed or driven off by a red heat. He showed that this something possessed the characters of an acid; for when Prussian blue is boiled with caustic potash it loses its blue colour, and if the potash be boiled with successive portions of Prussian blue, as long as it is capable of discolouring them, it loses the characters of an acid and assumes those of a neutral salt, and at the same time acquires the property of precipitating iron from the solutions of the sulphate at once of a blue colour. Macquer ascribed the green colour thrown down, by mixing the blood-lie and sulphate of iron to the potash in the blood-lie, not being saturated with the colouring matter of Prussian blue. Hence a portion of the iron is thrown down in the state of Prussian blue, and another portion in that of yellow oxide of iron : these two being mixed form a green. The muriatic acid dissolves the yellow oxide and leaves the Prussian blue untouched. Macquer, however, did not succeed in determining the nature of the colouring matter; a task reserved for Scheele, whose lot it was to take up the half-finished investigations of Macquer, and throw upon them a new and brilliant light. Macquer thought that this colouring matter was *phlogiston*. On that account the potash saturated with it, which was employed by chemists to detect the presence of iron by forming with it Prussian blue, was called *phlogisticated alkali*.

Macquer, conjointly with Baumé, subjected the grains of crude platinum, to which the attention of chemists had been newly drawn, to experiment. Their principle object was to examine its fusibility and ductility. They succeeded in fusing it imperfectly, by means of a burning mirror, and found that the grains thus treated were not destitute of ductility. But upon the whole the experiments of these chemists threw but little light upon the subject. Many years elapsed before chemists were able to work this refractory metal, and to make it into vessels fitted for the uses of the laboratory. For this important improvement, which constitutes an era in chemistry, the chemical world was chiefly indebted to Dr. Wollaston.

In the year 1750 M. Macquer was charged with a commission by the court. There existed at that time in Brittany a man, the Count de la Garaie, who, yielding to a passion for benevolence, had for forty years devoted himself to the service of suffering humanity. He had built an hospital by the side of a chemical laboratory: he took care of the patients in the hospital himself; and treated them with medicines prepared in his laboratory. Some of these were new, and, in his opinion, excellent medicines; and he offered to sell them to government for the service of his hospital. Macquer was charged by government with the examination of these medicines. The project of the Count de la Garaie was to extract the salutary parts of minerals, by a long maceration with neutral salts. Among other things he had prepared a mercurial tincture, by a process which lasted several months: but this tincture was merely a solution of corrosive sublimate in spirit of wine. Such is the history of most of those boasted secrets; sometimes they are chimerical, and sometimes known to all the world, except to those who purchase them.

M. Macquer had the fortune to live at a time when chemistry began to be freed from the reveries of alchymists; but methodical arrangement was a merit still unknown to the elementary chemical books, especially in France, where a residue of Cartesianism added to the natural obscurity of the science, by surcharging it with pretended mechanical explanations. Macquer was the first French chemist who gave to an elementary treatise the same clearness, simplicity, and method, which is to be found in the other branches of science. This was no small merit, and undoubtedly contributed considerably to the rapid improvement of the science which so speedily followed. His elements of chemistry were translated into different languages, especially into English; and long constituted the textbook employed in the different European universities. Dr. Black recommended it for many years in the University of Edinburgh. Indeed, it was only superseded in consequence of the new views introduced into chemistry by Lavoisier, which, requiring a new language to render them intelligible, naturally superseded all the elementary chemical books which had preceded the introduction of that language.

Macquer, during a number of years, delivered regular courses of chemical lectures, conjointly with Baumé. In these courses he preferred that arrangement which appeared to him to require the least preliminary knowledge of chemistry. He described the experiments, stated the facts with clearness and precision, and explained them in the way which appeared

to him most plausible, according to the opinions generally received; but without placing much confidence
in the accuracy of these explanations. He thought it
necessary to theorize a little, to enable his pupils the
better to connect the facts and to remember them;
and to put an end to that painful state of uncertainty
which always results from a collection of facts without
any theoretical links to bind them together. When
the discoveries of Lavoisier began to shake the foundation of the Stahlian theory, Macquer was old; and
it appears from a letter of his, published by Delametherie in the Journal de Physique, that he was
alarmed at the prophetic announcements of Lavoisier
in the academy that the reign of Phlogiston was
drawing towards an end. M. Condorcet assures us
that his attachment to theory, by which he means
phlogiston, was by no means strong;* but his own
letter to Delametherie rather shows that this statement was not quite correct. How, indeed, could he
fail to experience an attachment to opinions which it
had been the business of his whole life to inculcate?

Macquer also published a dictionary of chemistry,
which was very successful, and which was translated
into most of the European languages. This mode of
treating chemistry was well suited to a science still in
its infancy, and which did not yet constitute a complete whole. It enabled him to discuss the different
topics in succession, and independent of each other:
and thus to introduce much important matter which
could not easily have been introduced into a systematic
work on chemistry. The second edition of this dictionary was published just at the time when the gases
began to attract the attention of scientific men; when
facts began to multiply with prodigious rapidity, and to
shake the confidence of chemists in all received theories. He acquitted himself of the difficult task of

* Hist. de l'Acad. R. des Sciences, 1784, p. 24.

collecting and stating these new facts with considerable success; and doubtless communicated much new
information to his countrymen : for the discoveries
connected with the gases originated, and were chiefly
made, in England, from which, on account of the re-
volutionary American war, there was some difficulty
of obtaining early information.

M. Hellot, who was commissioner of the counsel
for dyeing, and chemist to the porcelain manufacture,
requested to have M. Macquer for an associate. This
request did much honour to Hellot, as he was conscious
that the reputation of Macquer as a chemist was superior to his own. Macquer endeavoured, in the first
place, to lay down the true principles of the art of
dyeing, as the best method of dissipating the obscurity
which still hung over it. A great part of his treatise
on the art of dyeing silk, published in the collection
of the Academy of Sciences, has these principles for
its object. He gave processes also for dyeing silk
with Prussian blue, and for giving to silk, by means
of cochineal, as brilliant a scarlet colour as can be
given to woollen cloth by the same dye-stuff. He
published nothing on the porcelain manufacture,
though he attended particularly to the processes, and
introduced several ameliorations. The beautiful porcelain earth at present used at Sevre, was discovered in
consequence of a premium which he offered to any
person who could point out a clay in every respect
proper for making porcelain.

Macquer passed a great part of his life with a brother, whom he affectionately loved : after his death
he devoted himself entirely to his wife and two children, whose education he superintended. He was
rather averse to society, but conducted himself while
in it with much sweetness and affability. He was
fond of tranquillity and independence. Though his
health had been injured a good many years before his
death, the calmness and serenity of his temper pre-

vented strangers from being aware that he was afflicted with any malady. He himself was sensible that his strength was gradually sinking; he predicted his approaching end to his wife, whom he thanked for the happiness which she had spread over his life. He left orders that his body should be opened after his decease, that the cause of his death might be discovered. He died on the 15th of February, 1784. An ossification of the aorta, and several calculous concretions found in the cavities of the heart, had been the cause of the disease under which he had suffered for several years before his death.

These four chemists, of whose lives a sketch has just been given, were the most eminent that France ever produced belonging to the Stahlian school of chemistry. Baron, Malouin, Rouelle senior, Tillet, Cadet, Baumé, Sage, and several others whose names I purposely omit, likewise cultivated chemistry, during that period, with assiduity and success; and were each of them the authors of papers which deserve attention, but which it would be impossible to particularize without swelling this work into a size greatly beyond its proper limits.

Hilaire-Marin Rouelle, who was born at Caen in 1718, was, however, too eminent a chemist to be passed over in silence. His elder brother, William Francis, was a member of the Academy of Sciences, and demonstrator to Macquer, who gave lectures in the Jardin du Roi. At the death of Macquer, in 1770, Hilaire-Marin Rouelle succeeded him. He devoted the whole of his time and money to this situation, and quite altered the nature of the experimental course of chemistry given in the Jardin du Roi. He was in some measure the author of the chemistry of animal bodies, at least in France. When he published his experiments on the salts of urine, and of blood, he had scarcely any model; and though he committed some considerable mistakes, he ascertained several essential

and important facts, which have been since fully con‑
firmed by more modern experimenters. He died on the
7th of April, 1779, aged sixty-one years. His temper
was peculiar, and he was too honest and too open for
the situation in which he was placed, and for a state
of society in which every thing was carried by intrigue
and finesse. This is the reason why, in France, his
reputation was lower than it ought to have been. It
accounts, too, for his never becoming a member of the
Academy of Sciences, nor of any of the other nume‑
rous academies which at that time swarmed in France.
Nothing is more common than to find these unjust
decisions raise or depress men of science far above
or far below their true standard. Romé de Lisle, the
first person who commenced the study of crystals, and
placed that study in a proper point of view, was a man
of the same stamp with the younger Rouelle, and
never on that account, became a member of any aca‑
demy, or acquired that reputation during his lifetime,
to which his laborious career justly entitled him. It
would be an easy, though an invidious task, to point
out various individuals, especially in France, whose
reputation, in consequence of accidental and adventi‑
tious circumstances, rose just as much above their
deserts, as those of Rouelle, and Romé de Lisle were
sunk below.

CHAPTER IX.

OF THE FOUNDATION AND PROGRESS OF SCIENTIFIC
CHEMISTRY IN GREAT BRITAIN.

THE spirit which Newton had infused for the mathematical science was so great, that during many years they drew within their vortex almost all the scientific men in Great Britain. Dr. Stephen Hales is almost the only remarkable exception, during the early part of the eighteenth century. His vegetable statics constituted a most ingenious and valuable contribution to vegetable physiology. His hæmastatics was a no less valuable contribution to iatro-mathematics, at that time the fashionable medical theory in Great Britain. While his *analysis of air*, and experiments on the animal calculus constituted, in all probability, the foundation-stone of the whole discoveries respecting the gases to which the great subsequent progress of chemistry is chiefly owing.

Dr. William Cullen, to whom medicine lies under deep obligations, and who afterwards raised the medical celebrity of the College of Edinburgh to so high a pitch, had the merit of first perceiving the importance of scientific chemistry, and the reputation which that man was likely to earn, who should devote himself to the cultivation of it. Hitherto chemistry in Great Britain, and on the continent also, was considered as a mere appendage to medicine, and useful only so far as it contributed to the formation of new and useful remedies. This was the reason why it came to constitute an essential part of the education of every medical man, and why a physician was considered as unfit for practice unless he was also a chemist. But Dr. Cullen viewed the science as far more important; as capable of throwing light on the constitution of bodies, and of improving and amending of those arts and manufactures [that are most useful to man. He resolved to devote himself to its cultivation and improvement; and he would undoubtedly have derived celebrity from this science, had not his fate led rather to the cultivation of medicine. But Dr. Cullen, as the true commencer of the study of scientific chemistry in Great Britain, claims a conspicuous place in this historical sketch.

William Cullen was born in Lanarkshire, in Scotland, in the year 1712, on the 11th of December. His father, though chief magistrate of Hamilton, was not in circumstances to lay out much money on his son. William, therefore, after serving an apprenticeship to a surgeon in Glasgow, went several voyages to the West Indies, as surgeon, in a trading-vessel from London; but tiring of this, he settled, when very young, in the parish of Shotts; and after residing for a short time among the farmers and country people, he went to Hamilton, with a view of practising as a physician.

While he resided near Shotts, it happened that Archibald, Duke of Argyle, who at that time bore the chief political sway in Scotland, paid a visit to a gentleman of rank in that neighbourhood. The duke was fond of science, and was at that time engaged in some chemical researches which required to be elucidated by experiment. Eager in these pursuits, while on his visit he found himself at a loss for some piece of chemical apparatus which his landlord could not furnish; but he mentioned young Cullen to the duke as a person fond of chemistry, and likely therefore to possess the required apparatus. He was accordingly invited to dine, and introduced to his Grace. The duke was so pleased with his knowledge, politeness, and address, that an acquaintance commenced, which laid the foundation of all Cullen's future advancement.

His residence in Hamilton naturally made his name known to the Duke of Hamilton, whose palace is situated in the immediate vicinity of that town. His Grace being taken with a sudden illness, sent for Cullen, and was highly delighted with the sprightly character, and ingenious conversation of the young physician. He found no difficulty, especially as young Cullen was already known to the Duke of Argyle, in getting him appointed to a place in the University of Glasgow, where his singular talents as a teacher soon became very conspicuous.

It was while Dr. Cullen was a practitioner in Shotts that he formed a connexion with William, afterwards Doctor Hunter, the famous lecturer on anatomy in London, who was a native of the same part of the country as Cullen. These two young men, stimulated by genius, though thwarted by the narrowness of their circumstances, entered into a copartnery business, as surgeons and apothecaries, in the country. The chief object of their contract was to furnish the parties with the means of carrying on their medical studies, which they were not able to do separately. It was stipulated that one of them, alternately, should be allowed to study in whatever college he preferred, during the winter, while the other carried on the common business in his absence. In consequence of this agreement, Cullen was first allowed to study in the University of Edinburgh, for a winter. When it came to Hunter's turn next winter, he rather chose to go to London. There his singular neatness in dissecting, and uncommon dexterity in making anatomical preparations, his assiduity in study, his mild manners, and easy temper, drew upon him the attention of Dr. Douglas, who at that time read lectures on anatomy and midwifery in the capital. He engaged him as his assistant, and he afterwards succeeded him in the same department with much honour to himself, and advantage to the public. Thus was dissolved a copartnership of perhaps as singular a kind as any that occurs in the annals of science. Cullen was not disposed to let any engagement with him prove a bar to his partner's advancement in the world. The articles were abandoned, and Cullen and Hunter kept up ever after a friendly correspondence; though there is reason to believe that they never afterwards met.

It was while a country practitioner that young Cullen married a Miss Johnston, daughter of a neighbouring clergyman. The connexion was fortunate and lasting. She brought her husband a numerous family, and continued his faithful companion through all the alterations of his fortune. She died in the summer of 1786.

In the year 1746 Cullen, who had now taken the degree of doctor of medicine, was appointed lecturer on chemistry in the University of Glasgow; and in the month of October began a course on that science. His singular talent for arrangement, his distinctness of enunciation, his vivacity of manner, and his knowledge of the science which he taught, rendered his lectures interesting to a degree which had been till then unknown in that university: he was adored by the students. The former professors were eclipsed by the brilliancy of his reputation, and he had to encounter all those little rubs and insults that disappointed envy naturally threw in his way. But he proceeded in his career regardless of these petty mortifications; and supported by the public, he was more than consoled for the contumely heaped upon him by the illnature and pitiful malignity of his colleagues. His practice as a physician increased every day, and a vacancy occurring in the chair in 1751, he was appointed by the crown professor of medicine, which put him on a footing of equality with his colleagues in the university. This new appointment called forth powers which he was not before known to possess, and thus served still further to increase his reputation.

At that time the patrons of the University of Edinburgh were eagerly bent on raising the reputation of their medical school, and were in consequence on the look out for men of abilities and reputation to fill their respective chairs. Their attention was soon drawn towards Cullen, and on the death of Dr. Plummer, in 1756, he was unanimously invited to fill the vacant *chemical chair*. He accepted the invitation, and began his academical career in the College of Edinburgh in October of that year, and here he continued during the remainder of his life.

The appearance of Dr. Cullen in the College of Edinburgh constitutes a memorable era in the progress of that celebrated school. Hitherto chemistry being reckoned of little importance, had been attended by very few students; when Cullen began to lecture it became a favourite study, almost all the students flocking to hear him, and the chemical class becoming immediately more numerous than any other in the college, anatomy alone excepted. The students in general spoke of the new professor with that rapturous ardour so natural to young men when highly pleased. These eulogiums were doubtless extravagant, and proved disgusting to his colleagues. A party was formed to oppose this new favourite of the public. His opinions were misrepresented, it was affirmed that he taught doctrines which excited the alarm of some of the most moderate and conscientious of his colleagues. Thus a violent ferment was excited, and some time elapsed before the malignant arts by which this flame had been blown up were discovered.

During this time of public ferment Cullen went steadily forward; he never gave an ear to the gossip brought him respecting the conduct of his colleagues, nor did he take any notice of the doctrines which they taught. Some of their unguarded strictures on himself might occasionally have come to his ears; but if it was so, he took no notice of them whatever; they seemed to have made no impression on him.

This futile attempt to lower his character being thus baffled, his fame as a professor, and his reputation as a physician, increased daily : nor could it be otherwise; his professional knowledge was always great, and his manner of lecturing singularly clear and intelligible, lively, and entertaining. To his patients his conduct was so pleasing, his address so affable and engaging, and his manner so open, so kind, and so little regulated by pecuniary considerations, that those who once applied to him for medical assistance could never afterwards dispense with it: he became the friend and companion of every family he visited, and his future acquaintance could not be dispensed with.

His private conduct to his students was admirable, and deservedly endeared him to every one of them. He was so uniformly attentive to them, and took so much interest in the concerns of those who applied to him for advice; was so cordial and so warm, that it was impossible for any one, who had a heart susceptible of generous emotions, not to be delighted with a conduct so uncommon and so kind. It was this which served more than any thing else to extend his reputation over every civilized quarter of the globe. Among ingenuous youth gratitude easily degenerates into rapture; hence the popularity which he enjoyed, and which to those who do not well weigh the causes which operated on the students must appear excessive.

The general conduct of Cullen to his students was this : with all such as he observed to be attentive and diligent he formed an early acquaintance, by inviting them by twos, by threes, and by fours at a time to sup with him; conversing with them at such times with the most engaging ease, entering freely with them into the subject of their studies, their amusements, their difficulties, their hopes and future prospects. In this way he usually invited the whole of his numerous class till he made himself acquainted with their private character, their abilities, and their objects of pursuit. Those of whom he formed the highest opinion were of course invited most frequently, till an intimacy was gradually formed which proved highly beneficial to them. To their doubts and difficulties he listened with the most obliging condescension, and he solved them to the utmost of his power. His library was at all times open for their accommodation : in short, he treated them as if they had been all his relatives and friends. Few men of distinction left the University of Edinburgh, in his time, with whom he did not keep up a correspondence till they were fairly established in business. This enabled him gradually to form an accurate knowledge of the state of medicine in every country, and the knowledge thus acquired put it in his power to direct students in the choice of places where they might have an opportunity of engaging in business with a reasonable prospect of success.

Nor was it in this way alone that he befriended the students in the University of Edinburgh. Remembering the difficulties with which he had himself to struggle in his younger days, he was at all times singularly attentive to the pecuniary wants of the students. From the general intimacy which he contracted with them he found no difficulty in discovering those whose circumstances were contracted, or who laboured under any pecuniary embarrassment, without being under the necessity of hurting their feelings by a direct inquiry. To such persons, when their habits of study admitted it, he was peculiarly attentive : they were more frequently invited to his house than others, they were treated with unusual kindness and familiarity, they were conducted to his library and encouraged by the most delicate address to borrow from it freely whatever books he thought they had occasion for; and as persons under such circumstances are often extremely shy, books were sometimes pressed upon them as a sort of task, the doctor insisting upon knowing their opinion of such and such passages which they had not read, and desiring them to carry the book home for that purpose : in short, he behaved to them as if he had courted their company. He thus raised them in the opinion of their acquaintances, which, to persons in their circumstances, was of no little consequence. They were inspired at the same time with a secret sense of dignity, which elevated their minds, and excited an uncommon ardour, instead of that desponding inactivity so natural to depressed circumstances. Nor was he less delicate in the manner of supplying their wants : he often found out some polite excuse for refusing to take money for a first course, and never was at a loss for one to an after course. Sometimes (as his lectures were never written) he would request the favour of a sight of their notes, if he knew that they were taken with care, in order to refresh his memory. Sometimes he would express a wish to have their opinion of a particular part of his course, and presented them with a ticket for the purpose. By such delicate pieces of address, in which he greatly excelled, he took care to anticipate their wants. Thus he not only gave them the benefit of his own lectures, but by refusing to take money enabled them to attend such others as were necessary for completing their course of medical study.

He introduced another general rule into the university dictated by the same spirit of disinterested benevolence. Before he came to Edinburgh, it was the custom of the medical professors to accept of fees for their medical attendance when wanted, even from medical students themselves, though they were perhaps attending the professor's lectures at the time. But Dr. Cullen never would take a fee from any student of the university, though he attended them, when called on as a physician, with the same assiduity and care as if they had been persons of the first rank who paid him most liberally. This gradually led others to follow his example; and it has now become a general rule for medical professors to decline taking any fees when their assistance is necessary to a student. For this useful reform, as well as for many others, the students in the University of Edinburgh are entirely indebted to Dr. Cullen.

The first lectures which Dr. Cullen delivered in Edinburgh were on chemistry; and for many years he also gave lectures on the cases that occurred in the infirmary. In the month of February, 1763, Dr. Alston died, after having begun his usual course of lectures on the materia medica. The magistrates of Edinburgh, who are the patrons of the university, appointed Dr. Cullen to that chair, requesting that he would finish the course of lectures that had been begun by his predecessor. This he agreed to do, and, though he had only a few days to prepare himself, he never once thought of reading the lectures of his predecessor, but resolved to deliver a new course, which should be entirely his own. Some idea may be formed of the popularity of Cullen, by the increase of students to a class nearly half finished: Dr. Alston had been lecturing to ten; as soon as Dr. Cullen began, a hundred new students enrolled themselves.

Some years after, on the death of Dr. Whytt, professor of the theory of medicine, Dr. Cullen was appointed to give lectures in his stead. It was then that he thought it requisite to resign the chemical chair in favour of Dr. Black, his former pupil, whose talents in that department of science were well known. Soon after, on the death of Dr. Rutherford, professor of the practice of medicine, Dr. John Gregory having become a candidate for this place, along with Dr. Cullen, a sort of compromise took place between them, by which they agreed to give lectures alternately, on the theory and practice of medicine, during their joint lives, the longest survivor being allowed to hold either of the classes he should incline. Unluckily this arrangement was soon destroyed, by the sudden and unexpected death of Dr. Gregory, in the flower of his age. Dr. Cullen thenceforth continued to give lectures on the practice of medicine till within a few months of his death, which happened on the 5th of February, 1790, when he was in the seventy-seventh year of his age.

It is not our business to follow Dr. Cullen's medical career, nor to point out the great benefits which he conferred on nosology and the practice of medicine. He taught four different classes in the University of Edinburgh, which we are not aware to have happened to any other individual, except to professor Dugald Stewart.

Notwithstanding the important impulse which he gave to chemistry, he published nothing upon that science, except a short paper on the cold produced by the evaporation of ether, which made its appearance in one of the volumes of the Edinburgh Physical and Literary Essays. Dr. Cullen employed Dr. Dobson of Liverpool, at that time his pupil, to make experiments on the heat and cold produced by mixing liquids and solids with each other. Dr. Dobson, in making these experiments, observed that the thermometer, when lifted out of many of the liquids, and suspended a short time in the air beside them, fell to a lower degree than indicated by another thermometer which had undergone no such process. After varying his observations on this phenomenon, he found reason to conclude that it was occasioned by the evaporation of the last drop of liquid which adhered to the bulb of the thermometer; the sinking of the thermometer being always greatest when this instrument was taken out of the most volatile liquids. Dr. Cullen had the curiosity to try whether the same phenomenon would appear on repeating these experiments under the exhausted receiver of an air-pump. To satisfy himself, he put on the plate of the air-pump a glass goblet containing water; and in the goblet he placed a wide-mouthed phial containing sulphuric ether. The whole was covered with an air-pump receiver, having at the upper end a collar of leathers in a brass socket, through which a thick smooth wire could be moved; and from the lower end of this wire, projecting into the receiver, was suspended a thermometer. By pushing down the wire, the thermometer could be dipped into the ether; by drawing it up it could be taken out, and suspended over the phial.

The apparatus being thus adjusted, the air-pump was worked to extract the air. An unexpected phenomenon immediately appeared, which prevented the experiment from being made in the way intended. The ether was thrown into a violent agitation, which Dr. Cullen ascribed to the extrication of a great quantity of air: in reality, however, it was boiling violently. What was still more remarkable, the ether, by this boiling or rapid evaporation, became all of a sudden so cold, as to freeze the water in the goblet around it; though the temperature of the air and of all the materials were at the fifty-fourth degree of Fahrenheit at the beginning of the experiment.

I have been particular in giving an account of this curious phenomenon, as it was the only direct contribution to the science of chemistry which Dr. Cullen communicated to the public. The nature of the phenomenon was afterwards explained by Dr. Black; in addition to Dr. Cullen, a philosopher, whom the grand stimulus which his lectures gave to the cultivation of scientific chemistry in this country, had the important merit of bringing forward.

Joseph Black was born in France, on the banks of the Garonne, in the year 1728: his father, Mr. John

Black, was a native of Belfast, but of a Scottish family which had been for some time settled there. Mr. Black resided for the most part at Bordeaux, where he was engaged in the wine trade. He married a daughter of Mr. Robert Gordon, of the family of Hilhead, in Aberdeenshire, who was also engaged in the same trade at Bordeaux. Mr. Black was a gentleman of most amiable manners, candid and liberal in his sentiments, and of no common information. These qualities, together with the warmth of his heart, appear very conspicuous in a series of letters to his son, which that son preserved with the nicest care. His good qualities did not escape the discerning eye of the great Montesquieu, one of the presidents of the court of justice in that province. This illustrious and excellent man honoured Mr. Black with a friendship and intimacy altogether rare; of which his descendants were justly proud.

Long before Mr. Black retired from business, his son Joseph was sent home to Belfast, that he might have the education of a British subject. This was in the year 1740, when he was twelve years of age. After the ordinary instruction at the grammar-school, he was sent, in 1746, to continue his education in the University of Glasgow. Here he studied with much assiduity and success: physical science, however, chiefly engrossed his attention. He was a favourite pupil of Dr. Robert Dick, professor of natural philosophy, and the intimate companion of his son and successor. This young professor was of a character peculiarly suited to Dr. Black's taste, having the clearest conception, and soundest judgment, accompanied by a modesty that was very uncommon. When he succeeded his father, in 1751, he became the delight of the students. He was carried off by a fever in 1757.

Young Black being required by his father to make choice of a profession, he preferred that of medicine as the most suitable to the general habits of his studies.

Fortunately Dr. Cullen had just begun his great career in the College of Glasgow, and having made choice of the field of philosophical chemistry which lay as yet unoccupied before him. Hitherto chemistry had been treated as a curious and useful art; but Cullen saw in it a vast department of the science of nature, depending on principles as immutable as the laws of mechanism, and capable of being formed into a system as comprehensive and as complete as astronomy itself. He conceived the resolution of attempting himself to explore this magnificent field, and expected much reputation from accomplishing his object. Nor was he altogether disappointed. He quickly took the science out of the hands of artists, and exhibited it as a study fit for a gentleman. Dr. Black attended his chemical lectures, and, from the character which has already been given of him, it is needless to say that he soon discovered the uncommon value of his pupil, and attached him to himself, rather as a co-operator and a friend, than a pupil. He was considered as his assistant in all his operations, and his experiments were frequently adduced in the lecture as good authority.

Young Black laid down a very comprehensive and serious plan of study. This appears from a number of note-books found among his papers. There are some in which he seems to have inserted every thing as it took his fancy, in medicine, chemistry, jurisprudence, or matters of taste. Into others, the same things are transferred, but distributed according to their scientific connexions. In short, he kept a journal and ledger of his studies, and has posted his books like a merchant. What particularly strikes one in looking over these books, is the steadiness with which he advanced in any path of knowledge. Things are inserted for the first time from some present impression of their singularity or importance, but without any allusion to their connexions. When a thing of the same kind is mentioned again, there is generally a reference back to its fellow; and thus the most isolated facts often acquired a connexion which gave them importance.

He went to Edinburgh to finish his medical studies in 1750 or 1751, where he lived with his cousin german, Mr. James Russel, professor of natural philosophy in that university.

It was the good fortune of chemical science, that at this very time the opinions of professors were divided concerning the manner in which certain lithontriptic medicines, particularly lime-water, acted in alleviating the excruciating pains of the stone and gravel. The students usually partake of such differences of opinion: they are thereby animated to more serious study, and science gains by their emulation. This was a subject quite to the taste of young Mr. Black, one of Dr. Cullen's most zealous and intelligent chemical pupils. It was, indeed, a most interesting subject, both to the chemist and the physician.

All the medicines which were then in vogue as solvents of urinary calculi had a greater or less resemblance to caustic potash or soda; substances so acrid, when in a concentrated state, that in a short time they reduce the fleshy parts of the animal body to a mere pulp. Thus, though they might possess lithontriptic properties, their exhibition was dangerous, if in unskilful hands. They all seemed to derive their efficacy from quicklime, which again derives its power from the fire. It was therefore very natural for them to ascribe its power to igneous matter imbibed from the fire, retained by the lime, and communicated by it to alkalies, which it renders powerfully acrid. Hence, undoubtedly, the term *caustic* applied to the alkalies in that state, and hence also the *acidum pingue* of Mayer, which was a peculiar state of fire. It appears from Dr. Black's note-books, that he originally entertained the opinion, that caustic alkalies acquired igneous matter from quicklime. In one of them he hints at some way of catching this matter as it escapes from lime, while it becomes mild by exposure to the air ; but on the opposite blank page is written, " Nothing escapes—the cup rises considerably by absorbing air." A few pages further on, he compares the loss of weight sustained by an ounce of chalk when calcined, with its loss while dissolved in muriatic acid. Immediately after this, a medical case is mentioned, which occurred in November, 1752. Hence it would appear, that he had before that time suspected the real cause of the difference between limestone and burnt lime. He had prosecuted his inquiry with vigour ; for the experiments with magnesia are soon after mentioned.

These experiments laid open the whole mystery, as appears by another memorandum. " When I precipitate lime by a common alkali there is no effervescence : the air quits the alkali for the lime; but it is lime no longer, but C. C. C. : it now effervesces, which good lime will not." What a multitude of important consequences naturally flowed from this discovery! He now knew to what the causticity of alkalies is owing, and how to induce it or remove it at pleasure. The common notion was entirely reversed. Lime imparts nothing to the alkalies; it only removes from them a peculiar kind of air (*carbonic acid gas*) with which they were combined, and which prevented their natural caustic properties from being developed. All the former mysteries disappear, and the greatest simplicity appears in those operations of nature which before appeared so intricate and obscure.

Dr. Black had fixed upon this subject for his inaugural dissertation, and was induced, in consequence, to defer applying for his degree till he had succeeded in establishing his doctrine beyond the possibility of contradiction. The inaugural essay was delivered at a moment peculiarly favourable to the advancement of science. Dr. Cullen had been just removed to Edinburgh, and there was a vacancy in the chemical chair in Glasgow: it could not be bestowed better than on such an *alumnus* of the university—on one who had distinguished himself both as a chemist and an excellent reasoner; for few finer models of inductive investigation exist than are displayed in Black's essay on quicklime and magnesia. He was appointed professor of anatomy and lecturer on chemistry in the University of Glasgow in 1756. It was a fortunate circumstance both for himself and for the public, that a situation thus presented itself, just at the time when he was under the necessity of settling in the world— a situation which allowed him to dedicate his talents chiefly to the cultivation of chemistry, his favourite science.

When Dr. Black took his degree in medicine, he sent some copies of his essay to his father at Bordeaux. A copy was given by the old gentleman to his friend, the President Montesquieu, who, after a few days called on Mr. Black, and said to him, " Mr. Black, my very good friend, I rejoice with you; your son will be the honour of your name and family." This anecdote was told Professor John Robison by the brother of Dr. Black.

Thus Dr. Black, while in Glasgow, taught at one and the same time two different classes. He did not consider himself very well qualified to teach anatomy, but determined to do his utmost; but he soon afterwards made arrangements with the professor of medicine, who, with the concurrence of the university, exchanged his own chair for that of Dr. Black.

Black's medical lectures constituted his chief task while in Glasgow. They gave the greatest satisfaction by their perspicuity and simplicity, and by the cautious moderation of all his general doctrines: and, indeed, all his perspicuity, and all his neatness of manner in exhibiting simple truths, were necessary to create a relish for moderation and caution, after the brilliant prospects of systematic knowledge to which the students had been accustomed by Dr. Cullen, his celebrated predecessor. But Dr. Black had no wish to form a medical school, distinguished by some all-comprehending doctrine: he satisfied himself with a clear account of as much of physiology as he thought founded on good principles, and a short sketch of such general doctrines as were maintained by the most eminent authors, though perhaps on a less firm foundation. He then endeavoured to deduce a few canons of medical practice, and concluded with certain rules founded on successful practice only, but not deducible from the principles of physiology previously laid down. With his medical lectures he does not appear to have been himself entirely satisfied: he did not encourage conversation on the different topics, and no remains of these lectures were to be found among his papers. The preceding account of them was given to Professor Robison by a surgeon in Glasgow, who attended the two last medical courses which Dr. Black ever delivered.

Dr. Black's reception at Glasgow by the university was in the highest degree encouraging. His former conduct as a student had not only done him credit in his classes, but had conciliated the affection of the professors to a very high degree. He became immediately connected in the strictest friendship with the celebrated Dr. Adam Smith—a friendship which continued intimate and confidential through the whole of their lives. Both were remarkable for a certain simplicity of character and the most incorruptible integrity. Dr. Smith used to say, that no one had less nonsense in his head than Dr. Black; and he often acknowledged himself obliged to him for setting him right in his judgment of character, confessing that he himself was too apt to form his opinion from a single feature.

It was during his residence in Glasgow, between the years 1759 and 1763, that he brought to maturity those speculations concerning the combination of *heat* with *matter*, which had frequently occupied a portion of his thoughts. It had long been known that ice has the property of continuing always at the temperature of 32° till it be melted. This happens equally though it be placed in contact with the warm hand or surrounded with bodies many degrees hotter than itself. The hotter the bodies are that surround it, the sooner is it melted; but its temperature during the whole process of melting, continues uniformly the same. Yet, during the whole process of melting, it is constantly robbing the surrounding bodies of heat; for it makes them colder, without acquiring itself any sensible heat.

Dr. Black had some vague notion that the heat so received by the ice, during its conversion into water, was not lost, but was contained in the water. This opinion was founded chiefly on a curious observation of Fahrenheit, recorded by Boerhaave; namely, that water might in some cases be made considerably colder than melting snow, without freezing. In such cases, when disturbed it would freeze in a moment, and in the act of freezing always gave out a quantity of heat. This opinion was confirmed by observing the slowness with which water is converted into ice, and ice into water. A fine winter-day of sunshine is never sufficient to clear the hills of snow; nor is one frosty night capable of covering the ponds with a thick coating of ice. The phenomena satisfied him that much heat was absorbed and fixed in the water which trickles from wreaths of snow, and that much heat emerged from it while water was slowly converted into ice; for during a thaw the melting snow is always colder than the air, and must, therefore, be always receiving heat from it; while, during a frost, the air is always colder than the freezing water, and must therefore be always receiving heat from it. These observations, and many others which it is needless to state, satisfied Dr. Black that when ice is converted into water it unites with a quantity of heat, without increasing in temperature; and that when water is frozen into ice it gives out a quantity of heat without diminishing in temperature. The heat thus combined is the cause of the fluidity of the water. As it is not sensible to the thermometer, Dr. Black called it *latent heat*. He made an experiment to determine the quantity of heat necessary to convert ice into water. This he estimated by the length of time necessary to melt a given weight of ice, and measuring how much heat entered into the same weight of water, reduced as nearly to the temperature of ice as possible during the first half-hour that the experiment lasted. As the ice continued during the whole of its melting at the same temperature as at first, he concluded that it would absorb, every half-hour that the process lasted, as much heat as the water did during the first half hour. The result of this experiment was, that the latent heat of water amounts to 140°; or, in other words, that this heat, if thrown into a quantity of water, equal in weight to that of the ice melted, would raise its temperature 140°.

Dr. Black, having established this discovery in the most incontrovertible manner by simple and decisive experiments, drew up an account of the whole investigation, and the doctrine which he founded upon it, and read it to a literary society which met every Friday in the faculty-room of the college, consisting of the members of the university and several gentlemen of the city, who had a relish for science and literature. This paper was read on the 23d of April, as appears by the registers of the society.

Dr. Black quickly perceived the vast importance of this discovery, and took a pleasure in laying before his students a view of the beneficial effects of this habitude of heat in the economy of nature. During the summer season a vast magazine of heat was accumulated in the water, which, by gradually emerging during congelation, serves to temper the cold of winter. Were it not for this accumulation of heat in water and other bodies, the sun would no sooner go a few degrees to the south of the equator, than we should feel all the horrors of winter. He did not confine his views to the congelation of water alone, but extended them to every case of congelation and liquefaction which he has ascribed equally to the evolution or fixation of latent heat. Even those bodies which change from solid to fluid, not all at once, but by slow degrees, as butter, tallow, resins, owe, he found, their gradual softening to the same absorption of heat, and the same combination of it with the substance undergoing liquefaction.

Another subject that engaged his attention at this time, was an examination of the scale of the thermometer, to learn whether equal differences of expansion corresponded to equal additions or abstractions of heat. His mode was to mix together equal weights of water of different temperatures, and to measure the temperature of the mixture by a thermometer. It is obvious that the temperature must be the exact mean of that of the two portions of water ; and that if the expansion or contraction of the mercury in the thermometer be an exact measure of the difference of temperature, a thermometer, so placed, will indicate the exact mean. Suppose one pound of water at 100° to be mixed with one pound of water at 200°, and the whole heat still to remain in the mixture, it is obvious that it would divide itself equally between the two portions of water. The water of 100° would become hotter, and the water of 200° would become colder: and the increase of temperature in the colder portion would be just as much as the diminution of temperature in the hotter portion. The colder portion would become hotter by 50°, while the hotter portion would become colder by 50°. Hence the real temperature, after mixture, would be 150° ; and a thermometer plunged into such a mixture, if a true measurer of heat, would indicate 150°. The result of his experi ments was, that as high up as he could try by mixing water of different temperatures, the mercurial thermometer is an accurate measurer of the alterations of temperature.

An account of his experiments on this subject was drawn up by him, and read to the literary society of the College of Glasgow, on the 28th of March, 1760. Dr. Black, at the time he made these experiments, did not know that he had been already anticipated in them by Dr. Brooke Taylor, the celebrated mathematician, who had obtained similar results, and had consigned his experiments to the Royal Society, in whose Transactions for 1723 they were published. It has been since found by Coulomb and Petit, that at higher temperatures than 212° the rate of the expansion of mercury begins to increase. Hence it happens that at high temperatures the expansion of mercury is no longer an accurate measurer of temperature. Fortunately, the expansion of glass very nearly equals the increment of that of mercury. The consequence is, that in a common glass-thermometer mercury measures the true increments of temperature very nearly up to its boiling point; for the boiling point of mercury measured by an air-thermometer is 662°: and if a glass mercurial thermometer be plunged into boiling mercury, it will indicate 660°, a difference of only 2° from the true point.

There is such an analogy between the cessation of thermometric expansion during the liquefaction of ice, and during the conversion of water into steam, that their could be no hesitation about explaining both in the same way. Dr. Black immediately concluded that as water is ice united to a certain quantity of *latent heat*, so steam is water united to a still greater quantity. The slow conversion of water into steam, notwithstanding the great quantity of heat constantly flowing into it from the fire, left no reasonable doubt about the accuracy of this conclusion. In short, all the phenomena are precisely similar to those of the conversion of ice into water; and so, of course, must the explanation be. So much was he convinced of this, that he taught the doctrine in his lectures in 1761, before he had made a single experiment on the subject; and he explained, with great felicity of argument, many phenomena of nature, which result from this vaporific combination of heat. From notes taken in his class during this session, it appears that nothing more was wanting to complete his views on this subject, than a set of experiments to determine the exact quantity of heat which was combined in steam in a state not indicated by the thermometer, and therefore *latent*, in the same sense that the heat of liquefaction in water is *latent*.

The requisite experiments were first attempted by Dr. Black, in 1764. They consisted merely in measuring the time requisite to convert a certain weight of water of a given temperature into steam. The water was put into a tin-plate wide-mouthed vessel, and laid upon a red-hot plate of iron, the initial tem· perature of the water was marked, and the time necessary to heat it from that point to the boiling point noted, and then the time requisite to boil the whole to dryness. It was taken for granted that as much heat would enter into the water during every minute that the experiment lasted, as did during the first minute. From this it was concluded that the latent heat of steam is not less than 810 degrees.

Mr. James Watt afterwards repeated these experiments with a better apparatus and very great care, and calculated from his results that the latent heat of steam is not under 950 degrees. Lavoisier and Laplace afterwards made experiments in a different way, and deduced 1000° as the result of their experiments. The subsequent experiments of Count Rumford, meda in a very ingenious manner, so as to obviate most of the sources of error, to which such researches are liable, come very nearly to those of Lavoisier. 1000° therefore, is usually now-a-days adopted as the number which denotes the true latent heat of steam.

Dr. Black continued in the University of Glasgow from 1756 to 1766, much esteemed as an eminent professor, much employed as an able and attentive physician, and much beloved as an amiable and accomplished man, happy in the enjoyment of a small but select society of friends. Meanwhile his reputation as a chemical philosopher was every day increasing, and pupils from foreign countries carried home with them the peculiar doctrines of his courses—so that *fixed air* and *latent heat* began to be spoken of among the naturalists of the continent. In 1766 Dr. Cullen, at that time professor of chemistry in Edinburgh, was appointed professor of medicine, and thus a vacancy was made in the chemical chair of that university. There was but one wish with regard to a successor. Indeed, when the vacancy happened in 1756, on the death of Dr. Plummer, the reputation of Dr. Black, who had just taken his degree, was so high, both as a chemist and an accurate thinker and reasoner, that, had the choice depended on the university, he would have been the new professor of chemistry. He had now, in 1766, greatly added to his claim of merit by his important discovery of latent heat ; and he had acquired the esteem of all by the singular moderation and scrupulous caution which marked all his researches.

Dr. Black was appointed to the chemical chair in Edinburgh in 1766, to the general satisfaction of the public, but the University of Glasgow suffered an irreparable loss. In this new situation his talents were more conspicuous and more extensively useful. He saw that the case was so, and while he could not but be gratified by the number of students whom the high reputation of Edinburgh, as a medical school, brought together, his mind was forcibly struck by the importance of his duties as a teacher. This led him to form the resolution of devoting the whole of his study to the improvement of his pupils in the elementary knowledge of chemistry. Many of them came to his class with a very scanty stock of previous knowledge. Many from the workshop of the manufacturer had little or none. He was conscious that the number of this kind of pupils must increase with the increasing activity and prosperity of the country; and they appeared to him by no means the least important part of his auditory. To engage the attention of such pupils, and to be perfectly understood by the most illiterate of his audience, Dr. Black considered as a sacred duty: he resolved, therefore, that plain doctrines taught in the plainest manner, should henceforth employ his chief study. To render his lectures perfectly intelligible they were illustrated by suitable experiments, by the exhibition of specimens, and by the repetition of chemical processes.

To this method of lecturing Dr. Black rigidly adhered, endeavouring every year to make his courses more plain and familiar, and illustrating them by a greater variety of examples in the way of experiment. No man could perform these more neatly or successfully; they were always ingeniously and judiciously contrived, clearly establishing the point in view, and were never more complicated than was sufficient for the purpose. Nothing that had the least appearance of quackery; nothing calculated to surprise and astonish his audience; nothing savouring of a showman or sleight-of-hand man was ever permitted in his lecture-room. Every thing was simple, neat, and elegant, calculated equally to please and inform: indeed simplicity and neatness stamped his character. It was this that constituted the charm of his lectures, and rendered them so delightful to his pupils. I can speak of them from experience, for I was fortunate enough to hear the last course of lectures which he ever delivered. I can say with perfect truth that I never listened to any lectures with so much pleasure as to his : and it was the elegant simplicity of his manner, the perfect clearness of his statements, and the vast quantity of information which he contrived in this way to communicate, that delighted me. I was all at once transported into a new world—my views were suddenly enlarged, and I looked down from a height which I had never before reached ; and all this knowledge was communicated without any apparent effort either on the part of the professor or his pupils. His illustrations were just sufficient to answer completely the object in view, and nothing more. No quackery, no trickery, no love of mere dazzle and glitter, ever had the least influence upon his conduct. He constituted the most complete model of a perfect chemical lecturer that I have ever had an opportunity of witnessing.

The discovery which Dr. Black had made that marble is a combination of lime and a peculiar substance, to which he gave the name of *fixed air*, began gradually to attract the attention of chemists in other parts of the world. It was natural in the first place to examine the nature and properties of this fixed air, and the circumstances under which it is generated. It may seem strange and unaccountable that Dr. Black did not enter with ardour into this new career which he had himself opened, and that he allowed others to reap the corn after having himself sown the grain. Yet he did take some steps towards ascertaining the properties of *fixed air;* though I am not certain what progress he made. He knew that a candle would not burn in it, and that it is destructive to life, when any living animal attempts to breathe it. He knew that it was formed in the lungs during the breathing of animals, and that it is generated during the fermentation of wine and beer. Whether he was aware that it possesses the properties of an

acid I do not know ; though with the knowledge which
he possessed that it combines with alkalies and alkaline
earths, and neutralizes them, or at least blunts and di-
minishes their alkaline properties, the conclusion that
it partook of alkaline properties was scarcely avoidable.
All these, and probably some other properties of *fixed
air* he was in the constant habit of stating in his lectures
from the very commencement of his academical career;
though, as he never published any thing on the subject
himself, it is not possible to know exactly how far his
knowledge of the properties of *fixed air* extended. The
oldest manuscript copy of his lectures that I have seen
was taken down in writing in the year 1773 ; and
before that time Mr. Cavendish had published his
paper on *fixed air* and *hydrogen gas*, and had detailed
the properties of each.   It was impossible from the
manuscript of Dr. Black's lectures to know which of the
properties of *fixed air* stated by him were discovered
by himself, and which were taken from Mr. Cavendish.

This languor and listlessness, on the part of Dr.
Black, is chiefly to be ascribed to the delicate state of
his health, which precluded much exertion, and was
particularly inconsistent with any attempt at putting
his thoughts down upon paper.   Hence, probably, that
carelessness about posthumous fame, and that regard-
lessness of reputation, which, however it may be accounted for from bodily ailment, must still be considered as a blemish. How differently did Paschal act in
a similar state of health !   With what energy did he
exert himself in spite of bodily ailment! But the tone
of his mind was quite different from that of Dr. Black.
Gentleness, diffidence, and perhaps even slowness
of apprehension, were the characteristic features by
which the latter was distinguished.

There is an anecdote of Black which I was told by
the late Mr. Benjamin Bell, of Edinburgh, author of
a well-known system of surgery, and he assured me
that he had it from the late Sir George Clarke, of

Pennicuik, who was a witness of the circumstance related. Soon after the appearance of Mr. Cavendish's paper on hydrogen gas, in which he made an approximation to the specific gravity of that body, showing that it was at least ten times lighter than common air, Dr. Black invited a party of his friends to supper, informing them that he had a curiosity to show them. Dr. Hutton, Mr. Clarke of Elden, and Sir George Clarke of Pennicuik, were of the number. When the company invited had assembled, he took them into a room. He had the allentois of a calf filled with hydrogen gas, and upon setting it at liberty, it immediately ascended, and adhered to the ceiling. The phenomenon was easily accounted for: it was taken for granted that a small black thread had been attached to the allentois, that this thread passed through the ceiling, and that some one in the apartment above, by pulling the thread, elevated it to the ceiling, and kept it in this position. This explanation was so probable, that it was acceded to by the whole company; though, like many other plausible theories, it turned out wholly unfounded; for when the allentois was brought down no thread whatever was found attached to it. Dr. Black explained the cause of the ascent to his admiring friends; but such was his carelessness of his own reputation, and of the information of the public, that he never gave the least account of this curious experiment even to his class; and more than twelve years elapsed before this obvious property of hydrogen gas was applied to the elevation of air-balloons, by M. Charles, in Paris.

The constitution of Dr. Black had always been exceedingly delicate. The slightest cold, the most trifling approach to repletion, immediately affected his chest, occasioned feverishness, and if the disorder continued for two or three days, brought on a spitting of blood. In this situation, nothing restored him to ease, but relaxation of thought, and gentle exercise.

The sedentary life to which study confined him, was manifestly hurtful ; and he never allowed himself to indulge in any investigation that required intense thought, without finding these complaints increased.

Thus situated, Dr. Black was obliged to be a contented spectator of the rapid progress which chemistry was making, without venturing himself to engage in any of the numerous investigations which presented themselves on every side. Such indeed was the eagerness with which chemistry was at that time prosecuted, and such the passion for discovery, that there was some risk that his undoubted claim to originality and priority in his own great discoveries, might be called in question, and even rendered doubtful. His friends at least were afraid of this, and often urged him to do justice to himself, by publishing an account of his own discoveries. He more than once began the task ; but was so nice in his notions of the manner in which it should be executed, that the pains he took in forming a plan of the work never failed to affect his health, and oblige him to desist. It is known that he felt hurt at the publication of several of Lavoisier's papers, in the Mémoires de l'Académie, without any allusion whatever to what he himself had previously done on the same subject. How far Lavoisier was really culpable, and whether he did not intend to do full justice to all the claims of his predecessors, cannot now be known ; as he was cut off in the midst of his career, while so many of his scientific projects remained unexecuted. From the posthumous works of Lavoisier, there is some reason for believing that if he had lived, he would have done justice to all parties ; but there is no doubt that Dr. Black, in the mean time, thought himself aggrieved, and that he formed the intention of doing himself justice, by publishing an account of his own discoveries ; however this intention was thwarted and prevented by bad health.

No one contributed more largely to establish, to support, and to increase, the high character of the medical school in the University of Edinburgh than Dr. Black. His talent for communicating knowledge was not less eminent than his faculty of observation. He soon became one of the principal ornaments of the university; and his lectures were attended by an audience which continued increasing from year to year for more than thirty years. His personal appearance and manners were those of a gentleman, and peculiarly pleasing: his voice, in lecturing, was low, but fine; and his articulation so distinct, that he was perfectly well heard by an audience consisting of several hundreds. While in Glasgow, he had practised extensively as a physician; but in Edinburgh he declined general practice, and confined his attendance to a few families of intimate and respected friends. He was, however, a physician of good repute in a place where the character of a physician implied no common degree of liberality, propriety, and dignity of manners, as well as of learning and skill.

Such was Dr. Black as a public man. While young, his countenance was comely and interesting; and as he advanced in years, it continued to preserve that pleasing expression of inward satisfaction which, by giving ease to the beholder, never fails to please. His manners were simple, unaffected, and graceful; he was of the most easy approach, affable, and readily entered into conversation, whether serious or trivial: for he was not merely a man of science, but was well acquainted with the elegant accomplishments. He had an accurate musical ear, and a voice which would obey it in the most perfect manner; he sang and performed on the flute with great taste and feeling; and could sing a plain air at sight, which many instrumental performers cannot do. Music was his amusement in Glasgow; after his removal to Edinburgh he gave it up entirely. Without having studied drawing he had acquired a considerable power of expression with his pencil, both in figures and in landscape. He was peculiarly happy in expressing the passions, and seemed in this respect to have the talents of a historical painter. Figure indeed, of every kind, attracted his attention ; in architecture, furniture, ornament of every sort, it was never a matter of indifference to him. Even a retort, or a crucible, was to his eye an example of beauty, or deformity. These are not indifferent things ; they are features of an elegant mind, and they account for some part of that satisfaction and pleasure which persons of different habits and pursuits felt in Dr. Black's company and conversation.

Those circumstances of form, and in which Dr. Black perceived or sought for beauty, were suitableness or propriety : something that rendered them will adapted for the purposes for which they were intended. This love of propriety constituted the leading feature in Dr. Black's mind ; it was the standard to which he constantly appealed, and which he endeavoured to make the directing principle of his conduct.

Dr. Black was fond of society, and felt himself beloved in it. His chief companions, in the earlier part of his residence in Edinburgh, were Dr. Adam Smith, Mr. David Hume, Dr. Adam Ferguson, Mr. John Home, Dr. Alexander Carlisle, and a few others. Mr. Clarke of Elden, and his brother Sir George, Dr. Roebuck, and Dr. James Hutton, particularly the latter, were affectionately attached to him, and in their society he could indulge in his professional studies. Dr. Hutton was the only person near him to whom Dr. Black imparted every speculation in chemical science, and who knew all his literary labours : seldom were the two friends asunder for two days together.

Towards the close of the eighteenth century, the infirmities of advanced life began to bear more heavily on his feeble constitution. Those hours of walking and gentle exercise, which had hitherto been necessary for his ease, were gradually curtailed. Company and conversation began to fatigue : he went less abroad, and was visited only by his intimate friends. His duty at college became too heavy for him, and he got an assistant, who took a share of the lectures, and relieved him from the fatigue of the experiments. The last course of lectures which he delivered was in the winter of 1796-7. After this, even lecturing was too much for his diminished strength, and he was obliged to absent himself from the class altogether; but he still retained his usual affability of temper, and his habitual cheerfulness, and even to the very last was accustomed to walk out and take occasional exercise. As his strength declined, his constitution became more and more delicate. Every cold he caught occasioned some degree of spitting of blood ; yet he seemed to have this unfortunate disposition of body almost under command, so that he never allowed it to proceed far, or to occasion any distressing illness. He spun his thread of life to the very last fibre. He guarded against illness by restricting himself to an abstemious diet; and he met his increasing infirmities with a proportional increase of attention and care, regulating his food and exercise by the measure of his strength. Thus he made the most of a feeble constitution, by preventing the access of disease from abroad. And enjoyed a state of health which was feeeble,indeed, but scarcely interrupted ; as well as a mind undisturbed in the calm and cheerful use of its faculties. His only apprehension was that of a long-continued sick-bed — from the humane consideration of the trouble and distress that he might thus occasion to attending friends; and never was such generous wish more completely gratified.

On the 10th of November, 1799, in the seventy-first year of his age, he expired without any convulsion, shock, or stupor, to announce or retard the approach of death. Being at table with his usual fare, some bread, a few prunes, and a measured quantity of milk,

diluted with water, and having the cup in his hand
when the last stroke of his pulse was to be given, he
set it down on his knees, which were joined together,
and kept it steady with his hand in the manner of a
person perfectly at ease ; and in this attitude expired
without spilling a drop, and without a writhe in his
countenance ; as if an experiment had been required to show to his friends the facility with which
he departed. His servant opened the door to tell him
that some one had left his name ; but getting no answer, stepped about halfway to him; and seeing him
sitting in that easy posture, supporting his basin of
milk with one hand, he thought that he had dropped
asleep, which was sometimes wont to happen after
meals. He went back and shut the door ; but before
he got down stairs some anxiety, which he could not
account for, made him return and look again at his
master. Even then he was satisfied, after coming
pretty near him, and turned to go away ; but he again
returned, and coming close up to him, he found him
without life. His very near neighbour, Mr. Benjamin
Bell, the surgeon, was immediately sent for ; but nothing whatever could be done.*

Dr. Black's writings are exceedingly few, consisting
altogether of no more than three papers. The first,
entitled " Experiments upon Magnesia alba, Quicklime, and other Alkaline Substances," constituted the
subject of his inaugural dissertation. It afterwards
appeared in an English dress in one of the volumes of
The Edinburgh Physical and Literary Essays, in the
year 1755. Mr. Creech, the bookseller, published it
in a separate pamphlet, together with Dr. Cullen's
little essay on the " cold produced by evaporating

* The preceding character of Dr. Black is from Professor
Robison, who knew him intimately ; and from Dr. Adam Ferguson, who was his next relation. See the preface to Dr. Black's
lectures. The portrait of Dr. Black prefixed to these lectures
is an excellent likeness.

fluids," in the year 1796. This essay exhibits one of the very finest examples of inductive reasoning to be found in the English language. The author shows that magnesia is a peculiar earthy body, possessed of properties very different from lime. He gives the properties of lime in a pure state, and proves that it differs from limestone merely by the absence of the carbonic acid, which is a constituent of limestone. Limestone is a *carbonate of lime;* quicklime is the pure uncombined earth. He shows that magnesia has also the property of combining with carbonic acid ; that caustic potash, or soda, is merely these bodies in a pure or isolated state ; while the mild alkalies are combinations of these bodies with carbonic acid. The reason why quicklime converts mild into caustic alkali is, that the lime has a stronger affinity for the carbonic acid than the alkali; hence the lime is converted into carbonate of lime, and the alkali, deprived of its carbonic acid, becomes caustic. Mild potash is a carbonate of potash ; caustic potash, is potash freed from carbonic acid.—The publication of this essay occasioned a controversy in Germany, which was finally settled by Jacquin and Lavoisier, who repeated Dr. Black's experiments and showed them to be correct.

Dr. Black's second paper was published in the Philosophical Transactions for 1775. It is entitled " The supposed Effect of boiling on Water, in disposing it to freeze more readily, ascertained by Experiments." He shows, that when water that has been recently boiled is exposed to cold air, it begins to freeze as soon as it reaches the freezing point; while water that has not been boiled may be cooled some degrees below the freezing point before it begins to congeal. But if the unboiled water be constantly stirred during the whole time of its exposure, it begins to freeze when cooled down to the freezing point as well as the other. He shows that the difference between the two waters consists in this, that the boiled water is constantly absorbing air, which disturbs it, whereas the other water remains in a state of rest.

His last paper was "An Analysis of the Water of some boiling Springs in Iceland," published in the Transactions of the Royal Society of Edinburgh. This was the water of the Geyser spring, brought from Iceland by Sir J. Stanley. Dr. Black found it to contain a great deal of silica, held in solution in the water by caustic soda.

The tempting career which Dr. Black opened, and which he was unable to prosecute for want of health, soon attracted the attention of one of the ablest men that Great Britain has produced—I mean Mr. Cavendish.

The Honourable Henry Cavendish was born in London on the 10th of October, 1731: his father was Lord Charles Cavendish, a cadet of the house of Devonshire, one of the oldest families in England. During his father's lifetime he was kept in rather narrow circumstances, being allowed an annuity of £500 only; while his apartments were a set of stables, fitted up for his accommodation. It was during this period that he acquired those habits of economy, and those singular oddities of character, which he exhibited ever after in so striking a manner. At his father's death he was left a very considerable fortune; and an aunt who died at a later period bequeathed him a very handsome addition to it; but, in consequence of the habits of economy which he had acquired, it was not in his power to spend the greater part of his annual income. This occasioned a yearly increase to his capital, till at last it accumulated so much, without any care on his part, that at the period of his death he left behind him nearly £1,300,000; and he was at that time the greatest proprietor of stock in the Bank of England.

On one occasion, the money in the hands of his bankers had accumulated to the amount of £70,000. These gentlemen thinking it improper to keep so large a sum in their hands, sent one of the partners to wait upon him, in order to learn how he desired it disposed of. This gentleman was admitted; and, after employing the necessary precautions to a man of Mr. Cavendish's peculiar disposition, stated the circumstance, and begged to know whether it would not be proper to lay out the money at interest. Mr. Cavendish dryly answered, " You may lay it out if you please," and left the room.

He hardly ever went into any other society than that of his scientific friends : he never was absent from the weekly dinner of the Royal Society club at the Crown and Anchor Tavern in the Strand. At these dinners, when he happened to be seated near those that he liked, he often conversed a great deal; though at other times he was very silent. He was likewise a constant attendant at Sir Joseph Banks's Sunday evening meetings. He had a house in London, which he only visited once or twice a-week at stated times, and without ever speaking to the servants : it contained an excellent library, to which he gave all literary men the freest and most unrestrained access. But he lived in a house on Clapham Common, where he scarcely ever received any visiters. His relation, Lord George Cavendish, to whom he left by will the greatest part of his fortune, visited him only once a-year, and the visit hardly ever exceeded ten or twelve minutes.

He was shy and bashful to a degree bordering on disease ; he could not bear to have any person introduced to him, or to be pointed out in any way as a remarkable man. One Sunday evening he was standing at Sir Joseph Banks's in a crowded room, conversing with Mr. Hatchett, when Dr. Ingenhousz, who had a good deal of pomposity of manner, came up with an Austrian gentleman in his hand, and introduced him formally to Mr. Cavendish. He mentioned the titles and qualifications of his friend at great length, and said that he had been peculiarly anxious to be introduced to a philosopher so profound and so universally known and celebrated as Mr. Cavendish. As soon as Dr. Ingenhousz had finished, the Austrian gentleman began, and assured Mr. Cavendish that his principal reason for coming to London was to see and converse with one of the greatest ornaments of the age, and one of the most illustrious philosophers that ever existed. To all these high-flown speeches Mr. Cavendish answered not a word, but stood with his eyes cast down quite abashed and confounded. At last, spying an opening in the crowd, he darted through it with all the speed of which he was master; nor did he stop till he reached his carriage, which drove him directly home.

Of a man, whose habits were so retired, and whose intercourse with society was so small, there is nothing else to relate except his scientific labours: the current of his life passed on with the utmost regularity; the description of a single day would convey a correct idea of his whole existence. At one time he was in the habit of keeping an individual to assist him in his experiments. This place was for some time filled by Sir Charles Blagden; but they did not agree well together, and after some time Sir Charles left him. Mr. Cavendish died on the 4th of February, 1810, aged seventy-eight years, four months, and six days. When he found himself dying, he gave directions to his servant to leave him alone, and not to return till a certain time which he specified, and by which period he expected to be no longer alive. The servant, however, who was aware of the state of his master, and was anxious about him, opened the door of the room before the time specified, and approached the bed to take a look at the dying man. Mr. Cavendish, who was still sensible, was offended at the intrusion, and ordered him out of the room with a voice of displeasure, commanding him not by any means to return till the time specified.   When he did come back at that time, he found his master dead.   What a contrast between the characters of Mr. Cavendish and Dr. Black!

The appearance of Mr. Cavendish did not much prepossess strangers in his favour; he was somewhat above the middle size, his body rather thick, and his neck rather short.   He stuttered a little in his speech, which gave him an air of awkwardness: his countenance was not strongly marked, so as to indicate the profound abilities which he possessed.   This was probably owing to the total absence of all the violent passions.   His education seems to have been very complete; he was an excellent mathematician, a profound electrician, and a most acute and ingenious chemist. He never ventured to give an opinion on any subject, unless he had studied it to the bottom.   He appeared before the world first as a chemist, and afterwards as an electrician.   The whole of his literary labours consist of eighteen papers, published in the Philosophical Transactions, which, though they occupy only a few pages, are full of the most important discoveries and the most profound investigations.   Of these papers, there are ten which treat of chemical subjects, two treat of electricity, two of meteorology, three are connected with astronomy, and there is one, the last which he wrote, which gives his method of dividing astronomical instruments.   Of the papers in question, those alone which treat of Chemistry can be analyzed in a work like this.

1. His first paper, entitled, "Experiments on fictitious Air," was published in the year 1766, when Mr. Cavendish was thirty-five years of age.   Dr. Hales had demonstrated (as had previously been done by Van Helmont and Glauber) that *air* is given out by a vast number of bodies in peculiar circumstances.   But he never suspected that any of the *airs* which he obtained differed from common air.   Indeed common air had always been considered as an elementary substance to

which every elastic fluid was referred.   Dr. Black had
shown that the mild alkalies and limestone, and carbonate of magnesia, were combinations of these bodies
with a gaseous substance, to which he had given the
name of *fixed air*; and he had pointed out various
methods of collecting this fixed air; though he himself had not made much progress in investigating its
properties.   This paper of Mr. Cavendish may be considered as a continuation of the investigations begun
by Dr. Black. He shows that there exist two species of
air quite different in their properties from common air :
and he calls them *inflammable air* and *fixed air.*

Inflammable air (hydrogen gas) is evolved when
iron, zinc, or tin, are dissolved in dilute sulphuric or
muriatic acid.   Iron yielded about 1-22d part of its
weight, of inflammable air, zinc about 1-23d or
1-24th of its weight, and tin about 1-44th of its
weight.   The properties of the inflammable air were
the same, whichever of the three metals was used
to procure it, and whether they were dissolved in sulphuric or muriatic acids. When the sulphuric acid was
concentrated, iron and zinc dissolved in it with difficulty and only by the assistance of heat. The air given
out was not inflammable, but consisted of sulphurous
acid. These facts induced Mr. Cavendish to conclude
that the inflammable air evolved in the first case was
the unaltered *phlogiston* of the metals, while the sulphurous acid evolved in the second case, was a compound of the same phlogiston and a portion of the
acid, which deprived it of its inflammability.   This
opinion was very different from that of Stahl, who considered combustible bodies as compounds of phlogiston
with acids or calces.

Cavendish found the specific gravity of his inflammable air about eleven times less than that of common air.
This determination is under the truth ; but the error is, at
least in part, owing to the quantity of water held in
solution by the air, and which, as Mr. Cavendish showed,

amounted to about 1-9th of the weight of the air. He tried the combustibility of the inflammable air, when mixed with various proportions of common air, and found that it exploded with the greatest violence when mixed with rather more than its bulk of common air.

Copper he found, when dissolved in muriatic acid by the assistance of heat, yielded no inflammable air, but an air which lost its elasticity when it came in contact with water. This *air*, the nature of which Mr. Cavendish did not examine, was *muriatic acid gas*, the properties of which were afterwards investigated by Dr. Priestley.

The *fixed air* (*carbonic acid gas*) on which Mr. Cavendish made his experiments was obtained by dissolving marble in muriatic acid. He found that it might be kept over mercury for any length of time without undergoing any alteration; that it was gradually absorbed by cold water; and that 100 measures of water of the temperature 55° absorbed 103·8 measures of fixed air. The whole of the air thus absorbed was separated again by exposing the water to a boiling heat, or by leaving it for some time in an open vessel. Alcohol (the specific gravity not mentioned) absorbed $2\frac{1}{4}$ times its bulk of this air, and olive-oil about 1-3d of its bulk.

The specific gravity of fixed air he found 1·57, that of common air being 1.* Fixed air is incapable of supporting combustion, and common air, when mixed with it, supports combustion a much shorter time than when pure. A small wax taper burnt eighty seconds in a receiver which held 180 ounce measures, when filled with common air only. The same taper burnt fifty-one seconds in the same receiver when filled with a mixture of one volume fixed air, and nineteen volumes of common air. When the fixed air was 3-40ths of

* This I apprehend to be a little above the truth, the true specific gravity of carbonic acid gas being 1·5277, that of air being unity.

the whole volume the taper burnt twenty-three seconds. When the fixed air was 1-10th, the taper burnt eleven seconds. When it was 6-55ths or 1-9·16 of the whole mixture, the taper would not burn at all.

Mr. Cavendish was of opinion that more than one kind of fixed air was given out by marble; in other words, that the elastic fluid emitted, consisted of two different airs, one more absorbable by water than the other. He drew his conclusion from the circumstance that after a solution of potash had been exposed to a quantity of fixed air for some time, it ceased to absorb any more; yet, if the residual portion of air were thrown away and new fixed air substituted in its place, it began to absorb again; but Mr. Dalton has since given a satisfactory explanation of this seeming anomaly by showing that the absorbability of fixed air in water is proportional to its purity, and that when mixed with a great quantity of common air or any other gas not soluble in water, it ceases to be sensibly absorbed.

Mr. Cavendish ascertained the quantity of fixed air contained in marble, carbonate of ammonia, common pearlashes, and carbonate of potash: but notwithstanding the care with which these experiments were made they are of little value; because the proper precautions could not be taken, in that infant state of chemical science, to have these salts in a state of purity. The following were the results obtained by Mr. Cavendish:

1000 grains of marble contained 408 grs. fixed air.
1000 — carb. of ammonia 533 —
1000 — pearlashes . . 284 —
1000 — carb. of potash 423 —

Supposing the marble, carbonate of ammonia, and carbonate of potash, to have been pure anhydrous simple salts, their composition would be

1000 grains of marble contain 440 grs. fixed air.
1000 — carb. of ammonia 709·6 —
1000 — carb. of potash 314·2 —

Bicarbonate of potash was first obtained by Dr. Black. Mr. Cavendish formed the salt by dissolving pearlashes in water, and passing a current of carbonic acid gas through the solution till it deposited crystals. These crystals were not altered by exposure to the air, did not deliquesce, and were soluble in about four times their weight of cold water.

Dr. M'Bride had already ascertained that vegetable and animal substances yield fixed air by putrefaction and fermentation. Mr. Cavendish found by experiment that sugar when dissolved in water and fermented, gives out 57-100ths of its weight of fixed air, possessing exactly the properties of fixed air from marble. During the fermentation no air was absorbed, nor was any change induced on the common air, at the surface of the fermenting liquor. Apple-juice fermented much faster than sugar; but the phenomena were the same, and the fixed air emitted amounted to $\frac{381}{1000}$ of the weight of the solid extract of apples. Gravy and raw meat yielded inflammable air during their putrefaction, the former in much greater quantity than the latter. This air, as far as Mr. Cavendish's experiments went, he found the same as the inflammable air from zinc by dilute sulphuric acid; but its specific gravity was a little higher.

This paper of Mr. Cavendish was the first attempt by chemists to collect the different kinds of air, and endeavour to ascertain their nature. Hence all his processes were in some measure new : they served as a model to future experimenters, and were gradually brought to their present state of simplicity and perfection. He was the first person who attempted to determine the specific gravity of airs, by comparing their weight with that of the same bulk of common air; and though his apparatus was defective, yet the principle was good, and is the very same which is still employed to accomplish the same object. Mr. Cavendish then first began the true investigation of gases, and in his first paper he determined the peculiar nature of two very remarkable gases, *carbonic* and *hydrogen*.

2. Mineral waters have at all times attracted the attention of the faculty in consequence of their peculiar properties and medical virtues. Some faint steps towards their investigation were taken by Boyle. Du Clos attempted a chemical analysis of the mineral waters in France; and Hierne made a similar investigation of the mineral waters of Sweden. Though these experiments were rude and inaccurate, they led to the knowledge of several facts respecting mineral waters which chemists were unable to explain. One of these was the existence of a considerable quantity of *calcareous earth* in some mineral waters, which was precipitated by boiling. Nobody could conceive in what way this insoluble substance *(carbonate of lime)* was held in solution, nor why it was thrown down when the water was raised to a boiling heat. It was to determine this point that Mr. Cavendish made his experiments on Rathbone-place water, which were published in the year 1767, and which may be considered as the first analysis of a mineral water that possessed tolerable accuracy. Rathbone-place water was raised by a pump, and supplied the portion of London in its immediate neighbourhood. Mr. Cavendish found that when boiled, it deposited a quantity of earthy matter, consisting chiefly of lime, but containing also a little magnesia. This he showed was held in solution by fixed air; and he proved experimentally, that when an excess of this gas is present, it has the property of holding lime and magnesia in solution.* Besides these earthy carbonates, the water was found to contain a little ammonia, some sulphate of lime, and some common salt. Mr. Cavendish examined, likewise, some

---

* The salts held in solution are in the state of bicarbonates of lime and magnesia. Boiling drives off half the carbonic acid, and the simple carbonates being insoluble are precipitated.

other pump-water in London, and showed that it contained lime, held in solution by carbonic acid.

3. Dr. Priestley, at a pretty early period of his chemical career, had discovered that when nitrous gas is mixed with common air over water, a diminution of bulk takes place; that there is a still greater diminution of bulk when oxygen gas is employed instead of common air; and that the diminution is always proportional to the quantity of oxygen gas present in the gas mixed with the nitrous gas. This discovery induced him to employ nitrous gas as a test of the quantity of oxygen present in common air; and various instruments were contrived to facilitate the mixture of the gases, and the measurement of the diminution of volume which took place. As the goodness of air, or its fitness to support combustion, and maintain animal life, was conceived to depend upon the proportion of oxygen gas which it contained, these instruments were distinguished by the name of *eudiometers;* the simplest of them was contrived by Fontana, and is usually distinguished by the name of the *eudiometer of Fontana.* Philosophers, in examining air by means of this instrument, at various seasons, and in various places, had found considerable differences in the diminution of bulk : hence they inferred that the proportion of oxygen varies in different places; and to this variation they ascribed the healthiness or noxiousness of particular situations. For example, Dr. Ingenhousz had found a greater proportion of oxygen in the air above the sea, and on the sea-coast; and to this he ascribed the healthiness of maritime situations. Mr. Cavendish examined this important point with his usual patient industry and acute discernment, and published the result in the Philosophical Transactions for 1783. He ascertained that the apparent variations were owing to inaccuracies in making the experiment; and that when the requisite precautions are taken, the proportion of oxygen in air is found constant in all

places, and at all seasons. This conclusion has since been confirmed by numerous observations in every part of the globe. Mr. Cavendish also analyzed common air, and found it to consist of

79·16 volumes azotic gas,
20·84 volumes oxygen gas.
_____

100·00

4. For many years it was the opinion of chemists that mercury is essentially liquid, and that no degree of cold is capable of congealing it. Professor Braun's accidental discovery that it may be frozen by cold, like other liquids, was at first doubted ; and when it was finally established by the most conclusive experiments, it was inferred from the observations of Braun that the freezing point of mercury is several hundred degrees below zero on Fahrenheit's scale. It became an object of great importance to determine the exact point of the congelation of this metal by accurate experiments. This was done at Hudson's Bay, by Mr. Hutchins, who followed a set of directions given him by Mr. Cavendish, and from his experiments Mr. Cavendish, in a paper inserted in the Philosophical Transactions for 1783, deduced that the freezing point of mercury is 38·66 degrees below the zero of Fahrenheit's thermometer.

5. These experiments naturally drew the attention of Mr. Cavendish to the phenomena of freezing, to the action of freezing mixtures, and the congelation of acids. He employed Mr. M'Nab, who was settled in the neighbourhood of Hudson's Bay, to make the requisite experiments; and he published two very curious and important papers on these subjects in the Philosophical Transactions for 1786 and 1788. He explained the phenomena of congelation exactly according to the theory of Dr. Black, but rejecting the hypothesis that heat is a *substance* sui generis, and

thinking it more probable, with Sir Isaac Newton, that it is owing to the rapid internal motion of the particles of the hot body. The latent heat of water, he found to be 150°. The observations on the congelation of nitric and sulphuric acids are highly interesting: he showed that their freezing points vary considerably, according to the strength of each; and drew up tables indicating the freezing points of acids, of various degrees of strength.

6. But the most splendid and valuable of Mr. Cavendish's chemical experiments were published in two papers, entitled, "Experiments on Air," in the Transactions of the Royal Society for 1784 and 1785. The object of these experiments was to determine what happened during the *phlogistication of air*, as it was at that time termed; that is, the change which air underwent when metals were calcined in contact with it, when sulphur or phosphorus was burnt in it, and in several similar processes. He showed, in the first place, that there was no reason for supposing that carbonic acid was formed, except when some animal or vegetable substance was present; that when *hydrogen gas* was burnt in contact with air or oxygen gas, it *combined* with that gas, and formed *water;* that *nitrous gas*, by combining with the oxygen of the atmosphere, formed *nitrous acid;* and that when *oxygen* and *azotic* gas are mixed in the requisite proportions, and electric sparks passed through the mixture, they *combine*, and form *nitric* acid.

The first of these opinions occasioned a controversy between Mr. Cavendish, and Mr. Kirwan, who maintained that carbonic acid is always produced when air is phlogisticated. Two papers on this subject by Kirwan, and one by Cavendish, are inserted in the Philosophical Transactions for 1784, each remarkable examples of the peculiar manner of the respective writers. All the arguments of Kirwan are founded on the experiments of others. He displays great reading, and a strong memory; but does not discriminate between the merits of the chemists on whose authority he founds his opinions. Mr. Cavendish, on the other hand, never advances a single opinion, which he has not put to the test of experiment; and never suffers himself to go any further than his experiment will warrant. Whatever is not accurately determined by unexceptionable trials, is merely stated as a conjecture on which little stress is laid.

In the first of these celebrated papers, Mr. Cavendish has drawn a comparison between the phlogistic and antiphlogistic theories of chemistry; he has shown that each of them is capable of explaining the phenomena in a satisfactory manner; though it is impossible to demonstrate the truth of either; and he has given the reasons which induced him to prefer the phlogistic theory—reasons which the French chemists were unable to refute, and which they were wise enough not to notice. There cannot be a more striking proof of the influence of fashion, even in science, and of the unwarrantable precipitation with which opinions are rejected or embraced by philosophers, than the total inattention paid by the chemical world to this admirable dissertation. Had Mr. Kirwan adopted the opinions of Mr. Cavendish, when he undertook the defence of phlogiston, instead of trusting to the vague experiments of inaccurate chemists, he would not have been obliged to yield to his French antagonists, and the antiphlogistic theory would not so speedily have gained ground.

Such is an epitome of the chemical papers of Mr. Cavendish. They contain five notable discoveries; namely, 1. The nature and properties of hydrogen gas. 2. The solubility of bicarbonates of lime and magnesia in water. 3. The exact proportion of the constituents of common air. 4. The composition of water. 5. The composition of nitric acid. It is to him also that we are indebted for our knowledge of the freezing point of mercury; and he was likewise the first person who showed that potash has a stronger affinity for acids than soda has. His experiments on the subject are to be found in a paper on Mineral Waters, published in the Philosophical Transactions, by Dr. Donald Monro.

THE END.

C. WHITING, BEAUFORT HOUSE, STRAND.

HISTORY OF CHEMISTRY.

CHAPTER I.

OF THE FOUNDATION AND PROGRESS OF SCIENTIFIC CHEMISTRY IN GREAT BRITAIN.

WHILE Mr. Cavendish was extending the bounds of pneumatic chemistry, with the caution and precision of a Newton, Dr. Priestley, who had entered on the same career, was proceeding with a degree of rapidity quite unexampled; while from his happy talents and inventive faculties, he contributed no less essentially to the progress of the science, and certainly more than any other British chemist to its popularity.

Joseph Priestley was born in 1733, at Fieldhead, about six miles from Leeds in Yorkshire. His father, Jonas Priestley, was a maker and dresser of woollen cloth, and his mother, the only child of Joseph Swift a farmer in the neighbourhood. Dr. Priestley was the eldest child; and, his mother having children very fast, he was soon committed to the care of his maternal grandfather. He lost his mother when he was only six years of age, and was soon after taken home by his father and sent to

school in the neighbourhood. His father being but
poor, and encumbered with a large family, his sister,
Mrs. Keighley, a woman in good circumstances,
and without children, relieved him of all care of his
eldest son, by taking him and bringing him up as
her own. She was a dissenter, and her house was
the resort of all the dissenting clergy in the country.
Young Joseph was sent to a public school in the
neighbourhood, and, at sixteen, had made considerable progress in Latin, Greek, and Hebrew.
Having shown a passion for books and for learning at
a very early age, his aunt conceived hopes that he
would one day become a dissenting clergyman,
which she considered as the first of all professions;
and he entered eagerly into her views: but his
health declining about this period, and something
like phthisical symptoms having come on, he was
advised to turn his thoughts to trade, and to settle
as a merchant in Lisbon. This induced him to apply
to the modern languages; and he learned French,
Italian, and German, without a master. Recovering his health, he abandoned his new scheme and
resumed his former plan of becoming a clergyman.
In 1752 he was sent to the academy of Daventry,
to study under Dr. Ashworth, the successor of Dr.
Doddridge. He had already made some progress
in mechanical philosophy and metaphysics, and
dipped into Chaldee, Syriac, and Arabic. At Daventry he spent three years, engaged keenly in studies
connected with divinity, and wrote some of his
earliest theological tracts. Freedom of discussion
was admitted to its full extent in this academy.
The two masters espoused different sides upon most
controversial subjects, and the scholars were divided
into two parties, nearly equally balanced. The discussions, however, were conducted with perfect good
humour on both sides; and Dr. Priestley, as he tells

us himself, usually supported the heterodox opinion; but he never at any time, as he assures us, advanced arguments which he did not believe to be good, or supported an opinion which he did not consider as true. When he left the academy, he settled at Needham in Suffolk, as an assistant in a small obscure dissenting meeting-house, where his income never exceeded 30*l*. a-year. His hearers fell off, in consequence of their dislike of his theological opinions; and his income underwent a corresponding diminution. He attempted a school; but his scheme failed of success, owing to the bad opinion which his neighbours entertained of his orthodoxy. His situation would have been desperate, had he not been occasionally relieved by sums out of charitable funds, procured by means of Dr. Benson, and Dr. Kippis.

Several vacancies occurred in his vicinity; but he was treated with contempt, and thought unworthy to fill any of them. Even the dissenting clergy in the neighbourhood thought it a degradation to associate with him, and durst not ask him to preach: not from any dislike to his theological opinions; for several of them thought as freely as he did; but because the genteeler part of their audience always absented themselves when he appeared in the pulpit. A good many years afterwards, as he informs us himself, when his reputation was very high, he preached in the same place, and multitudes flocked to hear the very same sermons, which they had formerly listened to with contempt and dislike.

His friends being aware of the disagreeable nature of his situation at Needham, were upon the alert to procure him a better. In 1758, in consequence o the interest of Mr. Gill, he was invited to appear as a candidate for a meeting-house in Sheffield, vacant by the resignation of Mr. Wadsworth. He appear-

ed accordingly and preached, but was not approved
of.   Mr. Haynes, the other minister, offered to procure him a meeting-house at Nantwich in Cheshire.
This situation he accepted, and, to save expenses, he
went from Needham to London by sea.   At Nantwich he continued three years, and spent his time
much more agreeably than he had done at Needham.
His opinions were not obnoxious to his hearers, and
controversial discussions were never introduced.
Here he established a school, and found the business
of teaching, contrary to his expectation, an agreeable
and even interesting employment.   He taught from
seven in the morning, till four in the afternoon; and
after the school was dismissed, he went to the house
of Mr. Tomlinson, an eminent attorney in the neighbourhood, where he taught privately till seven in
the evening.   Being thus engaged twelve hours
every day in teaching, he had little time for private
study.   It is, indeed, scarcely conceivable how,
under such circumstances, he could prepare himself
for Sunday.   Here, however, his circumstances
began to mend.   At Needham it required the utmost economy to keep out of debt; but at Nantwich, he was able to purchase a few books and some
philosophical instruments, as a small air-pump, an
electrical machine, &c.   These he taught his eldest
scholars to keep in order and manage : and by
entertaining their parents and friends with experiments, in which the scholars were generally the
operators, and sometimes the lecturers too, he considerably extended the reputation of his school.   It
was at Nantwich that he wrote his grammar for the
use of his school, a book of considerable merit,
though its circulation was never extensive.   This
latter circumstance was probably owing to the
superior reputation of Dr. Lowth, who published
his well-known grammar about two years afterwards.

Being boarded in the house of Mr. Eddowes, a very sociable and sensible man, and a lover of music, Dr. Priestley was induced to play a little on the English flute; and though he never was a proficient, he informs us that it contributed more or less to his amusement for many years. He recommends the knowledge and practice of music to all studious persons, and thinks it rather an advantage for them if they have no fine ear or exquisite taste, as they will, in consequence, be more easily pleased, and less apt to be offended when the performances they hear are but indifferent.

The academy at Warrington was instituted while Dr. Priestley was at Needham, and he was recommended by Mr. Clark, Dr. Benson, and Dr. Taylor, as tutor in the languages; but Dr. Aiken, whose qualifications were considered as superior, was preferred before him. However, on the death of Dr. Taylor, and the advancement of Dr. Aiken to be tutor in divinity, he was invited to succeed him: this offer he accepted, though his school at Nantwich was likely to be more gainful; for the employment at Warrington was more liberal and less painful. In this situation he continued six years, actively employed in teaching and in literary pursuits. Here he wrote a variety of works, particularly his History of Electricity, which first brought him into notice as an experimental philosopher, and procured him celebrity. After the publication of this work, Dr. Percival of Manchester, then a student at Edinburgh, procured him the title of doctor in laws, from that university. Here he married a daughter of Mr. Isaac Wilkinson, an ironmonger in Wales; a woman whose qualities he has highly extolled, and who died after he went to America.

In the academy he spent his time very happily, but it did not flourish. A quarrel had broken out

between Dr. Taylor and the trustees, in consequence
of which all the friends of that gentleman were hostile to the institution. This, together with the smallness of his income, 100*l.* a-year, and 15*l.* for each
boarder, which precluded him from making any pro-
vision for his family, induced him to accept an
invitation to take charge of Millhill chapel, at
Leeds, where he had a considerable acquaintance,
and to which he removed in 1767.

Here he engaged keenly in the study of theology,
and produced a great number of works, many of
them controversial. Here, too, he commenced his
great chemical career, and published his first tract
on· *air.* He was led accidentally to think of pneumatic chemistry, by living in the immediate vicinity
of a brewery. Here, too, he published his history
of the Discoveries relative to Light and Colours, as
the first part of a general history of experimental
philosophy; but the expense of this book was so
great, and its sale so limited, that he did not venture
to prosecute the undertaking. Here, likewise, he
commenced and published three volumes of a periodical work, entitled "The Theological Repository,"
which he continued after he settled in Birmingham.

After he had been six years at Leeds, the Earl of
Shelburne (afterwards Marquis of Lansdowne),
engaged him, on the recommendation of Dr. Price,
to live with him as a kind of librarian and literary
companion, at a salary of 250*l.* a-year, with a house.
With his lordship he travelled through Holland,
France, and a part of Germany, and spent some
time in Paris. He was delighted with this excur-
sion, and expressed himself thoroughly convinced
of the great advantages to be derived from foreign
travel. The men of science and politicians in
Paris were unbelievers, and even professed atheists,
and as Dr. Priestley chose to appear before them as

a Christian, they told him that he was the first person they had met with, of whose understanding they had any opinion, who was a believer of Christianity; but, upon interrogating them closely, he found that none of them had any knowledge either of the nature or principles of the Christian religion.—While with Lord Shelburne, he published the first three volumes of his Experiments on Air, and had collected materials for a fourth, which he published soon after settling in Birmingham. At this time also he published his attack upon Drs. Reid, Beattie, and Oswald; a book which, he tells us, he finished in a fortnight: but of which he afterwards, in some measure, disapproved. Indeed, it was impossible for any person of candour to approve of the style of that work, and the way in which he treated Dr. Reid, a philosopher certainly much more deeply skilled than himself in metaphysics.

After some years Lord Shelburne began to be weary of his associate, and, on his expressing a wish to settle him in Ireland, Dr. Priestley of his own accord proposed a separation, to which his lordship consented, after settling on him an annuity of 150*l*., according to a previous stipulation. This annuity he continued regularly to pay during the remainder of the life of Dr. Priestley.

His income being much diminished by his separation from Lord Shelburne, and his family increasing, he found it now difficult to support himself. At this time Mrs. Rayner made him very considerable presents, particularly at one period a sum of 400*l*.; and she continued her contributions to him almost annually. Dr. Fothergill had proposed a subscription, in order that he might prosecute his experiments to their utmost extent, and be enabled to live without sacrificing his time to his pupils. This he accepted. It amounted at first to 40*l*. per annum, and was afterwards much increased. Dr. Watson,

Mr Wedgewood, Mr. Galton, and four or five more, were the gentlemen who joined with Dr. Fothergill in this generous subscription.

Soon after, he settled in a meeting-house in Birmingham, and continued for several years engaged in theological and chemical investigations. His apparatus, by the liberality of his friends, had become excellent, and his income was so good that he could prosecute his researches to their full extent. Here he published the three last volumes of his Experiments on Air, and various papers on the same subject in the Philosophical Transactions. Here, too, he continued his Theological Repository, and published a variety of tracts on his peculiar opinions in religion, and upon the history of the primitive church. He now unluckily engaged in controversy with the established clergy of the place; and expressed his opinions on political subjects with a degree of freedom, which, though it would have been of no consequence at any former period, was ill suited to the peculiar circumstances that were introduced into this country by the French revolution, and to the political maxims of Mr. Pitt and his administration. His answer to Mr. Burke's book on the French revolution excited the violent indignation of that extraordinary man, who inveighed against his character repeatedly, and with peculiar virulence, in the house of commons. The clergy of the church of England, too, who began about this time to be alarmed for their establishment, of which Dr. Priestley was the open enemy, were particularly active; the press teemed with their productions against him, and the minds of their hearers seem to have been artificially excited; indeed some of the anecdotes told of the conduct of the clergy of Birmingham, were highly unbecoming their character. Unfortunately, Dr. Priestley did not seem to be aware of the state of the nation, and of the plan of conduct laid down by Mr. Pitt and his political friends; and he was too fond of controversial discussions to yield tamely to the attacks of his antagonists.

These circumstances seem in some measure to explain the disgraceful riots which took place in Birmingham in 1791, on the day of the anniversary of the French revolution. Dr. Priestley's meeting-house and his dwelling-house were burnt; his library and apparatus destroyed, and many manuscripts, the fruits of several years of industry, were consumed in the conflagration. The houses of several of his friends shared the same fate, and his son narrowly escaped death, by the care of a friend who forcibly concealed him for several days. Dr. Priestley was obliged to make his escape to London, and a seat was taken for him in the mail-coach under a borrowed name. Such was the ferment against him that it was believed he would not have been safe any where else; and his friends would not allow him, for several weeks, to walk through the streets.

He was invited to Hackney, to succeed Dr. Price in the meeting-house of that place. He accepted the office, but such was the dread of his unpopularity, that nobody would let him a house, from an apprehension that it would be burnt by the populace as soon as it was known that he inhabited it. He was obliged to get a friend to take a lease of a house in another name; and it was with the utmost difficulty that he could prevail with the landlord to allow the lease to be transferred to him. The members of the Royal Society, of which he was a fellow, declined admitting him into their company; and he was obliged to withdraw his name from the society.

When we look back upon this treatment of a man of Dr. Priestley's character, after an interval of forty

years, it cannot fail to strike us with astonishment;
and it must be owned, I think, that it reflects an
indelible stain upon that period of the history of
Great Britain. To suppose that he was in the least
degree formidable to so powerful a body as the
church of England, backed as it was- by the
aristocracy, by the ministry, and by the opinions
of the people, is perfectly ridiculous. His theological sentiments, indeed, were very different from
those of the established church; but so were those
of Milton, Locke, and Newton. Nay, some of the
members of the church itself entertained opinions,
not indeed so decided or so openly expressed as
those of Dr. Priestley, but certainly having the same
tendency. To be satisfied of this it is only necessary to recollect the book which Dr. Clarke published on the Trinity. Nay, some of the bishops,
unless they are very much belied, entertained
opinions similar to those of Dr. Clarke. The same
observation applies to Dr. Lardner, Dr. Price, and
many others of the dissenters. Yet, the church of
England never attempted to persecute these respectable and meritorious men, nor did they consider their opinions as at all likely to endanger the
stability of the church. Besides, Dr. Horsley had
taken up the pen against Dr. Priestley's theological
opinions, and had refuted them so completely in the
opinion of the members of the church, that it was
thought right to reward his meritorious services by
a bishopric.

It could hardly, therefore, be the dread of Dr.
Priestley's theological opinions that induced the
clergy of the church of England to bestir themselves against him with such alacrity. Erroneous
opinions advanced and refuted, so far from being
injurious, have a powerful tendency to support and
strengthen the cause which they were meant to

overturn. Or, if there existed any latent suspicion that the refutation of Horsley was not so complete as had been alleged, surely persecution was not the best means of supporting weak arguments ; and indeed it was rather calculated to draw the attention of mankind to the theological opinions of Priestley ; as has in fact been the consequence.

Neither can the persecutions which Dr. Priestley was subjected to be accounted for by his political opinions, even supposing it not to be true, that in a free country like Great Britain, any man is at liberty to maintain whatever theoretic opinions of government he thinks proper, provided he be a peaceable subject and obey rigorously all the laws of his country.

Dr. Priestley was an advocate for the perfectibility of the human species, or at least its continually increasing tendency to improvement—a doctrine extremely pleasing in itself, and warmly supported by Franklin and Price ; but which the wild principles of Condorcet, Godwin, and Beddoes at last brought into discredit. This doctrine was taught by Priestley in the outset of his Treatise on Civil Government, first published in 1768. It is a speculation of so very agreeable a nature, so congenial to our warmest wishes, and so flattering to the prejudices of humanity, that one feels much pain at being obliged to give it up. Perhaps it may be true, and I am willing to hope so, that improvements once made are never entirely lost, unless they are superseded by something much more advantageous, and that therefore the knowledge of the human race, upon the whole, is progressive. But political establishments, at least if we are to judge from the past history of mankind, have their uniform periods of progress and decay. Nations seem incapable of profiting by experience. Every nation seems destined to run the same career, and the history may be comprehended under the following heads : Poverty, liberty, industry, wealth, power, dissipation, anarchy, destruction. We have no example in history of a nation running through this career and again recovering its energy and importance. Greece ran through it more than two thousand years ago : she has been in a state of slavery ever since. An opportunity is now at last given her of recovering her importance : posterity will ascertain whether she wil lembrace it.

Dr. Priestley's short Essay on the First Principles of Civil Government was published in 1768. In it he lays down as the foundation of his reasoning, that " it must be understood, whether it be expressed or not, that all people live in society for their mutual advantage ; so that the good and happiness of the members, that is the majority of the members of any state, is the great standard by which every thing relating to that state must be finally determined ; and though it may be supposed that a body of people may be bound by a voluntary resignation of all their rights to a single person or to a few, it can never be supposed that the resignation is obligatory on their posterity, because it is manifestly contrary to the good of the whole that it should be so." From this first principle he deduces all his political maxims. Kings, senators, and nobles, are merely the servants of the public ; and when they abuse their power, in the people lies the right of deposing and consequently of punishing them. He examines the expediency of hereditary sovereignty, of hereditary rank and privileges, of the duration of parliament, and of the right of voting, with an evident tendency to democratical principles, though he does not express himself very clearly on the subject.

Such were his political principles in 1768, when his book was published. They excited no alarm and drew but little attention ; these principles he maintained ever after, or indeed he may be said to have become more moderate instead of violent. Though he approved of a republic in the abstract ; yet, considering the prejudices and habits of the people of Great Britain, he laid it down as a principle that their present form of government was best suited to them. He thought, however, that there should be. a reform in parliament; and that parliaments should be triennial instead of septennial. He was an enemy to all violent reforms, and thought that the change ought to be brought about gradually and peaceably. When the French revolution broke out he took the side of the patriots, as he had done during the American war ; and he wrote a refutation of Mr. Burke's extraordinary performance. Being a dissenter, it is needless to say that he was an advocate for complete religious freedom. He was ever hostile to all religious establishments, and an open enemy to the church of England.

How far these opinions were just and right this is not the place to inquire ; but that they were perfectly harmless, and that many other persons in this country during the last century, and even at present, have adopted similar opinions without incurring any odium whatever, and without exciting the jealousy or even the attention of government, is well known to every person. It comes then to be a question of some curiosity at least, to what we are to ascribe the violent persecutions raised against Dr. Priestley. It seems to have been owing chiefly to the alarm caught by the clergy of the established church that their establishment was in danger;—and, considering the ferment excited soon after the breaking out of the French revolution, and the rage for reform, which pervaded all ranks, the almost general alarm of the aristocracy, at least, was not entirely without foundation. I cannot, however, admit that there was occasion for the violent alarm caught by Mr. Pitt and his political friends, and for the very despotic measures which they adopted in consequence. The disease would probably have subsided of itself, or it would have been cured by a much gentler treatment. As Dr. Priestley was an open enemy to the establishment, its clergy naturally conceived a prejudice against him, and this prejudice was violently inflamed by the danger to which they thought themselves exposed; their influence with the ministry was very great, and Mr. Pitt and his friends naturally caught their prejudices and opinions. Mr. Burke, too, who had changed his political principles, and who was inflamed with the burning zeal which distinguishes all converts, was provoked at Dr. Priestley's answer to his book on the French revolution, and took every opportunity to inveigh against him in the house of commons. The conduct of the French, likewise, who made Dr. Priestley a citizen of France, and chose him a member of their assembly, though intended as a compliment, was injurious to him in Great Britain. It was laid hold of by his antagonists to convince the people that he was an enemy to his country; that he had abjured his rights as an Englishman; and that he had adopted the principles of the hereditary enemies of Great Britain. These causes, and not his political opinions, appear to me to account for the persecution which was raised against him.

His sons, disgusted with this persecution of their father, had renounced their native country and gone over to France; and, on the breaking out of the war between this country and the French republic, they emigrated to America. It was this circumstance, joined to the state of insulation in which

he lived, that induced Dr. Priestley, after much
consideration, to form the resolution of following
his sons and emigrating to America. He published
his reasons in the preface to a Fast-day Sermon,
printed in 1794, one of the gravest and most forcible
pieces of composition I have ever read. He left England in April, 1795, and reached New York in June.
In America he was received with much respect by
persons of all ranks; and was immediately offered
the situation of professor of chemistry in the
College of Philadelphia; which, however, he declined, as his circumstances, by the liberality of
his friends in England, continued independent.
He settled, finally, in Northumberland, about 130
miles from Philadelphia, where he built a house,
and re-established his library and laboratory, as
well as circumstances permitted. Here he published a considerable number of chemical papers,
some of them under the form of pamphlets, and
the rest in the American Transactions, the New
York Medical Repository, and Nicholson's Journal
of Natural Philosophy and Chemistry. Here, also,
he continued keenly engaged in theological pursuits; and published, or republished, a great
variety of books on theological subjects. Here he
lost his wife and his youngest and favourite son,
who, he had flattered himself, was to succeed him in
his literary career :—and here he died, in 1804, after
having been confined only two days to bed, and but
a few hours after having arranged his literary concerns, inspected some proof-sheets of his last theological work, and given instructions to his son how
it should be printed.

During the latter end of the presidency of Mr.
Adams, the same kind of odium which had banished
Dr. Priestley from England began to prevail in
America. He was threatened with being sent out of

the country as an alien. Notwithstanding this, he declined being naturalized; resolving, as he said, to die as he had lived, an Englishman. When his friend Mr. Jefferson, whose political opinions coincided with his own, became president, the odium against him wore off, and he became as much respected as ever.

As to the character of Dr. Priestley, it is so well marked by his life and writings, that it is difficult to conceive how it could have been mistaken by many eminent men in this kingdom. Industry was his great characteristic; and this quality, together with a facility of composition, acquired, as he tells us, by a constant habit while young of drawing out an abstract of the sermons which he had preached, and writing a good deal in verse, enabled him to do so much: yet, he informs us that he never was an intense student, and that his evenings were usually passed in amusement or company. He was an early riser, and always lighted his own fire before any one else was stirring: it was then that he composed all his works. It is obvious, from merely glancing into his books, that he was precipitate; and indeed, from the way he went on thinking as he wrote, and writing only one copy, it was impossible he could be otherwise: but, as he was perfectly sincere and anxious to obtain the truth, he freely acknowledged his mistakes as soon as he became sensible of them. This candour is very visible in his philosophical speculations; but in his theological writings it was not so much to be expected. He was generally engaged in controversy in theology; and his antagonists were often insolent, and almost always angry. We all know the effect of such opposition; and need not be surprised that it operated upon Dr. Priestley, as it would do upon any other man. By all accounts his powers of conversation were very great, and his manners in every respect very agreeable. That this must have been the case is obvious from the great number of his friends, and the zeal and ardour with which they continued to serve him, notwithstanding the obloquy under which he lay, and even the danger that might be incurred by appearing to befriend him. As for his moral character, even his worst enemies have been obliged to allow that it was unexceptionable. Many of my readers will perhaps smile, when I say that he was not only a sincere, but a zealous Christian, and would willingly have died a martyr to the cause. Yet I think the fact is of easy proof; and his conduct through life, and especially at his death, affords irrefragable proofs of it. His tenets, indeed, did not coincide with those of the majority of his countrymen; but though he rejected many of the doctrines, he admitted the whole of the sublime morality and the divine origin of the Christian religion; which may charitably be deemed sufficient to constitute a true Christian. Of vanity he seems to have possessed rather more than a usual share; but perhaps he was deficient in pride.

His writings were exceedingly numerous, and treated of science, theology, metaphysics, and politics. Of his theological, metaphysical, and political writings it is not our business in this work to take any notice. His scientific works treat of *electricity*, *optics*, and *chemistry*. As an electrician he was respectable; as an optician, a compiler; as a chemist, a discoverer. He wrote also a book on perspective which I have never had an opportunity of perusing.

It is to his chemical labours that he is chiefly indebted for the great reputation which he acquired. No man ever entered upon any undertaking with less apparent means of success than Dr. Priestley did on the chemical investigation of *airs*. He was unacquainted with chemistry, excepting that he had, some years before, attended an elementary course delivered by Mr. Turner, of Liverpool. He was not in possession of any apparatus, nor acquainted with the method of making chemical experiments; and his circumstances were such, that he could neither lay out a great deal of money on experiments, nor could he hope, without a great deal of expense, to make any material progress in his investigations. These circumstances, which, at first sight, seem so adverse, were, I believe, of considerable service to him, and contributed very much to his ultimate success. The branch of chemistry which he selected was new: an apparatus was to be invented before any thing of importance could be effected; and, as simplicity is essential in every apparatus, *he* was most likely to contrive the best, whose circumstances obliged him to attend to economical considerations.

Pneumatic chemistry had been begun by Mr. Cavendish in his valuable paper on carbonic acid and hydrogen gases, published in the Philosophical Transactions for 1766. The apparatus which he employed was similar to that used about a century before by Dr. Mayow of Oxford. Dr. Priestley contrived the apparatus still used by chemists in pneumatic investigations; it is greatly superior to that of Mr. Cavendish, and, indeed, as convenient as can be desired. Were we indebted to him for nothing else than this apparatus, it would deservedly give him high consideration as a pneumatic chemist.

His discoveries in pneumatic chemistry are so numerous, that I must satisfy myself with a bare outline; to enumerate every thing, would be to transcribe his three volumes, into which he digested his discoveries. His first paper was published in 1772, and was on the method of impregnating water

with carbonic acid gas; the experiments contained
in it were the consequence of his residing near a
brewery in Leeds. This pamphlet was immediately
translated into French; and, at a meeting of the
College of Physicians in London, they addressed
the Lords of the Treasury, pointing out the advantage that might result from water impregnated with
carbonic acid gas in cases of scurvy at sea. His
next essay was published in the Philosophical Transactions, and procured him the Copleyan medal.
His different volumes on air were published in succession, while he lived with Lord Shelburne, and
while he was settled at Birmingham. They drew the
attention of all Europe, and raised the reputation of
this country to a great height.

The first of his discoveries was *nitrous gas,* now
called *deutoxide of azote,* which had, indeed, been
formed by Dr. Hales; but that philosopher had not
attempted to investigate its properties. Dr. Priestley
ascertained its properties with much sagacity, and
almost immediately applied it to the analysis of air.
It contributed very much to all subsequent investigations in pneumatic chemistry, and may be said
to have led to our present knowledge of the constitution of the atmosphere.

The next great discovery was *oxygen gas,* which was
made by him on the 1st of August, 1774, by heating the red oxide of mercury, and collecting the
gaseous matter given out by it. He almost immediately detected the remarkable property which this
gas has of supporting combustion better, and animal
life longer, than the same volume of common air;
and likewise the property which it has of condensing
into red fumes when mixed with nitrous gas. Lavoisier, likewise, laid claim to the discovery of
oxygen gas; but his claim is entitled to no attention
whatever; as Dr. Priestley informs us that he pre-

pared this gas in M. Lavoisier's house, in Paris, and showed him the method of procuring it in the year 1774, which is a considerable time before the date assigned by Lavoisier for his pretended discovery. Scheele, however, actually obtained this gas without any previous knowledge of what Priestley had done; but the book containing this discovery was not published till three years after Priestley's process had become known to the public.

Dr. Priestley first made known sulphurous acid, fluosilicic acid, muriatic acid, and ammonia in the gaseous form; and pointed out easy methods of procuring them: he describes with exactness the most remarkable properties of each. He likewise pointed out the existence of carburetted hydrogen gas; though he made but few experiments to determine its nature. His discovery of protoxide of azote affords a beautiful example of the advantages resulting from his method of investigation, and the sagacity which enabled him to follow out any remarkable appearances which occurred. Carbonic oxide gas was discovered by him while in America, and it was brought forward by him as an incontrovertible refutation of the antiphlogistic theory.

Though he was not strictly the discoverer of hydrogen gas, yet his experiments on it were highly interesting, and contributed essentially to the revolution which chemistry soon after underwent. Nothing, for example, could be more striking, than the reduction of oxide of iron, and the disappearance of the hydrogen when the oxide is heated sufficiently in contact with hydrogen gas. Azotic gas was known before he began his career; but we are indebted to him for most of the properties of it yet known. To him, also, we owe the knowledge of the fact, that an acid is formed when electric sparks are made to pass for some time through a given bulk of common air;

a fact which led afterwards to Mr. Cavendish's great discovery of the composition of nitric acid.

He first discovered the great increase of bulk which takes place when electric sparks are made to pass through ammoniacal gas—a fact which led Berthollet to the analysis of this gas. He merely repeated Priestley's experiment, determined the augmentation of bulk, and the nature of the gases evolved by the action of the electricity. His experiments on the amelioration of atmospherical air by the vegetation of plants, on the oxygen gas given out by their leaves, and on the respiration of animals, are not less curious and interesting.

Such is a short view of the most material facts for which chemistry is indebted to Dr. Priestley. As a discoverer of new substances, his name must always stand very high in the science ; but as a reasoner or theorist his position will not be so favourable. It will be observed that almost all his researches and discoveries related to gaseous bodies. He determined the different processes, by means of which the different gases can be procured, the substances which yield them, and the effects which they are capable of producing on other bodies. Of the other departments of chemistry he could hardly be said to know any thing. As a pneumatic chemist he stands high; as an analytical chemist he can scarcely claim any rank whatever. In his famous experiments on the formation of water by detonating mixtures of oxygen and hydrogen in a copper globe, the copper was found acted upon, and a blue liquid was obtained, the nature of which he was unable to ascertain; but Mr. Keir, whose assistance he solicited, determined it to be a solution of nitrate of copper in water. This formation of nitric acid induced him to deny that water was a compound of oxygen and hydrogen. The same acid was formed

in the experiments of Mr. Cavendish ; but he investigated the circumstances of the formation, and
showed that it depended upon the presence of azotic
gas in the gaseous mixture. Whenever azotic gas
is present, nitric acid is formed, and the quantity of
this acid depends upon the relative proportion of the
azotic and hydrogen gases in the mixture. When no
hydrogen gas is present, nothing is formed but nitric
acid : when no azotic gas is present, nothing is
formed but water. These facts, determined by
Cavendish, invalidate the reasoning of Priestley altogether ; and had he possessed the skill, like Cavendish, to determine with sufficient accuracy the proportions of the different gases in his mixtures, and
the relative quantities of nitric acid formed, he
would have seen the inaccuracy of his own conclusions.

He was a firm believer in the existence of phlogiston ; but he seems, at least ultimately, to have
adopted the view of Scheele, and many other eminent contemporary chemists—indeed, the view of
Cavendish himself—that hydrogen gas is phlogiston
in a separate and pure state. Common air he considered as a compound of oxygen and phlogiston.
Oxygen, in his opinion, was air quite free from
phlogiston, or air in a simple and pure state ; while
azotic gas (the other constituent of common air)
was air saturated with phlogiston. Hence he called
oxygen *dephlogisticated*, and azote *phlogisticated
air*. The facts that when common air is converted
into azotic gas its bulk is diminished about one-fifth
part, and that azotic gas is lighter than common air
or oxygen gas, though not quite unknown to him,
do not seem to have drawn much of his attention.
He was not accustomed to use a balance in his experiments, nor to attend much to the alterations
which took place in the weight of bodies. Had he

done so, most of his theoretical opinions would have fallen to the ground.

When a body is allowed to burn in a given quantity of common air, it is known that the quality of the common air is deteriorated ; it becomes, in his language, more phlogisticated. This, in his opinion, was owing to an affinity which existed between phlogiston and air. The presence of air is necessary to combustion, in consequence of the affinity which it has for phlogiston. It draws phlogiston out of the burning body, in order to combine with it. When a given bulk of air is saturated with phlogiston, it is converted into azotic gas, or *phlogisticated air*, as he called it ; and this air, having no longer any affinity for phlogiston, can no longer attract that principle, and consequently combustion cannot go on in such air.

All combustible bodies, in his opinion, contain hydrogen. Of course the metals contain it as a constituent. The calces of metals are those bodies deprived of phlogiston. To prove the truth of this opinion, he showed that when the oxide of iron is heated in hydrogen gas, that gas is absorbed, while the calx is reduced to the metallic state. Finery cinder, which he employed in these experiments, is, in his opinion, iron not quite free from phlogiston. Hence it still retains a quantity of hydrogen. To prove this, he mixed together finery cinder and carbonates of lime, barytes and strontian, and exposed the mixture to a strong heat; and by this process obtained inflammable gas in abundance. In his opinion every inflammable gas contains hydrogen in abundance. Hence this experiment was adduced by him as a demonstration that hydrogen is a constituent of finery cinder.

All these processes of reasoning, which appear so plausible as Dr. Priestley states them, vanish into nothing, when his experiments are made, and the weights of every thing determined by means of a balance : it is then established that a burning body becomes heavier during its combustion, and that the surrounding air loses just as much weight as the burning body gains. Scheele and Lavoisier showed clearly that the loss of weight sustained by the air is owing to a quantity of oxygen absorbed from it, and condensed in the burning body. Cruikshank first elucidated the nature of the inflammable gas, produced by the heating a mixture of finery cinder and carbonate of lime, or other earthy carbonate. He found that iron filings would answer better than finery cinder. The gas was found to contain no hydrogen, and to be in fact a compound of oxygen and carbon. It was shown to be derived from the carbonic acid of the earthy carbonate, which was deprived of half its oxygen by the iron filings or finery cinder. Thus altered, it no longer preserved its affinity for the lime, but made its escape in the gaseous form, constituting the gas now known by the name of carbonic oxide.

Though the consequence of the Birmingham riots, which obliged Dr. Priestley to leave England and repair to America, is deeply to be lamented, as fixing an indelible disgrace upon the country ; perhaps it was not in reality so injurious to Dr. Priestley as may at first sight appear. He had carried his peculiar researches nearly as far as they could go. To arrange and methodize, and deduce from them the legitimate consequences, required the application of a different branch of chemical science, which he had not cultivated, and which his characteristic rapidity, and the time of life to which he had arrived, would have rendered it almost impossible for him to acquire. In all probability, therefore, had he been allowed to prosecute his researches unmolested, his reputation, instead of an increase, might have suffered a diminution, and he might have lost that eminent situation as a man of science which he had so long occupied.

With Dr. Priestley closes this period of the History of British Chemistry—for Mr. Cavendish, though he had not lost his activity, had abandoned that branch of science, and turned his attention to other pursuits.

CHAPTER II.

___

OF THE PROGRESS OF PHILOSOPHICAL CHEMISTRY IN
SWEDEN.

THOUGH Sweden, partly in consequence of her
scanty population, and the consequent limited sale
of books in that country, and partly from the pro-
pensity of her writers to imitate the French, which
has prevented that originality in her poets and historians that is requisite for acquiring much eminence
—though Sweden, for these reasons, has never
reached a very high rank in literature ; yet the case
has been very different in science. She has produced men of the very first eminence, and has contributed more than her full share in almost every
department of science, and in none has she shone
with greater lustre than in the department of Chemistry. Even in the latter part of the seventeenth
century, before chemistry had, properly speaking,
assumed the rank of a science, we find Hierne in
Sweden, whose name deserves to be mentioned with
respect. Moreover, in the earlier part of the eighteenth
century, Brandt, Scheffer, and Wallerius, had distinguished themselves by their writings. Cronstedt,
about the middle of the eighteenth century, may be
said to have laid the foundation of systematic mi-

neralogy upon chemical principles, by the publication of his System of Mineralogy. But Bergman is entitled to the merit of being the first person who prosecuted chemistry in Sweden on truly philosophical principles, and raised it to that high estimation to which its importance justly entitles it.

Torbern Bergman was born at Catherinberg, in West Gothland, on the 20th of March, 1735. His father, Barthold Bergman, was receiver of the revenues of that district, and his mother, Sara Hägg, the daughter of a Gotheborg merchant. A receiver of the revenues was at that time, in Sweden, a post both disagreeable and hazardous. The creatures of a party which had had the ascendancy in one diet, they were exposed to the persecution of the diet next following, in which an opposite party usually had the predominance. This circumstance induced Bergman to advise his son to turn his attention to the professions of law or divinity, which were at that time the most lucrative in Sweden. After having spent the usual time at school, and acquired those branches of learning commonly taught in Sweden, in the public schools and academies to which Bergman was sent, he went to the University of Upsala, in the autumn of 1752, where he was placed under the guidance of a relation, whose province it was to superintend his studies, and direct them to those pursuits that were likely to lead young Bergman to wealth and distinction. Our young student showed at once a decided predilection for mathematics, and those branches of physics which were connected with mathematics, or depended upon them. But these were precisely the branches of study which his relation was anxious to prevent his indulging in. Bergman attempted at once to indulge his own inclination, and to gratify the wishes of his relation. This obliged him to study with a degree of ardour and perseverance which has few examples. His mathematical and physical studies claimed the first share of his attention ; and, after having made such progress in them as would alone have been sufficient to occupy the whole time of an ordinary student—to satisfy his relation, Jonas Victorin, who was at that time a *magister docens* in Upsala, he thought it requisite to study some law books besides, that he might be able to show that he had not neglected his advice, nor abandoned the views which he had held out.

He was in the habit of rising to his studies every morning at four o'clock, and he never went to bed till eleven at night. The first year of his residence at Upsala, he had made himself master of Wolf's Logic, of Wallerius's System of Chemistry, and of twelve books of Euclid's Elements : for he had already studied the first book of that work in the Gymnasium before he went to college. He likewise perused Keil's Lectures on Astronomy, which at that time were considered as the best introduction to physics and astronomy. His relative disapproved of his mathematical and physical studies altogether ; but, not being able to put a stop to them, he interdicted the books, and left his young charge merely the choice between law and divinity. Bergman got a small box made, with a drawer, into which he put his mathematical and physical books, and over this box he piled the law books which his relative had urged him to study. At the time of the daily visits of his relative, the mathematical and physical books were carefully locked up in the drawer, and the law books spread upon the table ; but no sooner was his presence removed, than the drawer was opened, and the mathematical studies resumed.

This incessant study ; this necessity under which

he found himself to consult his own inclinations and those of his relative; this double portion of labour, without time for relaxation, exercise, or amusement, proved at last injurious to young Bergman's health.   He fell ill, and was obliged to leave the university and return home to his father's house in a state of bad health.   There constant and moderate exercise was prescribed him, as the only means of restoring his health.   That his time here might not be altogether lost to him, he formed the plan of making his walks subservient to the study of botany and entomology.

At this time Linnæus, after having surmounted obstacles which would have crushed a man of ordinary energy, was in the height of his glory; and was professor of botany and natural history in the University of Upsala.   His lectures were attended by crowds of students from every country in Europe : he was enthusiastically admired and adored by his students.   This influence on the minds of his pupils was almost unbounded; and at Upsala, every student was a natural historian.   Bergman had studied botany before he went to college, and he had acquired a taste for entomology from the lectures of Linnæus himself.   Both of these pursuits he continued to follow after his return home to West Gothland; and he made a collection of plants and of insects.   Grasses and mosses were the plants to which he turned the most of his attention, and of which he collected the greatest number.   But he felt a predilection for the study of insects, which was a field much less explored than the study of plants.

Among the insects which he collected were several not to be found in the *Fauna Suecica*.   Of these he sent specimens to Linnæus at Upsala, who was lelighted with the present.   All of them were till

then unknown as Swedish insects, and several of them
were quite new. The following were the insects at
this time collected by Bergman, and sent to Upsala,
as they were named by Linnæus:

Phalæna. Bombyx monacha, camelina.
     Noctua Parthenias, conspicillaris.
     Perspicillaris, flavicornis, Plebeia.
     Geometra pennaria.
     Tortrix Bergmanniana, Lediana.
     Tinea Harrisella, Pedella, Punctella.
Tenthredo. Vitellina, ustulata.
Ichneumon. Jaculator niger.
Tipula. Tremula.

When Bergman's health was re-established, he
returned to Upsala with full liberty to prosecute
his studies according to his own wishes, and to devote the whole of his time to mathematics, physics,
and natural history. His relations, finding it in vain
to combat his predilections for these studies, thought
it better to allow him to indulge them.

He had made himself known to Linnæus by the
collection of insects which he had sent him from
Catherinberg; and, drawn along by the glory with
which Linnæus was surrounded, and the zeal with
which his fellow-students prosecuted such studies,
he devoted a great deal of his attention to natural
history. The first paper which he wrote upon the
subject contained a discovery. There was a substance observed in some ponds not far from Upsala,
to which the name of *coccus aquaticus* was given,
but its nature was unknown. Linnæus had conjectured that it might be the *ovarium* of some insect; but he left the point to be determined by
future observations. Bergman ascertained that it
was the ovum of a species of leech, and that it con-

tained from ten to twelve young animals. When he stated what he had ascertained to Linnæus, that great naturalist refused to believe it; but Bergman satisfied him of the truth of his discovery by actual observation. Linnæus, thus satisfied, wrote under the paper of Bergman, *Vidi et obstupui*, and sent it to the academy of Stockholm with this flattering panegyric. It was printed in the Memoirs of that learned body for 1756 (p. 199), and was the first paper of Bergman's that was committed to the press.

He continued to prosecute the study of natural history as an amusement; though mathematics and natural philosophy occupied by far the greatest part of his time. Various useful papers of his, connected with entomology, appeared from time to time in the Memoirs of the Stockholm Academy; in particular, a paper on the history of insects which attack fruit-trees, and on the methods of guarding against their ravages : on the method of classing these insects from the forms of their larvæ, a time when it would be most useful for the agriculturist to know, in order to destroy those that are hurtful: a great number of observations on this class of animals, so various in their shape and their organization, and so important for man to know —some of which he has been able to overcome, while others, defended by their small size, and powerful by their vast numbers, still continue their ravages ; and which offer so interesting a sight to the philosopher by their labours, their manners, and their foresight.—Bergman was fond of these pursuits, and looked back upon them in afterlife with pleasure. Long after, he used to mention with much satisfaction, that by the use of the method pointed out by him, no fewer than seven millions of destructive insects were destroyed in a single garden, and during the course of a single summer.

About the year 1757 he was appointed tutor to

the only son of Count Adolf Frederick Stackelberg,
a situation which he filled greatly to the satisfaction
both of the father and son, as long as the young
count stood in need of an instructor. He took his
master's degree in 1758, choosing for the subject of
his thesis on *astronomical interpolation.* Soon
after, he was appointed *magister docens* in natural
philosophy, a situation peculiar to the University of
Upsala, and constituting a kind of assistant to the
professor. For his promotion to this situation he
was obliged to M. Ferner, who saw how well qualified he was for it, and how beneficial his labours
would be to the University of Upsala. In 1761 he
was appointed *adjunct* in mathematics and physics,
which, I presume, means that he was raised to the
rank of an associate with the professor of these
branches of science. In this situation it was his
business to teach these sciences to the students of
Upsala, a task for which he was exceedingly well
fitted. During this period he published various
tracts on different branches of physical science,
particularly on the *rainbow,* the crepuscula, the
aurora-borealis, the electrical phenomena of Iceland
spar, and of the tourmalin. We find his name
among the astronomers who observed the first
transit of Venus over the sun, in 1761, whose results deserve the greatest confidence.[*] His observations on the electricity of the tourmalin are
important. It was he that first established the true
laws that regulate these curious phenomena.

During the whole of this period he had been si-
lently studying chemistry and mineralogy, though
nobody suspected that he was engaged in any such
pursuits. But in 1767 John Gottschalk Wallerius,
who had long filled the chair of chemistry in the

* See Phil. Trans., vol. lii. p. 227, and vol. lvi. p. 85.

University of Upsala, with high reputation, resigned his chair. Bergman immediately offered himself as a candidate for the vacant professorship: and, to show that he was qualified for the office, published two dissertations on the Manufacture of Alum, which probably he had previously drawn up, and had lying by him. Wallerius intended to resign his chair in favour of a pupil or relation of his own, whom he had destined to succeed him. He immediately formed a party to oppose the pretensions of Bergman; and his party was so powerful and so malignant, that few doubted of their success: for it was joined by all those who, despairing of equalling the industry and reputation of Bergman, set themselves to oppose and obstruct his success. Such men unhappily exist in all colleges, and the more eminent a professor is, the more is he exposed to their malignant activity. Many of those who cannot themselves rise to any eminence, derive pleasure from the attempt to pull down the eminent to their own level. In these attempts, however, they seldom succeed, unless from some want of prudence and steadiness in the individual whom they assail. Bergman's Dissertations on Alum were severely handled by Wallerius and his party: and such was the influence of the ex-professor, that every body thought Bergman would be crushed by him.

Fortunately, Gustavus III. of Sweden, at that time crown prince, was chancellor of the university. He took up the cause of Bergman, influenced, it is said, by the recommendation of Von Swab, who pledged himself for his qualifications, and was so keen on the subject that he pleaded his cause in person before the senate. Wallerius and his party were of course baffled, and Bergman got the chair.

For this situation his previous studies had fitted

him in a peculiar manner. His mathematical, physical, and natural-historical knowledge, so far from being useless, contributed to free him from prejudices, and to emancipate him from that spirit of routine under which chemistry had hitherto suffered. They gave to his ideas a greater degree of precision, and made his views more correct. He saw that mathematics and chemistry divided between them the whole extent of natural science, and that its bounds required to be enlarged, to enable it to embrace all the different branches of science with which it was naturally connected, or which depended upon it. He saw the necessity of banishing from chemistry all vague hypotheses and explanations, and of establishing the science on the firm basis of experiment. He was equally convinced of the necessity of reforming the nomenclature of chemistry, and of bringing it to the same degree of precision that characterized the language of the other branches of natural philosophy.

His first care, after getting the chair, was to make as complete a collection as he could of mineral substances, and to arrange them in order according to the nature of their constituents, as far as they had been determined by experiment. To another cabinet he assigned the Swedish minerals, ranged in a geographical manner according to the different provinces which furnished them.

When I was at Upsala, in 1812, the first of these collections still remained, greatly augmented by his nephew and successor, Afzelius. But no remains existed of the geographical collection. However, there was a very considerable collection of this kind in the apartments of the Swedish school of mines at Stockholm, under the care of Mr. Hjelm, which I had an opportunity of inspecting. It is not improbable that Bergman's collection might have formed the nucleus of this. A geographical collection of minerals, to be of much utility, should exhibit all the different formations which exist in the kingdom: and in a country so uniform in its nature as Sweden, the minerals of one county are very nearly similar to those of the other counties ; with the exception of certain peculiarities derived from the mines, or from some formations which may belong exclusively to certain parts of the country, as, for example, the coal formations in the south corner of Sweden, near Helsinburg, and the porphyry rocks, in Elfsdale.

Bergman attempted also to make a collection of models of the apparatus employed in the different chemical manufactories, to be enabled to explain these manufactures with greater clearness to his students. I was informed by M. Ekeberg, who, in 1812, was *magister docens* in chemistry at Upsala, that these models were never numerous. Nor is it likely that they should be, as Sweden cannot boast of any great number of chemical manufactories, and as, in Bergman's time, the processes followed in most of the chemical manufactories of Europe were kept as secret as possible.

Thus it was Bergman's object to exhibit to his pupils specimens of all the different substances which the earth furnishes, with the order in which these productions are arranged on the globe—to show them the uses made of all these different productions—how practice had preceded theory and had succeeded in solving many chemical problems of the most complicated nature.

His lectures are said to have been particularly valuable. He drew around him a considerable number of pupils, who afterwards figured as chemical discoverers themselves. Of all these Assessor Gahn, of Fahlun, was undoubtedly the most remarkable ; but Hjelm, Gadolin, the Elhuyarts, and various other individuals, likewise distinguished themselves as chemists.

After his appointment to the chemical chair at Upsala, the remainder of his life passed with very little variety; his whole time was occupied with his favourite studies, and not a year passed that he did not publish some dissertation or other upon some more or less important branch of chemistry. His reputation gradually extended itself over Europe, and he was enrolled among the number of the members of most scientific academies. Among other honourable testimonies of the esteem in which he was held, he was elected rector of the University of Upsala. This university is not merely a literary body, but owns extensive estates, over which it possesses great authority, and, having considerable control over its students, and enjoying considerable immunities and privileges (conferred in former times as an encouragement to learning, though, in reality, they serve only to cramp its energies, and throw barriers in the way of its progress), constitutes, therefore, a kind of republic in the midst of Sweden : the professors being its chiefs. But while, in literary establishments, all the institutions ought to have for an object to maintain peace, and free their members from every occupation unconnected with letters, the constitution of that university obliges its professors to attend to things very inconsistent with their usual functions ; while it gives men of influence and ambition a desire to possess the power and patronage, though they may not be qualified to perform the duties, of a professor. Such temptations are very injurious to the true cause of science ; and it were to be wished, that no literary body, in any part of the world, were possessed of such powers and privileges. When Bergman was rector, the university was divided into two great parties, the one consisting of the theological and law faculties, and the other of the scientific professors. Bergman's object was to preserve peace and agreement between these two parties, and to convince them that it was the interest of all to unite for the good of the university and the promotion of letters. The period of his magistracy is remarkable in the annals of the university for the small number of deliberations, and the little business recorded in the registers; and for the good sense and good behaviour of the students. The students in Upsala are numerous, and most of them are young men. They had been accustomed frequently to brave or elude the severity of the regulations; but during Bergman's rectorship they were restrained effectually by their respect for his genius, and their admiration of his character and conduct.

When the reputation of Bergman was at its height, in the year 1776, Frederick the Great of Prussia formed the wish to attach him to the Academy of Sciences of Berlin, and made him offers of such a nature that our professor hesitated for a short time as to whether he ought not to accept them. His health had been injured by the assiduity with which he had devoted himself to the double duty of teaching and experimenting. He might look for an alleviation of his ailments, if not a complete recovery, in the milder climate of Prussia, and he would be able to devote himself entirely to his academical duties; but other considerations prevented him from acceding to this proposal, tempting as it was. The King of Sweden had been his benefactor, and it was intimated to him that his leaving the kingdom would afflict that monarch. This information induced him, without further hesitation, to refuse the proposals of the King of Prussia. He requested of the king, his master, not to make him lose the merit of his sacrifice by augmenting his income; but to this demand the King of Sweden very properly refused to accede.

In the year 1771, Professor Bergman married a widow lady, Margaretha Catharina Trast, daughter of a clergyman in the neighbourhood of Upsala. By her he had two sons; but both of them died when infants. This lady survived her husband. The King of Sweden settled on her an annuity of 200 rix dollars, on condition that she gave up the library and apparatus of her late husband to the Royal Society of Upsala.

Bergman's health had been always delicate; indeed he seems never to have completely recovered the effects of his first year's too intense study at Upsala. He struggled on, however, with his ailments; and, by way of relaxation, was accustomed sometimes, in summer, to repair to the waters of Medevi—a celebrated mineral spring in Sweden, situated near the banks of the great inland lake, Wetter. One of these visits seems to have restored him to health for the time. But his malady returned in 1784 with redoubled violence. He was afflicted with hemorrhoids, and his daily loss of blood amounted to about six ounces. This constant drain soon exhausted him, and on the 8th of July, 1784, he died at the baths of Medevi, to which he had repaired in hopes of again benefiting by these waters.

The different tracts which he published, as they have been enumerated by Hjelm, who gave an interesting account of Bergman to the Stockholm Academy in the year 1785, amount to 106. They have been all collected into six octavo volumes entitled "Opuscula Torberni Bergman Physica et Chemica"—with the exception of his notes on Scheffer, his Sciagraphia, and his chapter on Physical Geography, which was translated into French, and published in the Journal des Mines (vol. iii. No. 15, p. 55). His Sciagraphia, which is an attempt to arrange minerals according to their composition, was translated into English by Dr. Withering. His notes on Scheffer were interspersed in an edition of the " Chemiske Föreläsningar" of that chemist, published in 1774, which he seems to have employed as a text-book in his lectures: or, at all events, the work was published for the use of the students of chemistry at Upsala. There was a new edition of it published, after Bergman's death, in the year 1796, to which are appended Bergman's Tables of Affinities.

The most important of Bergman's chemical papers were collected by himself, and constitute the three first volumes of his Opuscula. The three last volumes of that work were published after his death. The fourth volume was published at Leipsic, in 1787, by Hebenstreit, and contains the rest of his chemical papers. The fifth volume was given to the world in 1788, by the same editor. It contains three chemical papers, and the rest of it is made up with papers on natural history, electricity, and other branches of physics, which Bergman had published in the earlier part of his life. The same indefatigable editor published the sixth volume in 1790. It contains three astronomical papers, two chemical, and a long paper on the means of preventing any injurious effects from lightning. This was an oration, delivered before the Royal Academy of Sciences of Stockholm, in 1764, probably at the time of his admission into the academy.

It would serve little purpose in the present state of chemical knowledge, to give a minute analysis of Bergman's papers. To judge of their value, it would be necessary to compare them, not with our present chemical knowledge, but with the state of the science when his papers were published.

A very short general view of his labours will be sufficient to convey an idea of the benefits which the
science derived from them.

1. His first paper, entitled " On the Aerial Acid,"
that is, *carbonic acid*, was published in 1774.   In
it he gives the properties of this substance in considerable detail, shows that it possesses acid qualities, and that it is capable of combining with the
bases, and forming salts.   What is very extraordinary, in giving an account of carbonate of lime and
carbonate of magnesia, he never mentions the name
of Dr. Black; though it is very unlikely that a controversy, which had for years occupied the attention
of chemists, should have been unknown to him.
Mr. Cavendish's name never once appears in the
whole paper; though that philosopher had preceded
him by seven or eight years.   He informs us, that he
had made known his opinions respecting the nature
of this substance, to various foreign correspondents,
among others to Dr. Priestley, as early as the year
1770, and that Dr. Priestley had mentioned his
views on the subject, in a paper inserted in the Philosophical Transactions for 1772.   Bergman found
the specific gravity of carbonic acid gas rather higher than 1·5, that of air being 1.   His result is not
far from the truth.   He obtained his gas, by mixing calcareous spar with dilute sulphuric acid.   He
shows that this gas has a sour taste, that it reddens
the infusion of litmus, and that it combines with
bases.   He gives figures of the apparatus which he
used.   This apparatus demands attention.   Though
far inferior to the contrivances of Priestley, it answered pretty well, enabling him to collect the gas,
and examine its properties.

It is unnecessary to enter into any further details
respecting this paper.   Whoever will take the trouble to compare it with Cavendish's paper on the same

subject, will find that he had been anticipated by that philosopher in a great many of his most important facts. Under these circumstances, I consider as singular his not taking any notice of Cavendish's previous labours.

2. His next paper, " On the Analyses of Mineral Waters," was first published in 1778, being the subject of a thesis, supported by J. P. Scharenberg. This dissertation, which is of great length, is entitled to much praise. He lays therein the foundation of the mode of analyzing waters, such as is followed at present. He points out the use of different reagents, for detecting the presence of the various constituents in mineral water, and then shows how the quantity of each is to be determined. It would be doing great injustice to Bergman, to compare his analyses with those of any modern experimenter. At that time, the science was not in possession of any accurate analyses of the neutral salts, which exist in mineral waters. Bergman undertook these necessary analyses, without which, the determination of the saline constituents of mineral waters was out of the question. His determinations were not indeed accurate, but they were so much better than those that preceded them, and Bergman's character as an experimenter stood so high, that they were long referred to as a standard by chemists. The first attempt to correct them was by Kirwan. But Bergman's superior reputation as a chemist enabled his results still to keep their ground, till his character for accuracy was finally destroyed by the very accurate experiments which the discovery of the atomic theory rendered it necessary to make. These, when once they became generally known, were of course preferred, and Bergman's analyses were laid aside.

It is a curious and humiliating fact, as it shows

how much chemical reputation depends upon situation, or accidental circumstances, that Wenzel had, in 1766, in his book on *affinity*, published much more accurate analyses of all these salts, than Bergman's—analyses indeed which were almost perfectly correct, and which have scarcely been surpassed, by the most careful ones of the present day. Yet these admirable experiments scarcely drew the attention of chemists; while the very inferior ones of Bergman were held up as models of perfection.

3. Bergman, not satisfied with pointing out the mode of analyzing mineral waters, attempted to imitate them artificially by chemical processes, and published two essays on the subject; in the first he showed the processes by which cold mineral waters might be imitated, and in the other, the mode of imitating hot mineral waters. The attempt was valuable, and served to extend greatly the chemical knowledge of mineral waters, and of the salts which they contain; but it was made at too early a period of the analytical art, to approach perfection. A similar remark applies to his analysis of sea-water. The water examined was brought by Sparmann from a depth of eighty fathoms, near the latitude of the Canaries: Bergman found in it only common salt, muriate of magnesia, and sulphate of lime. His not having discovered the presence of sulphate of magnesia is a sufficient proof of the imperfection of his analytical methods; the other constituents exist in such small quantity in sea-water that they might easily have been overlooked, but the quantity of sulphate of magnesia in sea-water is considerable.

4. I shall pass over the paper on oxalic acid, which constituted the subject of a thesis, supported in 1776, by John Afzelius Arfvedson. It is now known that oxalic acid was discovered by Scheele, not by Bergman. It is impossible to say how many

of the numerous facts stated in this thesis were ascertained by Scheele, and how many by Afzelius. For, as Afzelius was already a *magister docens* in chemistry, there can be little doubt that he would himself ascertain the facts which were to constitute the foundation of his thesis. It is indeed now known that Bergman himself intrusted all the details of his experiments to his pupils. He was the contriver, while his pupils executed his plans. That Scheele has nowhere laid claim to a discovery of so much importance as that of oxalic acid, and that he allowed Bergman peaceably to bear away the whole credit, constitutes one of the most remarkable facts in the history of chemistry. Moreover, while it reflects so much credit on Scheele for modesty and forbearance, it seems to bear a little hard upon the character of Bergman. When he published the essay in the first volume of his Opuscula, in 1779, why did he not in a note inform the world that Scheele was the true discoverer of this acid? Why did he allow the discovery to be universally assigned to him, without ever mentioning the true state of the case? All this appeared so contrary to the character of Bergman, that I was disposed to doubt the truth of the statement, that Scheele was the discoverer of oxalic acid. When I was at Fahlun, in the year 1812, I took an opportunity of putting the question to Assessor Gahn, who had been the intimate friend of Scheele, and the pupil, and afterwards the friend of Bergman. He assured me that Scheele really was the discoverer of oxalic acid, and ascribed the omission of Bergman to inadvertence. Assessor Gahn showed me a volume of Scheele's letters to him, which he had bound up : they contained the history of all his chemical labours. I have little doubt that an account of oxalic acid would be found in these letters. If the son of Assessor Gahn, in whose possession these letters must now be, would take the trouble to inspect the volume in question, and to publish any notices respecting this acid which they may contain, he would confer an important favour on every person interested in the history of chemistry.

5. The dissertation on the manufacture of alum has been mentioned before. Bergman shows himself well acquainted with the processes followed, at least in Sweden, for making alum. He had no notion of the true constitution of alum; nor was that to be expected, as the discovery was thereby years later in being made. He thought that the reason why alum leys did not crystallize well was, that they contained an excess of acid, and that the addition of potash gave them the property of crystallizing readily, merely by saturating that excess of acid. Alum is a double salt, composed of three integrant particles of sulphate of alumina, and one integrant particle of sulphate of potash, or sulphate of ammonia. In some cases, the alum ore contains all the requisite ingredients. This is the case with the ore at Tolfa, in the neighbourhood of Rome. It seems, also, to be the case with respect to some of the alum ores in Sweden; particularly at Hœnsœter on Kinnekulle, in West Gothland, which I visited in 1812. If any confidence can be put in the statements of the manager of those works, no alkaline salt whatever is added; at least, I understood him to say so when I put the question.

6. In his dissertation on tartar-emetic, he gives an interesting historical account of this salt and its uses. His notions respecting the antimonial preparations best fitted to form it, are not accurate: nor, indeed, could they be expected to be so, till the nature and properties of the different oxides of antimony were accurately known. Antimony forms

three *oxides :* now it is the protoxide alone that is useful in medicine, and that enters into the composition of tartar-emetic; the other two oxides are inert, or nearly so.   Bergman was aware that tartaremetic is a double salt, and that its constituents are tartaric acid, potash, and oxide of antimony; but it was not possible, in 1773, when his dissertation was published, to have determined the true constituents of this salt by analysis.

7. Bergman's paper on magnesia was also a thesis defended in 1775, by Charles Norell, of West Gothland, who in all probability made the experiments described in the essay.   In the introduction we have a history of the discovery of magnesia, and he mentions Dr. Black as the person who first accurately made out its peculiar chemical characters, and demonstrated that it differs from lime.   This essay contains a pretty full and accurate account of the salts of magnesia, considering the state of chemistry at the time when it was published.   There is no attempt to analyze any of the magnesian salts; but, in his treatise on the analysis of mineral waters, he had stated the quantity of magnesia contained in one hundred parts of several of them.

8. His paper on the *shapes of crystals*, published in 1773, contains the germ of the whole theory of crystallization afterwards developed by M. Hauy.   He shows how, from a very simple primary form of a mineral, other shapes may proceed, which seem to have no connexion with, or resemblance to the primary form.   His view of the subject, so far as it goes, is the very same afterwards adopted by Hauy : and, what is very curious, Hauy and Bergman formed their theory from the very same crystalline shape of calcareous spar—from which, by mechanical divisions, the same rhombic nucleus was extracted by both.   Nothing prevented

Bergman from anticipating Hauy but a sufficient quantity of crystals to apply his theory to.*

9. In his paper on silica he gives us a history of the progress of chemical knowledge respecting this substance. Its nature was first accurately pointed out by Pott; though Glauber, and before him Van Helmont, were acquainted with the *liquor silicus*, or the combination of silica and potash, which is soluble in water. Bergman gives a detailed account of its properties; but he does not suspect it to possess acid properties. This great discovery, which has thrown a new light upon mineral bodies, and shown them all to be chemical combinations, was reserved for Mr. Smithson.

10. Bergman's experiments on the precious stones constitute the first rudiments of the method of analyzing stony bodies. His processes are very imperfect, and his apparatus but ill adapted to the purpose. We need not be surprised, therefore, that the results of his analyses are extremely wide of the truth. Yet, if we study his processes, we shall find in them the rudiments of the very methods which we follow at present. The superiority of the modern analyses over those of Bergman must in a great measure be ascribed to the platinum vessels which we now employ, and to the superior purity of the substances which we use as reagents in our analyses. The methods, too, are simplified and perfected. But we must not forget that this paper of Bergman's, imperfect as it is, constitutes the commencement of the art, and that fully as much genius and invention may be requisite to contrive the first rude processes, how imperfect soever they may be, as are required to bring these processes when once invented to a

---

* I shall mention afterwards that the real discoverer of this fact was Assessor Gahn, of Fahlun.

state of comparative perfection. The great step in analyzing minerals is to render them soluble in acids. Bergman first thought of the method for accomplishing this which is still followed, namely, fusing them or heating them to redness with an alkali or alkaline carbonate.

11. The paper on fulminating gold goes a great way to explain the nature of that curious compound He describes the properties of this substance, and the effects of alkaline and acid bodies on it. He shows that it cannot be formed without ammonia, and infers from his experiments that it is a compound of oxide of gold and ammonia. He explains the fulmination by the elastic fluid suddenly generated by the decomposition of the ammonia.

12. The papers on platinum, carbonate of iron, nickel, arsenic, and zinc, do not require many remarks. They add considerably to the knowledge which chemists at that time possessed of these bodies ; though the modes of analysis are not such as would be approved of by a modern chemist ; nor were the results obtained possessed of much precision.

13. The Essay on the Analysis of Metallic Ores by the wet way, or by solution, constitutes the first attempt to establish a regular method of analyzing metallic ores. The processes are all imperfect, as might be expected from the then existing state of analytical chemistry, and the imperfect knowledge possessed, of the different metallic ores. But this essay constituted a first beginning, for which the author is entitled to great praise. The subject was taken up by Klaproth, and speedily brought to a great degree of improvement by the labours of modern chemists.

14. The experiments on the way in which minerals behave before the blowpipe, which Bergman pub-

lished, were made at Bergman's request by Assessor
Gahn, of Fahlun, who was then his pupil. They
constitute the first results obtained by that very
ingenious and amiable man. He afterwards continued the investigation, and added many improvements, simplifying the reagents and the manner of
using them. But he was too indolent a man to
commit the results of his investigations to writing.
Berzelius, however, had the good sense to see the
importance of the facts which Gahn had ascertained.
He committed them to writing, and published them
for the use of mineralogists. They constitute the
book entitled " Berzelius on the Blowpipe," which
has been translated into English.

15. The object of the Essay on Metallic Precipitates is to determine the quantity of phlogiston
which each metal contains, deduced from the quantity of one metal necessary to precipitate a given
weight of another. The experiments are obviously
made with little accuracy : indeed they are not
susceptible of very great precision. Lavoisier afterwards made use of the same method to determine
the quantity of oxygen in the different metallic
oxides; but his results were not more successful
than those of Bergman.

16. Bergman's paper on iron is one of the most
important in his whole works, and contributed very
materially to advance the knowledge of the cause of
the difference between iron and steel. He employed
his pupils to collect specimens of iron from the different Swedish forges, and gave them directions
how to select the proper pieces. All these specimens, to the number of eighty-nine, he subjected to
a chemical examination, by dissolving them in dilute
sulphuric acid. He measured the volume of hydrogen gas, which he obtained by dissolving a given
weight of each, and noted the quantity and the

nature of the undissolved residue.   The general result of the whole investigation was that pure malleable iron yielded most hydrogen gas; steel less, and cast-iron least of all.   Pure malleable iron left the smallest quantity of insoluble matter, steel a greater quantity, and cast-iron the greatest of all. From these experiments he drew conclusions with respect to the difference between iron, steel, and cast-iron.   Nothing more was necessary than to apply the antiphlogistic theory to these experiments, (as was done soon after by the French chemists,) in order to draw important conclusions respecting the nature of these bodies.   Iron is a simple body; steel is a compound of iron and carbon ; and cast-iron of iron and a still greater proportion of carbon. The defective part of the experiments of Bergman in this important paper is his method of determining the quantity of *manganese* in iron.   In some specimens he makes the manganese amount to considerably more than a third part of the weight of the whole.   Now we know that a mixture of two parts iron and one part manganese is brittle and useless. We are sure, therefore, that no malleable iron whatever can contain any such proportion of manganese. The fact is, that Bergman's mode of separating manganese from iron was defective.   What he considered as manganese was chiefly, and might be in many cases altogether, oxide of iron.   Many years elapsed before a good process for separating iron from manganese was discovered.

17. Bergman's experiments to ascertain the cause of the brittleness of cold-short iron need not occupy much of our attention.   He extracted from it a white powder, by dissolving the cold-short iron in dilute sulphuric acid.   This white powder he succeeded in reducing to the state of a white brittle metal, by fusing it with a flux and charcoal.

Klaproth soon after ascertained that this metal was a phosphuret of iron, and that the white powder was a phosphate of iron : and Scheele, with his usual sagacity, hit on a method of analyzing this phosphate, and thus demonstrating its nature. Thus Bergman's experiments led to the knowledge of the fact that cold-short iron owes its brittleness to a quantity of phosphorus which it contains. It ought to be mentioned that Meyer, of Stettin, ascertained the same fact, and made it known to chemists at about the same time with Bergman.

18. The dissertation on the products of volcanoes, first published in 1777, is one of the most striking examples of the sagacity of Bergman which we possess. He takes a view of all the substances certainly known to have been thrown out of volcanoes, attempts to subject them to a chemical analysis, and compares them with the basalt, and greenstone or trap-rocks, the origin of which constituted at that time a keen matter of dispute among geologists. He shows the identity between lavas and basalt and greenstone, and therefore infers the identity of formation. This is obviously the true mode of proceeding, and, had it been adopted at an earlier period, many of those disputes respecting the nature of trap-rocks, which occupied geologists for so long a period, would never have been agitated ; or, at least, would have been speedily decided. The whole dissertation is filled with valuable matter, still well entitled to the attention of geologists. His observations on *zeolites*, which he considered as unconnected with volcanic products, were very natural at the time when he wrote : though the subsequent experiments of Sir James Hall, and Mr. Gregory Watt, and, above all, an accurate attention to the scoriæ from different smelting-houses, have thrown a new light on the subject, and have shown the way in

which zeolitic crystals might easily have been formed in melted lava, provided circumstances were favourable. In fact, we find abundant cavities in real lava from Vesuvius, filled with zeolitic crystals.

19. The last of the labours of Bergman which I shall notice here is his Essay on Elective Attractions, which was originally published in 1775, but was much augmented and improved in the third volume of his Opuscula, published in 1783. An English translation of this last edition of the Essay was made by Dr. Beddoes, and was long familiar to the British chemical world. The object of this essay was to elucidate and explain the nature of chemical affinity, and to account for all the apparent anomalies that had been observed. He laid it down as a first principle, that all bodies capable of combining chemically with each other, have an attraction for each other, and that this attraction is a definite and fixed force which may be represented by a number. Now the bodies which have the property of uniting together are chiefly the acids and the alkalies, or bases. Every acid has an attraction for each of the alkalies or bases; but the force of this attraction differs in each. Some bases have a strong attraction for acids, and others a weak; but the attractive force of each may be expressed by numbers.

Now, suppose that an acid a is united with a base m with a certain force, if we mix the compound $a\ m$ with a certain quantity of the base n, which has a stronger attraction for a than m has, the consequence will be, that a will leave m and unite with n;—n having a stronger attraction for a than m has, will disengage it and take its place. In consequence of this property, which Bergman considered as the foundation of the whole of the science, the strength of affinity of one body for another is

E 2

determined by these decompositions and combinations. If n has a stronger affinity for a than m has, then if we mix together a, m, and n in the requisite proportions, a and n will unite together, leaving m uncombined : or if we mix n with the compound a m, m will be disengaged. Tables, therefore, may be drawn up, exhibiting the strength of these affinities. At the top of a column is put the name of an *acid* or a *base*, and below it are put the names of all the *bases* or *acids* in the order of their affinity. The following little table will exhibit a specimen of these columns :

> *Sulphuric Acid.*
> Barytes
> Strontian
> Potash
> Soda
> Lime
> Magnesia.

Here sulphuric acid is the substance placed at the head of the column, and under it are the names of the bases capable of uniting with it in the order of their affinity. Barytes, which is highest up, has the strongest affinity, and magnesia, which is lowest down, has the weakest affinity. If sulphuric acid and magnesia were combined together, all the bases whose names occur in the table above magnesia would be able to separate the sulphuric acid from it. Potash would be disengaged from sulphuric acid by barytes and strontian, but not by soda, lime, and magnesia.

Such tables then exhibited to the eye the strength of affinity of all the different bodies that are capable of uniting with one and the same substance, and the order in which decompositions are effected. Bergman drew up tables of affinity according to these views in fifty-nine columns. Each column contained the name of a particular substance, and under it was arranged all the bodies capable of uniting with it, each in the order of its affinity. Now bodies may be made to unite, either by mixing them together, and then exposing them to heat, or by dissolving them in water and mixing the respective solutions together. The first of these ways is usually called the *dry way*, the second the *moist way*. The order of decompositions often varies with the mode employed. On this account, Bergman divided each of his fifty-nine columns into two. In the first, he exhibited the order of decompositions in the moist way, in the second in the dry. He explained also the cases of double decomposition, by means of these unvarying forces acting together or opposing each other—and gave sixty-four cases of such double decompositions.

These views of Bergman's were immediately acceded to by the chemical world, and continued to regulate their processes till Berthollet published his Chemical Statics in 1802. He there called in question the whole doctrine of Bergman, and endeavoured to establish one of the very opposite kind. I shall have occasion to return to the subject when I come to give an account of the services which Berthollet conferred upon chemistry.

I have already observed, that we are under obligations to Bergman, not merely for the improvements which he himself introduced into chemistry, but for the pupils whom he educated as chemists, and the discoveries which were made by those persons, whose exertions he stimulated and encouraged. Among those individuals, whose chemical discoveries were chiefly made known to the world by his means, was Scheele, certainly one of the most extraordinary men, and most sagacious and industrious chemists that ever existed.

Charles William Scheele was born on the 19th of December, 1742, at Stralsund, the capital of Swedish Pomerania, where his father was a trades- man. He received the first part of his education at a private academy in Stralsund, and was after- wards removed to a public school. At a very early period he expressed a strong desire to study phar- macy, and obtained his father's consent to make choice of this profession. He was accordingly bound an apprentice for six years to Mr. Bouch, an apothecary in Gotheborg, and after his time was out, he remained with him still, two years longer.

It was here that he laid the groundwork of all his future celebrity, as we are informed by Mr. Grunberg, who was his fellow-apprentice, and af- terwards settled as an apothecary in Stralsund. He was at that time very reserved and serious, but un- commonly diligent. He attended minutely to all the processes, reflected upon them while alone, and studied the writings of Neumann, Lemery, Kun- kel, and Stahl, with indefatigable industry. He likewise exercised himself a good deal in draw- ing and painting, and acquired some proficiency in these accomplishments without a master. Kun- kel's Laboratorium was his favourite book, and he was in the habit of repeating experiments out of it secretly during the night-time. On one occasion, as he was employed in making pyrophorus, his fellow-apprentice was malicious enough to put a quantity of fulminating powder into the mixture. The consequence was a violent explosion, which, as it took place in the night, threw the whole fa- mily into confusion, and brought a very severe re- buke upon our young chemist. But this did not put a stop to his industry, which he pursued so constantly and judiciously, that, by the time his ap- prenticeship was ended, there were very few chemists indeed who excelled him in knowledge and practical skill. His fellow-apprentice, Mr. Grunberg, wrote to him in 1774, requesting to know by what means he had become such a proficient in chemistry, and received the following answer : " I look upon you, my dear friend, as my first instructor, and as the author of all I know on the subject, in consequence of your advising me to read Neumann's Chemistry. The perusal of this book first gave me a taste for experimenting, myself; and I very well remember, that upon mixing some oil of cloves and smoking spirit of nitre together, they took fire. However, I kept this matter secret. I have also before my eyes the unfortunate experiment which I made with pyrophorus. Such accidents only served to increase my passion for making experiments."

In 1765 Scheele went to Malmo, to the house of an apothecary, called Mr. Kalstrom. After spending two years in that place, he went to Stockholm, to superintend the apothecary's shop of Mr. Scharenberg. In 1773 he exchanged this situation for another at Upsala, in the house of Mr. Loock. It was here that he accidentally formed an acquaintance with Assessor Gahn, of Fahlun, who was at that time a student at Upsala, and a zealous chemist. Mr. Gahn happening to be one day in the shop of Mr. Loock, that gentleman mentioned to him a circumstance which had lately occurred to him, and of which he was anxious to obtain an explanation. If a quantity of saltpetre be put into a crucible and raised to such a temperature as shall not merely melt it, but occasion an agitation in it like boiling, and if, after a certain time, the crucible be taken out of the fire and allowed to cool, the saltpetre still continues neutral; but its properties are altered: for, if distilled vinegar be poured upon it, red fumes are given out, while vinegar produces no effect upon the saltpetre before it has been thus heated. Mr. Loock wished from Gahn an explanation of the cause of this phenomenon : Gahn was unable to explain it; but promised to put the question to Professor Bergman. He did so accordingly, but Bergman was as unable to find an explanation as himself. On returning a few days after to Mr. Loock's shop, Gahn was informed that there was a young man in the shop who had given an explanation of the phenomenon. This young man was Scheele, who had informed Mr. Loock that there were two species of acids confounded under the name of *spirit of nitre;* what we at present call *nitric* and *hyponitrous* acids. Nitric acid has a stronger affinity for potash than vinegar has; but hyponitrous acid has a weaker. The heat of the fire changes the *nitric* acid of the saltpetre to *hyponitrous:* hence the phenomenon.

Gahn was delighted with the information, and immediately formed an acquaintance with Scheele, which soon ripened into friendship. When he informed Bergman of Scheele's explanation, the professor was equally delighted, and expressed an eager desire to be made acquainted with Scheele; but when Gahn mentioned the circumstance to Scheele, and offered to introduce him to Bergman, our young chemist rejected the proposal with strong feelings of dislike.

It seems, that while Scheele was in Stockholm, he had made experiments on cream of tartar, and had succeeded in separating from it tartaric acid, in a state of purity. He had also determined a number of the properties of tartaric acid, and examined several of the tartrates. He drew up an account of these results, and sent it to Bergman. Bergman, seeing a paper subscribed by the name of a person who was unknown to him, laid it aside without looking at it, and forgot it altogether. Scheele was very much provoked at this contemptuous and unmerited treatment. He drew up another account of his experiments and gave it to Retzius, who sent it to the Stockholm Academy of Sciences (with some additions of his own), in whose Memoirs it was published in the year 1770.* It cost Assessor Gahn considerable trouble to satisfy Scheele that Bergman's conduct was merely the result of inadvertence, and that he had no intention whatever of treating him either with contempt or neglect. After much entreaty, he prevailed upon Scheele to allow him to introduce him to the professor of chemistry. The introduction took place accordingly, and ever after Bergman and Scheele continued steady friends—Bergman facilitating the researches of Scheele by every means in his power.

So high did the character of Scheele speedily rise in Upsala, that when the Duke of Sudermania visited the university soon after, in company with Prince Henry of Prussia, Scheele was appointed by the university to exhibit some chemical processes before him. He fulfilled his charge, and performed in different furnaces several curious and striking experiments. Prince Henry asked him various questions, and expressed satisfaction at the answers given. He was particularly pleased when informed that he was a native of Stralsund. These two princes afterwards stated to the professors that they would take it as a favour if Scheele could have free.access to the laboratory of the university whenever he wished to make experiments.

In the year 1775, on the death of Mr. Popler, apothecary at Köping (a small place on the north

* Konig. Vetensk. Acad. Handl. 1770, p. 207.

side of the lake Mæler), he was appointed by the
Medical College *provisor* of the apothecary's shop.
In Sweden all the apothecaries are under the control
of the Medical College, and no one can open a
shop without undergoing an examination and receiving licence from that learned body. In the course
of the examinations which he was obliged to undergo, Scheele gave great proofs of his abilities, and
obtained the appointment.   In 1777 the widow sold
him the shop and business, according to a written
agreement made between them ; but they still continued housekeeping at their joint expense.   He
had already distinguished himself by his discovery
of fluoric acid, and by his admirable paper on
manganese.   It is said, too, that it was he who
made the experiments on carbonic acid gas, which
constitute the substance of Bergman's paper on the
subject, and which confirmed and established Bergman's idea that it was an acid.   At Köping he continued his researches with unremitting perseverance, and
made more discoveries than all the chemists of his
time united together.   It was here that he made
the experiments on air and fire, which constitute the
materials of his celebrated work on these subjects.
The theory which he formed was indeed erroneous ;
but the numerous discoveries which the book contains must always excite the admiration of every
chemist.   His discovery of oxygen gas had been
anticipated by Priestley; but his analysis of atmospheric air was new and satisfactory—was peculiarly
his own.   The processes by means of which he procured oxygen gas were also new, simple, and easy,
and are still followed by chemists in general.   During
his residence at Köping he published a great number of chemical papers, and every one of them contained a discovery.   The whole of his time was
devoted to chemical investigations.   Every action

of his life had a tendency to forward the advancement of his favourite science; all his thoughts were turned to the same object; all his letters were devoted to chemical observations and chemical discussions. Crell's Annals was at that time the chief periodical work on chemistry in Germany. He got the numbers regularly as they were published, and was one of Crell's most constant and most valuable correspondents. Every one of his letters published in that work either contains some new chemical fact, or exposes the errors and mistakes of some one or other of Crell's numerous correspondents.

Scheele's outward appearance was by no means prepossessing. He seldom joined in the usual conversations and amusements of society, having neither leisure nor inclination for them. What little time he had to spare from the hurry of his profession was always employed in making experiments. It was only when he received visits from his friends, with whom he could converse on his favourite science, that he indulged himself in a little relaxation. For such intimate friends he had a sincere affection. This regard was extended to all the zealous cultivators of chemistry in every part of the world, whether personally known to him or not. He kept up a correspondence with several; though this correspondence was much limited by his ignorance of all languages except German; for at least he could not write fluently in any other language. His chemical papers were always written in German, and translated into Swedish, before they were inserted in the Memoirs of the Stockholm Academy, where most of them appeared.

He was kind and affable to all. Before he adopted an opinion in science, he reflected maturely on it; but, after he had once embraced it, his opinions were not easily shaken.　However, he did not hesitate to give up an opinion as soon as it had been proved to be erroneous.　Thus, he entirely renounced the notion which he once entertained that *silica* is a compound of *water* and *fluoric acid;* because it was demonstrated, by Meyer and others, that this *silica* was derived from the glass vessels in which the fluoric acid was prepared; that these glass vessels were speedily corroded into holes; and that, if fluoric acid was prepared in metallic vessels, and not allowed to come in contact with glass or any substance containing silica, it might be mixed with water without any deposition of silica whatever.

It appears also by a letter of his, published in Crell's Annals, that he was satisfied of the accuracy of Mr. Cavendish's experiments, showing that water was a compound of oxygen and hydrogen gases, and of Lavoisier's repetition of them.　He attempted to reconcile this fact with his own notion, that heat is a compound of oxygen and hydrogen.　But his arguments on that subject, though ingenious, are not satisfactory; and there is little doubt that if he had lived somewhat longer, and had been able to repeat his own experiments, and compare them with those of Cavendish and Lavoisier, he would have given up his own theory and adopted that of Lavoisier, or, at any rate, the explanation of Cavendish, which, being more conformable to his own preconceived notions, might have been embraced by him in preference.

It is said by Dr. Crell that Scheele was invited over to England, with an offer of an easy and advantageous situation; but that his love of quiet and retirement, and his partiality for Sweden, where he had spent the greatest part of his life, threw difficulties in the way of these overtures, and that a change in the English ministry put a stop to them

for the time. The invitation, Crell says, was renewed in 1786, with the offer of a salary of 300*l.*
a-year; but Scheele's death put a final stop to it.
I have very great doubts about the truth of this
statement; and, many years ago, during the lifetime
of Sir Joseph Banks, Mr. Cavendish, and Mr. Kirwan,
I made inquiry about the circumstance; but none
of the chemists in Great Britain, who were at that
time numerous and highly respectable, had ever
heard of any such negotiation. I am utterly at a
loss to conceive what one individual in any of the
ministries of George III. was either acquainted with
the science of chemistry, or at all interested in
its progress. They were all so intent upon accomplishing their own objects, or those of their sovereign,
that they had neither time nor inclination to think
of science, and certainly no money to devote to any
of its votaries. What minister in Great Britain ever
attempted to cherish the sciences, or to reward those
who cultivate them with success ? If we except Mr.
Montague, who procured the place of master of the
Mint for Sir Isaac Newton, I know of no one. While
in every other nation in Europe science is directly
promoted, and considerable sums are appropriated
for its cultivation, and for the support of a certain
number of individuals who have shown themselves
capable of extending its boundaries, not a single
farthing has been devoted to any such purpose in
Great Britain. Science has been left entirely to
itself; and whatever has been done by way of promoting it has been performed by the unaided exertions of private individuals. George III. himself
was a patron of literature and an encourager of
botany. He might have been disposed to reward the unrivalled eminence which Scheele had
attained; but this he could only have done by bestowing on him a pension out of his privy purse.
No situation which Scheele could fill was at his dis-

posal. The universities and the church were both shut against a Lutheran; and no pharmaceutical places exist in this country to which Scheele could have been appointed. If any such project ever existed, it must have been an idea which struck some man of science that such a proposal to a man of Scheele's eminence would redound to the credit of the country. But that such a project should have been broached by a British ministry, or by any man of great political influence, is an opinion that no person would adopt who has paid any attention to the history of Great Britain since the Revolution to the present time.

Scheele fell at last a sacrifice to his ardent love for his science. He was unable to abstain from experimenting, and many of his experiments were unavoidably made in his shop, where he was exposed during winter, in the ungenial climate of Sweden, to cold draughts of air. He caught rheumatism in consequence, and the disease was aggravated by his ardour and perseverance in his pursuits. When he purchased the apothecary's shop in which his business was carried on, he had formed the resolution of marrying the widow of his predecessor, and he had only delayed it from the honourable principle of acquiring, in the first place, sufficient property to render such an alliance desirable on her part. At length, in the month of March, 1786, he declared his intention of marrying her; but his disease at this time increased very fast, and his hopes of recovery daily diminished. He was sensible of this; but nevertheless he performed his promise, and married her on the 19th of May, at a time when he lay on his deathbed. On the 21st, he left her by his will the disposal of the whole of his property; and, the same day on which he so tenderly provided for her, he died.

I shall now endeavour to give the reader an idea

of the principal chemical discoveries for which we are indebted to Scheele : his papers, with the exception of his book on *air and fire*, which was published separately by Bergman, are all to be found either in the Memoirs of the Stockholm Academy of Science, or in Crell's Journal; they were collected, and a Latin translation of them, made by Godfrey Henry Schaefer, published at Leipsic, in 1788, by Henstreit, the editor of the three last volumes of Bergman's Opuscula. A French translation of them was made in consequence of the exertions of M. Morveau ; and an English translation of them, in 1786, by means of Dr. Beddoes, when he was a student in Edinburgh. There are also several German translations, but I have never had an opportunity of seeing them.

1. Scheele's first paper was published by Retzius, in 1770; it gives a method of obtaining pure tartaric acid : the process was to decompose cream of tartar by means of chalk. One half of the tartaric acid unites to the lime, and falls down in the state of a white insoluble powder, being *tartrate of lime*. The cream of tartar, thus deprived of half its acid, is converted into the neutral salt formerly distinguished by the name of *soluble tartar*, from its great solubility in water : it dissolves, and may be obtained in crystals, by the usual method of crystallizing salts. The tartrate of lime is washed with water, and then mixed with a quantity of dilute sulphuric acid, just capable of saturating the lime contained in the tartrate of lime ; the mixture is digested for some time; the sulphuric acid displaces the tartaric acid, and combines with the lime; and, as the sulphate of lime is but very little soluble in water, the greatest part of it precipitates, and the clear liquor is drawn off : it consists of tartaric acid, held in solution by water, but not quite free from sulphate of lime. By repeated

concentrations, all the sulphate of lime falls down, and at last the tartaric acid itself is obtained in large crystals. This process is still followed by the manufacturers of this country; for tartaric acid is used to a very considerable extent by the calico-printers, in various processes; for example, it is applied, thickened with gum, to different parts of cloth dyed Turkey red; the cloth is then passed through water containing the requisite quantity of chloride of lime: the tartaric acid, uniting with the lime, sets the chlorine at liberty, which immediately destroys the red colour wherever the tartaric acid has been applied, but leaves all the other parts of the cloth unchanged.

2. The paper on *fluoric acid* appeared in the Memoirs of the Stockholm Academy, for 1771, when Scheele was in Scharenberg's apothecary's shop in Stockholm, where, doubtless, the experiments were made. Three years before, Margraaf had attempted an analysis of fluor spar, but had discovered nothing. Scheele demonstrated that it is a compound of lime and a peculiar acid, to which he gave the name of *fluoric* acid. This acid he obtained in solution in water; it was separated from the fluor spar by sulphuric, muriatic, nitric, and phosphoric acids. When the fluoric acid came in contact with water, a white crust was formed, which proved, on examination, to be silica. Scheele at first thought that this silica was a compound of fluoric acid and water; but it was afterwards proved by Weigleb and by Meyer, that this notion is inaccurate, and that the silica was corroded from the retort into which the fluor spar and sulphuric acid were put. Bergman, who had adopted Scheele's theory of the nature of silica, was so satisfied by these experiments, that he gave it up, as Scheele himself did soon after.

Scheele did not obtain fluoric acid in a state of

purity, put only *fluosilicic acid;* nor were chemists
acquainted with the properties of fluoric acid till
Gay-Lussac and Thenard published their Recherches
Physico-chimiques, in 1811.

3. Scheele's experiments on *manganese* were undertaken at the request of Bergman, and occupied
him three years; they were published in the Memoirs
of the Stockholm Academy, for 1774, and constitute the most memorable and important of all his
essays, since they contain the discovery of two new
bodies, which have since acted so conspicuous a
part, both in promoting the progress of the science,
and in improving the manufactures of Europe. These
two substances are *chlorine* and *barytes,* the first
account of both of which occur in this paper.

The ore of manganese employed in these experiments was the *black oxide,* or *deutoxide,* of manganese, as it is now called.   Scheele's method of
proceeding was to try the effect of all the different
reagents on it.   It dissolved in sulphurous and nitrous
acids, and the solution was colourless.   Dilute sulphuric acid did not act upon it, nor nitric acid; but
concentrated sulphuric acid dissolved it by the assistance of heat.   The solution of sulphate of manganese in water was colourless and crystallized in very
oblique rhomboidal prisms, having a bitter taste.
Muriatic acid effervesced with it, when assisted by
heat, and the elastic fluid that passed off had a yellowish colour, and the smell of aqua regia.   He collected quantities of this elastic fluid *(chlorine)* in
bladders, and determined some of its most remarkable properties: it destroyed colours, and tinged the
bladder yellow, as nitric acid does.   This elastic
fluid, in Scheele's opinion, was muriatic acid deprived of phlogiston.   By phlogiston Scheele meant,
in this place, hydrogen gas.   He considered muriatic
acid as a compound of chlorine and hydrogen.   Now

this is the very theory that was established by Davy
in consequence of his own experiments and those of
Gay-Lussac and Thenard.   Scheele's mode of collecting chlorine gas in a bladder, did not enable him
to determine its characters with so much precision
as was afterwards done.   But his accuracy was so
great, that every thing which he stated respecting it
was correct so far as it went.

Most of the specimens of manganese ore which
Scheele examined, contained more or less barytes,
as has since been determined, in combination with
the oxide.   He separated this barytes, and determined its peculiar properties.   It dissolved in nitric
and muriatic acids, and formed salts capable of
crystallizing, and permanent in the air.   Neither
potash, soda, nor lime, nor any *base* whatever, was
capable of precipitating it from these acids.   But
the alkaline carbonates threw it down in the state of
a white powder, which dissolved with effervescence
in acids.   Sulphuric acid and all the sulphates threw
it down in the state of a white powder, which was
insoluble in water and in acids.   This sulphate cannot be decomposed by any acid or base whatever.
The only practicable mode of proceeding is to convert the sulphuric acid into sulphur, by heating the
salt with charcoal powder, along with a sufficient
quantity of potash, to bring the whole into fusion.
The fused mass, edulcorated, is soluble in nitric or
muriatic acid, and thus may be freed from charcoal,
and the barytes obtained in a state of purity.
Scheele detected barytes, also, in the potash made
from trees or other smaller vegetables; but at that
time he was unacquainted with *sulphate of barytes*,
which is so common in various parts of the earth,
especially in lead-mines.

To point out all the new facts contained in this
admirable essay, it would be necessary to transcribe

the whole of it. He shows the remarkable analogy between manganese and metallic oxides. Bergman, in an appendix affixed to Scheele's paper, states his reasons for being satisfied that it is really a metallic oxide. Some years afterwards, Assessor Gahn succeeded in reducing it to the metallic state, and thus dissipating all remaining doubts on the subject.

4. In 1775 he gave a new method of obtaining benzoic acid from benzoin. His method was, to digest the benzoin with pounded chalk and water, till the whole of the acid had combined with lime, and dissolved in the water. It is requisite to take care to prevent the benzoin from running into clots. The liquid thus containing benzoate of lime in solution is filtered, and muriatic acid added in sufficient quantity to saturate the lime. The benzoic acid is separated in white flocks, which may be easily collected and washed. This method, though sufficiently easy, is not followed by practical chemists, at least in this country. The acid when procured by precipitation is not so beautiful as what is procured by sublimation; nor is the process so cheap or so rapid. For these reasons, Scheele's process has not come into general use.

5. During the same year, 1775, his essay on arsenic and its acid was also published in the Memoirs of the Stockholm Academy. In this essay he shows various processes, by means of which white arsenic may be converted into an acid, having a very sour taste, and very soluble in water. This is the acid to which the name of *arsenic acid* has been since given. Scheele describes the properties of this acid, and the salts which it forms, with the different bases. He examines, also, the action of white arsenic upon different bodies, and throws light upon the arsenical salt of Macquer.

6. The object of the little paper on silica, clay,

and alum, published in the Memoirs of the Stockholm Academy, for 1776, is to prove that alumina
and silica are two perfectly distinct bodies, possessed
of different properties. This he does with his usual
felicity of experiment. He shows, also, that alumina
and lime are capable of combining together.

7. The same year, and in the same volume of the
Stockholm Memoirs, he published his experiments
on a urinary calculus. The calculus upon which
his experiments were made, happened to be composed of *uric acid*. He determined the properties
of this new acid, particularly the characteristic one
of dissolving in nitric acid, and leaving a beautiful
pink sediment when the solution is gently evaporated
to dryness.

8. In 1778 appeared his experiments on molybdena. What is now called *molybdena* is a soft
foliated mineral, having the metallic lustre, and
composed of two atoms sulphur united to one atom
of metallic molybdenum. It was known before,
from the experiments of Quest, that this substance
contains sulphur. Scheele extracted from it a white
powder, which he showed to possess acid properties,
though it was insoluble in water. He examined the
characters of this acid, called molybdic acid, and
the nature of the salts which it is capable of forming
by uniting with bases.

9. In the year 1777 was published the Experiments of Scheele on Air and Fire, with an introduction, by way of preface, from Bergman, who
seems to have superintended the publication. This
work is undoubtedly the most extraordinary production that Scheele has left us ; and is really wonderful, if we consider the circumstances under which
it was produced. Scheele ascertained that common
air is a mixture of two distinct elastic fluids, one of
which alone is capable of supporting combustion,

and which, therefore, he calls *empyreal air;* the other, being neither capable of maintaining combustion, nor of being breathed, he called *foul air.* These are the *oxygen* and *azote* of modern chemists. Oxygen he showed to be heavier than common air; bodies burnt in it with much greater splendour than in common air. Azote he found lighter than common air; bodies would not burn in it at all. He showed that metallic *calces,* or metallic *oxides,* as they are now called, contain oxygen as a constituent, and that when they are reduced to the metallic state, oxygen gas is disengaged. In his experiments on fulminating gold he shows, that during the fulmination a quantity of azotic gas is disengaged; and he deduces from a great many curious facts, which are stated at length, that ammonia is a compound of *azote* and *hydrogen.* His apparatus was not nice enough to enable him to determine the proportions of the various ingredients of the bodies which he analyzed: accordingly that is seldom attempted; and when it is, as was the case with common air, the results are very unsatisfactory. He deduces from his experiments, that the volume of oxygen gas, in common air, is between a third and a fourth: we now know that it is exactly a fifth.

In this book, also, we have the first account of sulphuretted hydrogen gas, and of its properties. He gives it the name of stinking sulphureous air.

The observations and new views respecting heat and light in this work are so numerous, that I am obliged to omit them: nor do I think it necessary to advert to his theory, which, when his book was published, was exceedingly plausible, and undoubtedly constituted a great step towards the improvements which soon after followed. His own experiments, had he attended a little more closely to the *weights,* and the alterations of them, would have been sufficient to have overturned the whole doctrine of phlogiston. Upon the whole it may be said, with confidence, that there is no chemical book in existence which contains a greater number of new and important facts than this work of Scheele, at the time it was published. Yet most of his discoveries were made, also, by others. Priestley and Lavoisier, from the superiority of their situations, and their greater means of making their labours speedily known to the public, deprived him of much of that reputation to which, in common circumstances, he would have been entitled. Priestley has been blamed for the rapidity of his publications, and the crude manner in which he ushered his discoveries to the world. But had he kept them by him till he had brought them to a sufficient degree of maturity, it is obvious that he would have been anticipated in the most important of them by Scheele.

10. In the Memoirs of the Stockholm Academy, for 1779, there is a short but curious paper of Scheele, giving an account of some results which he had obtained. If a plate of iron be moistened by a solution of common salt, or of sulphate of soda, and left for some weeks in a moist cellar, an afflorescence of carbonate of soda covers the surface of the plate. The same decomposition of common salt and evolution of soda takes place when unslacked quicklime is moistened with a solution of common salt, and left in a similar situation. These experiments led afterwards to various methods of decomposing common salt, and obtaining from it carbonate of soda. The phenomena themselves are still wrapped up in considerable obscurity. Berthollet attempted an explanation afterwards in his Chemical Statics; but founded on principles not easily admissible.

11. During the same year, his experiments on *plumbago* were published. This substance had been

long employed for making black-lead pencils; but nothing was known concerning its nature. Scheele, with his usual perseverance, tried the effect of all the different reagents, and showed that it consisted chiefly of *carbon*, but was mixed with a certain quantity of iron. It was concluded from these experiments, that plumbago is a carburet of iron. But the quantity of iron differs so enormously in different specimens, that this opinion cannot be admitted. Sometimes the iron amounts only to one-half per cent., and sometimes to thirty per cent. Plumbago, then, is carbon mixed with a variable proportion of iron, or carburet of iron.

12. In 1780 Scheele published his experiments on milk, and showed that sour milk contains a peculiar acid, to which the name of *lactic* acid has been given.

He found that when sugar of milk is dissolved in nitric acid, and the solution allowed to cool, small crystalline grains were deposited. These grains have an acid taste, and combine with bases: they have peculiar properties, and therefore constitute a particular acid, to which the name of *saclactic* was given. It is formed, also, when gum is dissolved in nitric acid; on this account it has been called, *mucic* acid.

13. In 1781 his experiments on a heavy mineral called by the Swedes *tungsten*, were published. This substance had been much noticed on account of its great weight; but nothing was known respecting its nature. Scheele, with his usual skill and perseverance, succeeded in proving that it was a compound of lime and a peculiar acid, to which the name of *tungstic acid* was given. Tungsten was, therefore, a tungstate of lime. Bergman, from its great weight, suspected that tungstic acid was in reality the oxide of a metal, and this conjecture was

afterwards confirmed by the Elhuyarts, who extracted the same acid from wolfram, and succeeded in reducing it to the metallic state.

14. In 1782 and 1783 appeared his experiments on *Prussian blue*, in order to discover the nature of the colouring matter. These experiments were exceedingly numerous, and display uncommon ingenuity and sagacity. He succeeded in demonstrating that *prussic acid*, the name at that time given to the colouring principle, was a compound of *carbon* and *azote*. He pointed out a process for obtaining prussic acid in a separate state, and determined its properties. This paper threw at once a ray of light on one of the obscurest parts of chemistry. If he did not succeed in elucidating this difficult department completely, the fault must not be ascribed to him, but to the state of chemistry when his experiments were made; in fact, it would have been impossible to have gone further, till the nature of the different elastic fluids at that time under investigation had been thoroughly established. Perhaps in 1783 there was scarcely any other individual who could have carried this very difficult investigation so far as it was carried by Scheele.

15. In 1783 appeared his observations on the *sweet principle of oils*. He observed, that when olive oil and litharge are combined together, a sweet substance separates from the oil and floats on the surface. This substance, when treated with nitric acid, yields *oxalic acid*. It was therefore closely connected with sugar in its nature. He obtained the same sweet matter from linseed oil, oil of almonds, of rape-seed, from hogs' lard, and from butter. He therefore concluded that it was a principle contained in all the expressed or fixed oils.

16. In 1784 he pointed out a method by which *citric acid* may be obtained in a state of purity from

lemon-juice. He likewise determined its characters, and showed that it was entitled to rank as a peculiar acid.

It was during the same year that he observed a white earthy matter, which may be obtained by washing rhubarb, in fine powder, with a sufficient quantity of water. This earthy matter he decomposed, and ascertained that it was a neutral salt, composed of oxalic acid, combined with lime. In a subsequent paper he showed, that the same oxalate of lime exists in a great number of roots of various plants.

17. In 1786 he showed that apples contain a peculiar acid, the properties of which he determined, and to which the name of *malic acid* has been given. In the same paper he examined all the common acid fruits of this country—gooseberries, currants, cherries, bilberries, &c., and determined the peculiar acids which they contain. Some owe their acidity to malic acid, some to citric acid, and some to tartaric acid; and not a few hold two, or even three, of these acids at the same time.

The same year he showed that the syderum of Bergman was phosphuret of iron, and the *acidum perlatum* of Proust *biphosphate of soda*.

The only other publication of Scheele, during 1785, was a short notice respecting a new mode of preparing *magnesia alba*. If sulphate of magnesia and common salt, both in solution, be mixed in the requisite proportions, a double decomposition takes place, and there will be formed sulphate of soda and muriate of magnesia. The greatest part of the former salt may be obtained out of the mixed ley by crystallization, and then the magnesia alba may be thrown down, from the muriate of magnesia, by means of an alkaline carbonate. The advantage of this new process is, the procuring of a considerable quantity of sulphate of soda in exchange for common salt, which is a much cheaper substance.

18. The last paper which Scheele published appeared in the Memoirs of the Stockholm Academy, for 1786 : in it he gave an account of the characters of gallic acid, and the method of obtaining that acid from nutgalls.

Such is an imperfect sketch of the principal discoveries of Scheele. I have left out of view his controversial papers, which have now lost their interest; and a few others of minor importance, that this notice might not be extended beyond its due length. It will be seen that Scheele extended greatly the number of acids; indeed, he more than doubled the number of these bodies known when he began his chemical labours. The following acids were discovered by him; or, at least, it was he that first accurately pointed out their characters :

Fluoric acid	Tartaric acid
Molybdic acid	Oxalic acid
Tungstic acid	Citric acid
Arsenic acid	Malic acid
Lactic acid	Saclactic
Gallic acid	Chlorine.

To him, also, we owe the first knowledge of barytes, and of the characters of manganese. He determined the nature of the constituents of ammonia and prussic acid : he first determined the compound nature of common air, and the properties of the two elastic fluids of which it is composed. What other chemist, either a contemporary or predecessor of Scheele, can be brought in competition with him as a discoverer? And all was performed under the most unpropitious circumstances, and during the continuance of a very short life, for he died in the 44th year of his age.

CHAPTER XI.

I HAVE already given an account of the state of chemistry in France, during the earlier part of the eighteenth century, as it was cultivated by the Stahlian school. But the new aspect which chemistry put on in Britain in consequence of the discoveries of Black, Cavendish, and Priestley, and the conspicuous part which the gases newly made known was likely to take in the future progress of the science, drew to the study of chemistry, sometime after the middle of the eighteenth century, a man who was destined to produce a complete revolution, and to introduce the same precision, and the same accuracy of deductive reasoning which distinguishes the other branches of natural science. This man was Lavoisier.

Antoine Laurent Lavoisier was born in Paris on the 26th of August, 1743. His father being a man of opulence spared no expense on his education. His taste for the physical sciences was early displayed, and the progress which he made in them was uncommonly rapid. In the year 1764 a prize was offered by the French government for the best and most economical method of lighting the streets of an extensive city. Young Lavoisier, though at that time only twenty-one years of age, drew up a memoir on the subject which obtained the gold medal. This essay was inserted in the Memoirs of the French Academy of Sciences, for 1768. It was during that year, when he was only twenty-five years of age that he became a member of that scientific body. By this time he was become fully conscious of his own strength; but he hesitated for some time to which of the sciences he should devote his attention. He tried pretty early to .determine, experimentally, some chemical questions which at that time drew the attention of practical chemists. For example: an elaborate paper of his appeared in the Memoirs of the French Academy, for 1768, on the composition of *gipsum*—a point at that time not settled; but which Lavoisier proved, as Margraaf had done before him, to be a compound of sulphuric acid and lime. In the Memoirs of the Academy, for 1770, two papers of his appeared, the object of which was to determine whether water could, as Margraaf had pretended, be converted into *silica* by long-continued digestion in glass vessels. Lavoisier found, as Margraaf stated, that when water is digested for a long time in a glass retort, a little silica makes its appearance; but he showed that this silica was wholly derived from the retort. Glass, it is well known, is a compound of silica and a fixed alkali. When water is long digested on it the glass is slightly corroded, a little alkali is dissolved in the water and a little silica separated in the form of a powder.

He turned a good deal of his attention also to geology, and made repeated journeys with Guettard into almost every part of France. The object in view was an accurate description of the mineralogical structure of France—an object accomplished to a considerable extent by the indefatigable exertions of Guettard, who published different papers on the subject in the Memoirs of the French Academy, accompanied with geological maps ; which were at that time rare.

The mathematical sciences also engrossed a considerable share of his attention. In short he displayed no great predilection for one study more than another, but seemed to grasp at every branch of science with equal avidity. While in this state of suspension he became acquainted with the new and unexpected discoveries of Black, Cavendish, and Priestley, respecting the gases. This opened a new creation to his view, and finally determined him to devote himself to scientific chemistry.

In the year 1774 he published a volume under the title of "Essays Physical and Chemical." It was divided into two parts. The first part contained an historical detail of every thing that had been done on the subject of airs, from the time of Paracelsus down to the year 1774. We have the opinions and experiments of Van Helmont, Boyle, Hales, Boerhaave, Stahl, Venel, Saluces, Black, Macbride, Cavendish, and Priestley. We have the history of Meyer's acidum pingue, and the controversy carried on in Germany, between Jacquin on the one hand, and Crans and Smeth on the other.

In the second part Lavoisier relates his own experiments upon gaseous substances. In the first four chapters he shows the truth of Dr. Black's theory of fixed air. In the 4th and 5th chapters he proves that when metallic calces are reduced, by heating them with charcoal, an elastic fluid is evolved, precisely of the same nature with carbonic acid gas. In the 6th chapter he shows that when metals are calcined their weight increases, and that a portion of air equal to their increase in weight is absorbed from the surrounding atmosphere. He observed that in a given bulk of air calcination goes on to a certain point and then stops altogether, and that air in which metals have been calcined does not support combustion so well as it did before any such process was performed in it. He also burned phosphorus in a given volume of air, observed the diminution of volume of the air and the increase of the weight of the phosphorus.

Nothing in these essays indicates the smallest suspicion that air was a mixture of two distinct fluids, and that only one of them was concerned in combustion and calcination; although this had been already deduced by Scheele from his own experiments, and though Priestley had already discovered the existence and peculiar properties of oxygen gas. It is obvious, however, that Lavoisier was on the way to make these discoveries, and had neither Scheele nor Priestley been fortunate enough to hit upon oxygen gas, it is exceedingly likely that he would himself have been able to have made that discovery.

Dr. Priestley, however, happened to be in Paris towards the end of 1774, and exhibited to Lavoisier, in his own laboratory in Paris, the method of procuring oxygen gas from red oxide of mercury. This discovery altered all his views, and speedily suggested not only the nature of atmospheric air, but also what happens during the calcination of metals and the combustion of burning bodies in general. These opinions when once formed he prosecuted with unwearied industry for more than twelve years, and after a vast number of experiments, conducted with a degree of precision hitherto unattempted in chemical investigations, he boldly undertook to disprove the existence of phlogiston altogether, and to explain all the phenomena hitherto supposed to depend upon that principle by the simple combination or separation of oxygen from bodies.

In these opinions he had for some years no coadjutors or followers, till, in 1785, Berthollet at a meeting of the Academy of Sciences, declared himself a convert. He was followed by M. Fourcroy, and soon after Guyton de Morveau, who was at that time the editor of the chemical department of the Encyclopédie Méthodique, was invited to Paris by Lavoisier and prevailed upon to join the same party. This was followed by a pretty vigorous controversy, in which Lavoisier and his associates gained a signal victory.

Lavoisier, after Buffon and Tillet, was treasurer to the academy, into the accounts of which he introduced both economy and order. He was consulted by the National Convention on the most eligible means of improving the manufacture of assignats, and of augmenting the difficulty of forging them. He turned his attention also to political economy, and between 1778 and 1785 he allotted 240 arpents in the Vendomois to experimental agriculture, and increased the ordinary produce by one-half. In 1791 the Constituent Assembly invited him to draw up a plan for rendering more simple the collection of the taxes, which produced an excellent report, printed under the title of " Territorial Riches of France."

In 1776 he was employed by Turgot to inspect the manufactory of gunpowder; which he made to carry 120 toises, instead of 90. It is pretty generally known, that during the war of the American revolution, the French gunpowder was much superior to the British; but it is perhaps not so generally understood, that for this superiority the French government were indebted to the abilities of Lavoisier. During the war of the French revolution, the quality of the powder of the two nations was reversed; the English being considerably superior to that of the French, and capable of carrying further. This was put to the test in a very remarkable way at Cadiz.

During the horrors of the dictatorship of Robespierre, Lavoisier began to suspect that he would be
stripped of his property, and informed Lalande that
he was extremely willing to work for his subsistence.
It was supposed that he meant to pursue the profession of an apothecary, as most congenial to his
studies : but he was accused, along with the other
farmers-general, of defrauding the revenue, and
thrown into prison. During that sanguinary period
imprisonment and condemnation were synonymous
terms. Accordingly, on the 8th of May, 1794, he
suffered on the scaffold, with twenty-eight farmers-
general, at the early age of fifty-one. It has been
alleged that Fourcroy, who at that time possessed
considerable influence, might have saved him had he
been disposed to have exerted himself. But this
accusation has never been supported by any evidence. Lavoisier was a man of too much eminence
to be overlooked, and no accused person at that
time could be saved unless he was forgotten. A
paper was presented to the tribunal, drawn up by
M. Hallé, giving a catalogue of the works, and a
recapitulation of the merits of Lavoisier; but it was
thrown aside without even being read, and M. Hallé
had reason to congratulate himself that his useless
attempts to save Lavoisier did not terminate in his
own destruction.

Lavoisier was tall, and possessed a countenance
full of benignity, through which his genius shone
forth conspicuous. He was mild, humane, sociable,
obliging, and he displayed an incredible degree of
activity. His influence was great, on account of his
fortune, his reputation, and the place which he held
in the treasury; but all the use which he made of it
was to do good. His wife, whom he married in
1771, was Marie-Anna-Pierette-Paulze, daughter of
a farmer-general, who was put to death at the same
time with her husband; she herself was imprisoned,

but saved by the fortunate destruction of the dictator himself, together with his abettors. It would appear that she was able to save a considerable part of her husband's fortune: she afterwards married Count Rumford, whom she survived.

Besides his volume of Physical and Chemical Essays, and his Elements of Chemistry, published in 1789, Lavoisier was the author of no fewer than sixty memoirs, which were published in the volumes of the Academy of Sciences, from 1772, to 1788, or in other periodical works of the time. I shall take a short review of the most important of these memoirs, dividing them into two parts: I. Those that are not connected with his peculiar chemical theory; II. Those which were intended to disprove the existence of phlogiston, and establish the anti-phlogistic theory.

I. I have already mentioned his paper on gypsum, published in the Memoirs of the Academy, for 1768. He proves, by very decisive experiments, that this salt is a compound of sulphuric acid, lime, and water. But this had been already done by Margraaf, in a paper inserted into the Memoirs of the Berlin Academy, for 1750, entitled "An Examination of the constituent parts of the Stones that become luminous." The most remarkable circumstance attending this paper is, that an interval of eighteen years should elapse without Lavoisier's having any knowledge of this important paper of Margraaf; yet he quotes Pott and Cronstedt, who had written on the same subject later than Margraaf, at least Cronstedt. What makes this still more singular and unaccountable is, that a French translation of Margraaf's Opuscula had been published in Paris, in the year 1762. That a man in Lavoisier's circumstances, who, as appears from his paper, had paid considerable attention to chemistry, should not have

perused the writings of one of the most eminent
chemists that had ever existed, when they were completely within his power, constitutes, I think, one
of the most extraordinary phenomena in the history
of science.

2. If a want of historical knowledge appears conspicuous in Lavoisier's first chemical paper, the same
remark cannot be applied to his second paper, " On
the Nature of Water, and the Experiments by which
it has been attempted to prove the possibility of changing it into Earth," which was inserted in the Memoirs
of the French Academy, for 1770. This memoir is
divided into two parts. In the first he gives a
history of the progress of opinions on the subject,
beginning with Van Helmont's celebrated experiment on the willow; then relating those of Boyle,
Triewald, Miller, Eller, Gleditch, Bonnet, Kraft,
Alston, Wallerius, Hales, Duhamel, Stahl, Boerhaave, Geoffroy, Margraaf, and Le Roy. This first
part is interesting, in an historical point of view,
and gives a very complete account of the progress
of opinions upon the subject from the very first
dawn of scientific chemistry down to his own time.
There is, it is true, a remarkable difference between
the opinions of his predecessors respecting the conversion of water into earth, and the experiments of
Margraaf on the composition of *selenite*. The former were inaccurate, and were recorded by him
that they might be refuted; but the experiments of
Margraaf were accurate, and of the same nature
with his own. The second part of this memoir contains his own experiments, made with much precision, which went to show that the earth was derived from the retort in which the experiments of
Margraaf were made, and that we have no proof
whatever that water may be converted into earth.

But these experiments of Lavoisier, though they

completely disproved the inferences that Margraaf drew from his observations, by no means demonstrated that water might not be converted into different animal and vegetable substances by the processes of digestion.   Indeed there can be no doubt that this is the case, and that the oxygen and hydrogen of which it is composed, enter into the composition of by far the greater number of animal and vegetable bodies produced by the action of the functions of living animals and vegetables.   We have no evidence that the carbon, another great constituent of vegetable bodies, and the carbon and azote which constitute so great a proportion of animal substances, have their origin from water.   They are probably derived from the food of plants and animals, and from the atmosphere which surrounds them, and which contains both of these principles in abundance.

Whether the silica, lime, alumina, magnesia, and iron, that exist in small quantity in plants, be derived from water and the atmosphere, is a question which we are still unable to answer.   But the experiments of Schrader, which gained the prize offered by the Berlin Academy, in the year 1800, for the best essay on the following subject : *To determine the earthy constituents of the different kinds of corn, and to ascertain whether these earthy parts are formed by the processes of vegetation,* show at least that we cannot account for their production in any other way.   Schrader analyzed the seeds of wheat, rye, barley, and oats, and ascertained the quantity of earthy matter which each contained. He then planted these different seeds in flowers of sulphur, and in oxides of antimony and zinc, watering them regularly with distilled water.   They vegetated very well.   He then dried the plants, and analyzed what had been the produce of a given weight of seed, and he found that the earthy matter in each was greater than it had been in the seeds from which they sprung. Now as the sulphur and oxides of zinc and antimony could furnish no earthy matter, no other source remains but the water with which the plants were fed, and the atmosphere with which they were surrounded. It may be said, indeed, that earthy matter is always floating about in the atmosphere, and that in this way they may have obtained all the addition of these principles which they contained. This is an objection not easily obviated, and yet it would require to be obviated before the question can be considered as answered.

3. Lavoisier's next paper, inserted in the Memoirs of the Academy, for 1771, was entitled " Calculations and Observations on the Project of the establishment of a Steam-engine to supply Paris with Water." This memoir, though long and valuable, not being strictly speaking chemical, I shall pass over. Mr. Watt's improvements seem to have been unknown to Lavoisier. Indeed as his patent was only taken out in 1769, and as several years elapsed before the merits of his new steam-engine became generally known, Lavoisier's acquaintance with it in 1771 could hardly be expected.

4. In 1772 we find a paper, by Lavoisier, in the Memoirs of the Academy, " On the Use of Spirit of Wine in the analysis of Mineral Waters." He shows how the earthy muriates may be separated from the sulphates by digesting the mixed mass in alcohol. This process no doubt facilitates the separation of the salts from each other: but it is doubtful whether the method does not occasion new inaccuracies that more than compensate the facility of such separations. When different salts are dissolved in water in small quantities, it may very well

happen that they do not decompose each other, being at too great a distance from each other to come within the sphere of mutual action. Thus it is possible that sulphate of soda and muriate of lime may exist together in the same water. But if we concentrate this water very much, and still more, if we evaporate to dryness, the two salts will gradually come into the sphere of mutual action, a double decomposition will take place, and there will be formed sulphate of lime and common salt. If upon the dry residue we pour as much distilled water as was driven off by the evaporation, we shall not be able to dissolve the saline matter deposited; a portion of sulphate of lime will remain in the state of a powder. Yet before the evaporation, all the saline contents of the water were in solution, and they continued in solution till the water was very much concentrated. This is sufficient to show that the nature of the salts was altered by the evaporation. If we digest the dry residue in spirit of wine, we may dissolve a portion of muriate of lime, if the quantity of that salt in the original water was greater than the sulphate of soda was capable of decomposing: but if the quantity was just what the sulphate of soda could decompose, the alcohol will dissolve nothing, if it be strong enough, or nothing but a little common salt, if its specific gravity was above 0·820. We cannot, therefore, depend upon the salts which we obtain after evaporating a mineral water to dryness, being the same as those which existed in the mineral water itself. The nature of the salts must always be determined some other way.

5. In the Memoirs of the Academy, for 1772 (published in 1776), are inserted two elaborate papers of Lavoisier, on the combustion of the diamond. The combustibility of the diamond was suspected by Newton, from its great refractive power. His sus-

picion was confirmed in 1694, by Cosmo III., Grand
Duke of Tuscany, who employed Averani and Tar-
gioni to try the effect of powerful burning-glasses
upon diamonds. They were completely dissipated
by the heat. Many years after, the Emperor Francis I. caused various diamonds to be exposed to the
heat of furnaces. They also were dissipated, without leaving any trace behind them. M. Darcet,
professor of chemistry at the Royal College of
Paris, being employed with Count Lauragais in a
set of experiments on the manufacture of porcelain,
took the opportunity of trying what effect the intense heat of the porcelain furnaces produced upon
various bodies. Diamonds were not forgotten. He
found that they were completely dissipated by the
heat of the furnace, without leaving any traces
behind them. Darcet found that a violent heat was
not necessary to volatilize diamonds. The heat of
an ordinary furnace was quite sufficient. In 1771
a diamond, belonging to M. Godefroi Villetaneuse,
was exposed to a strong heat by Macquer. It was
placed upon a cupel, and raised to a temperature
high enough to melt copper. It was observed to be
surrounded with a low red flame, and to be more
intensely red than the cupel. In short, it exhibited
unequivocal marks of undergoing real combustion.

These experiments were soon after repeated by
Lavoisier before a large company of men of rank and
science. The real combustion of the diamond was
established beyond doubt; and it was ascertained
also, that if it be completely excluded from the air,
it may be exposed to any temperature that can be
raised in a furnace without undergoing any alteration. Hence it is clear that the diamond is not a
volatile substance, and that it is dissipated by heat,
not by being volatilized, but by being burnt.

The object of Lavoisier in his experiments was to

determine the nature of the substance into which
the diamond was converted by burning.   In the first
part he gives as usual a history of every thing which
had been done previous to his own experiments on
the combustion of the diamond.   In the second par⁺
we have the result of his own experiments upon the
same subject.   He placed diamonds on porcelain
supports in glass jars standing inverted over water
and over mercury; and filled with common air and
with oxygen gas.*

The diamonds were consumed by means of burning-glasses.   No *water* or *smoke* or *soot* made their
appearance, and no alteration took place on the bulk
of the air when the experiments were made over mercury.   When they were made over water, the bulk of
the air was somewhat diminished.   It was obvious
from this that diamond when burnt in air or oxygen
gas, is converted into a gaseous substance, which is absorbed by water.   On exposing air in which diamond
had been burnt, to lime-water, a portion of it was
absorbed, and the lime-water was rendered milky.
From this it became evident, that when diamond
is burnt, *carbonic acid* is formed, and this was the
only product of the combustion that could be discovered.

Lavoisier made similar experiments with charcoal,
burning it in air and oxygen gas, by means of a
burning-glass.   The results were the same : carbonic
acid gas was formed in abundance, and nothing
else.   These experiments might have been employed
to support and confirm Lavoisier's peculiar theory,
and they were employed by him for that purpose
afterwards.   But when they were originally pub-

* The reader will bear in mind that though the memoir
was inserted in the Mem. de l'Acad., for 1772, it was in fact
published in 1776, and the experiments were made in 1775
and 1776.

lished, no such intention appeared evident; though doubtless he entertained it.

6. In the second volume of the Journal de Physique, for 1772, there is a short paper by Lavoisier on the conversion of water into ice. M. Desmarets had given the academy an account of Dr. Black's experiments, to determine the latent heat of water. This induced Lavoisier to relate his experiments on the same subject. He does not inform us whether they were made in consequence of his having become acquainted with Dr. Black's theory, though there can be no doubt that this must have been the case. The experiments related in this short paper are not of much consequence. But I have thought it worth while to notice it because it authenticates a date at which Lavoisier was acquainted with Dr. Black's theory of latent heat.

7. In the third volume of the Journal de Physique, there is an account of a set of experiments made by Bourdelin, Malouin, Macquer, Cadet, Lavoisier, and Baumé on the *white-lead ore* of Pullowen. The report is drawn up by Baumé. The nature of the ore is not made out by these experiments. They were mostly made in the dry way, and were chiefly intended to show that the ore was not a chloride of lead. It was most likely a phosphate of lead.

8. In the Memoirs of the Academy, for 1774, we have the experiments of Trudaine, de Montigny, Macquer, Cadet, Lavoisier, and Brisson, with the great burning-glass of M. Trudaine. The results obtained cannot be easily abridged, and are not of sufficient importance to be given in detail.

9. Analysis of some waters brought from Italy by M. Cassini, junior. This short paper appeared in the Memoirs of the Academy, for 1777. The waters in question were brought from alum-pits,

and were found to contain alum and sulphate of iron.

10. In the same volume of the Memoirs of the Academy, appeared his paper " On the Ash employed by the Saltpetre-makers of Paris, and on its use in the Manufacture of Saltpetre." This is a curious and valuable paper; but not sufficiently important to induce me to give an abstract of it here.

11. In the Memoirs of the Academy, for 1777, appeared an elaborate paper, by Lavoisier, " On the Combination of the matter of Fire, with Evaporable Fluids, and the Formation of Elastic aeriform Fluids." In this paper he adopts precisely the same theory as Dr. Black had long before established. It is remarkable that the name of Dr. Black never occurs in the whole paper, though we have seen that Lavoisier had become acquainted with the doctrine of latent heat, at least as early as the year 1772, as he mentioned the circumstance in a short paper inserted that year in the Journal de Physique, and previously read to the academy.

12. In the same volume of the Memoirs of the Academy, we have a paper entitled " Experiments made by Order of the Academy, on the Cold of the year 1775, by Messrs. Bezout, Lavoisier, and Vandermond." It is sufficiently known that the beginning of the year 1776 was distinguished in most parts of Europe by the weather. The object of this paper, however, is rather to determine the accuracy of the different thermometers at that time used in France, than to record the lowest temperature which had been observed. It has some resemblance to a paper drawn up about the same time by Mr. Cavendish, and published in the Philosophical Transactions.

13. In the Memoirs of the Academy, for 1778, appeared a paper entitled " Analysis of the Waters of the Lake Asphaltes, by Messrs. Macquer, Lavoi

sier, and Sage." This water is known to be saturated with *salt*. It is needless to state the result
of the analysis contained in this paper, because it is
quite inaccurate. Chemical analysis had not at that
time made sufficient progress to enable chemists to
analyze mineral waters with precision.

The observation of Lavoisier and Guettard, which
appeared at the same time, on a species of steatite,
which is converted by the fire into a fine biscuit of
porcelain, and on two coal-mines, the one in Franche-
Comté, the other in Alsace, do not require to be particularly noticed.

14. In the Mem. de l'Académie, for 1780 (published in 1784), we have a paper, by Lavoisier, " On
certain Fluids which may be obtained in an aeriform
State, at a degree of Heat not much higher than the
mean Temperature of the Earth." These fluids are
sulphuric ether, alcohol, and water. He points out
the boiling temperature of these liquids, and shows
that at that temperature the vapour of these bodies
possesses the elasticity of common air, and is permanent as long as the high temperature continues.
He burnt a mixture of vapour of ether and oxygen
gas, and showed that during the combustion carbonic acid gas is formed. Lavoisier's notions respecting these vapours, and what hindered the liquids
at the boiling temperature from being all converted
into vapour were not quite correct. Our opinions
respecting steam and vapours in general were first
rectified by Mr. Dalton.

15. In the Mem. de l'Académie, for 1780, appeared also the celebrated paper on *heat*, by Lavoisier and Laplace. The object of this paper was to
determine the specific heat of various bodies, and to
investigate the proposals that had been made by Dr.
Irvine for determining the point at which a thermometer would stand, if plunged into a body destitute
of heat. This point is usually called the real zero.

They begin by describing an instrument which they had contrived to measure the quantity of heat which leaves a body while it is cooling a certain number of degrees. To this instrument they gave the name of *calorimeter*. It consisted of a kind of hollow, surrounded on every side by ice. The hot body was put into the centre. The heat which it gave out while cooling was all expended in melting the ice, which was of the temperature of 32°, and the quantity of heat was proportional to the quantity of ice melted. Hence the quantity of ice melted, while equal weights of hot bodies were cooling a certain number of degrees, gave the direct ratios of the specific heats of each. In this way they obtained the following specific heats:

	Specific heat.
Water	1
Sheet-iron	0·109985
Glass without lead (crystal) .	0·1929
Mercury	0·029
Quicklime	0·21689
Mixture of 9 water with 16 lime	0·439116
Sulphuric acid of 1·87058 .	0·334597
4 sulphuric acid, 3 water .	0·603162
4 sulphuric acid, 5 water .	0·663102
Nitric acid of 1·29895 . .	0·661391
9⅓ nitric acid, 1 lime . .	0·61895
1 saltpetre, 8 water . . .	0·8167

Their experiments were inconsistent with the conclusions drawn by Dr. Irvine, respecting the real zero, from the diminution of the specific heat, and the heat evolved when sulphuric acid was mixed with various proportions of water, &c. If the experiments of Lavoisier and Laplace approached nearly to accuracy, or, indeed, unless they were quite inaccurate, it is obvious that the conclusions of Irvine must be quite erroneous. It is remarkable

that though the experiments of Crawford, and likewise those of Wilcke, and of several others, on specific heat had been published before this paper made
its appearance, no allusion whatever is made to
these publications. Were we to trust to the information communicated in the paper, the doctrine of
specific heat originated with Lavoisier and Laplace.
It is true that in the fourth part of the paper, which
treats of combustion and respiration, Dr. Crawford's
theory of animal heat is mentioned, showing clearly
that our authors were acquainted with his book on
the subject. And, as this theory is founded on the
different specific heats of bodies, there could be no
doubt that he was acquainted with that doctrine.

16. In the Mem. de l'Académie, for 1780, occur
the two following memoirs :

Report made to the Royal Academy of Sciences
on the Prisons. By Messrs. Duhamel, De Montigny, Le Roy, Tenon, Tillet, and Lavoisier.

Report on the Process for separating Gold and
Silver. By Messrs. Macquer, Cadet, Lavoisier,
Baumé, Cornette, and Berthollet.

17. In the Mem. de l'Académie, for 1781, we find
a memoir by Lavoisier and Laplace, on the electricity evolved when bodies are evaporated or sublimed. The result of these experiments was, that
when water was evaporated electricity was always
evolved. They concluded from these observations,
that whenever a body changes its state electricity is
always evolved. But when Saussure attempted to
repeat these observations, he could not succeed.
And, from the recent experiments of Pouillet, it
seems to follow that electricity is evolved only when
bodies undergo chemical decomposition or combination. Such experiments depend so much upon
very minute circumstances, which are apt to escape
the attention of the observer, that implicit confidence

cannot be put in them till they have been often repeated, and varied in every possible manner.

18. In the Memoires de l'Académie, for 1781,
there is a paper by Lavoisier on the comparative
value of the different substances employed as articles
of fuel. The substances compared to each other
are pit-coal, coke, charcoal, and wood. It would
serve no purpose to state the comparison here, as it
would not apply to this country; nor, indeed, would
it at present apply even to France.

We have, in the same volume, his paper on the
mode of illuminating theatres.

19. In the Memoires de l'Académie, for 1782
(printed in 1785), we have a paper by Lavoisier on
a method of augmenting considerably the action of
fire and of heat. The method which he proposes is
a jet of oxygen gas, striking against red-hot charcoal. He gives the result of some trials made in
this way. Platinum readily melted. Pieces of
ruby or sapphire were softened sufficiently to run
together into one stone. Hyacinth lost its colour,
and was also softened. Topaz lost its colour, and
melted into an opaque enamel. Emeralds and
garnets lost their colour, and melted into opaque
coloured glasses. Gold and silver were volatilized ;
all the other metals, and even the metallic oxides,
were found to burn. Barytes also burns when exposed to this violent heat. This led Lavoisier to
conclude, as Bergman had done before him, that
Barytes is a metallic oxide. This opinion has been
fully verified by modern chemists. Both silica and
alumina were melted. But he could not fuse lime
nor magnesia. We are now in possession of a still
more powerful source of heat in the oxygen and
hydrogen blowpipe, which is capable of fusing both
lime and magnesia, and, indeed, every substance which
can be raised to the requisite heat without burning

or being volatilized. This subject was prosecuted still further by Lavoisier in another paper inserted in a subsequent volume of the Memoires de l'Académie. He describes the effect on rock-crystal, quartz, sandstone, sand, phosphorescent quartz, milk quartz, agate, chalcedony, cornelian, flint, prase, nephrite, jasper, felspar, &c.

20. In the same volume is inserted a memoir " On the Nature of the aeriform elastic Fluids which are disengaged from certain animal Substances in a state of Fermentation." He found that a quantity of recent human fæces, amounting to about five cubic inches, when kept at a temperature approaching to 60° emitted, every day for a month, about half a cubic inch of gas. This gas was a mixture of eleven parts carbonic acid gas, and one part of an inflammable gas, which burnt with a blue flame, and was therefore probably carbonic oxide. Five cubic inches of old human fæces from a necessary kept in the same temperature, during the first fifteen days emitted about a third of a cubic inch of gas each day, and during each of the second fifteen days, about one fourth of a cubic inch. This gas was a mixture of thirty-eight volumes of carbonic acid gas, and sixty-two volumes of a combustible gas, burning with a blue flame, and probably carbonic oxide.

Fresh fæces do not effervesce with dilute sulphuric acid, but old moist fæces do, and emit about eight times their volume of carbonic acid gas. Quicklime, or caustic potash, mixed with fæces, puts a stop to the evolution of gas, doubtless by preventing all fermentation. During effervescence of fæcal matter the air surrounding it is deprived of a little of its oxygen, probably in consequence of its combining with the nascent inflammable gas which is slowly disengaged.

II. We come now to the new theory of combustion

of which Lavoisier was the author, and upon which his reputation with posterity will ultimately depend. Upon this subject, or at least upon matters more or less intimately connected with it, no fewer than twenty-seven memoirs of his, many of them of a very elaborate nature, and detailing expensive and difficult experiments, appeared in the different volumes of the academy between 1774 and 1788. The analogy between the combustion of bodies and the calcination of metals had been already observed by chemists, and all admitted that both processes were owing to the same cause; namely, the emission of *phlogiston* by the burning or calcining body. The opinion adopted by Lavoisier was, that during burning and calcination nothing whatever left the bodies, but that they simply united with a portion of the air of the atmosphere. When he first conceived this opinion he was ignorant of the nature of atmospheric air, and of the existence of oxygen gas. But after that principle had been discovered, and shown to be a constituent of atmospherical air, he soon recognised that it was the union of oxygen with the burning and calcining body that occasioned the phenomena. Such is the outline of the Lavoisierian theory stated in the simplest and fewest words. It will be requisite to make a few observations on the much-agitated question whether this theory originated with him.

It is now well known that John Rey, a physician at Bugue, in Perigord, published a book in 1630, in order to explain the cause of the increase of weight which lead and tin experience during their calcination. After refuting in succession all the different explanations of this increase of weight which had been advanced, he adds, " To this question, then, supported on the grounds already mentioned, I answer, and maintain with confidence, that the increase of weight arises from the air, which is condensed, rendered heavy and adhesive by the violent and long-continued heat of the furnace. This air mixes itself with the calx (frequent agitation conducing), and attaches itself to the minutest molecules, in the same manner as water renders heavy sand which is agitated with it, and moistens and adheres to the smallest grains." There cannot be the least doubt from this passage that Rey's opinion was precisely the same as the original one of Lavoisier, and had Lavoisier done nothing more than merely state in general terms that during calcination air unites with the calcining bodies, it might have been suspected that he had borrowed his notions from those of Rey. But the discovery of oxygen, and the numerous and decisive proofs which he brought forward that during burning and calcination oxygen unites with the burning and calcining body, and that this oxygen may be again separated and exhibited in its original elastic state oblige us to alter our opinion. And whether we admit that he borrowed his original notion from Rey, or that it suggested itself to his own mind, the case will not be materially altered. For it is not the man who forms the first vague notion of a thing that really adds to the stock of our knowledge, but he who demonstrates its truth and accurately determines its nature.

Rey's book and his opinions were little known. He had not brought over a single convert to his doctrine, a sufficient proof that he had not established it by satisfactory evidence. We may therefore believe Lavoisier's statement, when he assures us that when he first formed his theory he was ignorant of Rey, and never had heard that any such book had been published.

The theory of combustion advanced by Dr. Hook, in 1665, in his Micrographia, approaches still nearer to that of Lavoisier than the theory of Rey, and indeed, so far as he has explained it, the coincidence is exact. According to Hook there exists in common air a certain substance which is like, if not the very same with that which is fixed in saltpetre. This substance has the property of dissolving all combustibles; but only when their temperature is sufficiently raised. The solution takes place with such rapidity that it occasions fire, which in his opinion is mere *motion*. The dissolved substance may be in the state of air, or coagulated in a liquid or solid form. The quantity of this solvent in a given bulk of air is incomparably less than in the same bulk of saltpetre. Hence the reason why a combustible continues burning but a short time in a given bulk of air: the solvent is soon saturated, and then of course the combustion is at an end. This explains why combustion requires a constant supply of fresh air, and why it is promoted by forcing in air with bellows. Hook promised to develop this theory at greater length in a subsequent work; but he never fulfilled his promise; though in his Lampas, published about twelve years afterwards, he gives a beautiful chemical explanation of flame, founded on the very same theory.

From the very general terms in which Hook expresses himself, we cannot judge correctly of the extent of his knowledge. This theory, so far as it goes, coincides exactly with our present notions on the subject. His solvent is oxygen gas, which constitutes one-fifth part of the volume of the air, but exists in much greater quantity in saltpetre. It combines with the burning body, and the compound formed may either be a gas, a liquid, or a solid, according to the nature of the body subjected to combustion.

Lavoisier nowhere alludes to this theory of Hook nor gives the least hint that he had ever heard of it. This is the more surprising, because Hook was a man of great celebrity; and his Micrographia, as containing the original figures and descriptions of many natural objects, is well known, not merely in Great Britain, but on the continent. At the same time it must be recollected that Hook's theory is supported by no evidence; that it is a mere assertion, and that nobody adopted it. Even then, if we were to admit that Lavoisier was acquainted with this theory, it would derogate very little from his merit, which consisted in investigating the phenomena of combustion and calcination, and in showing that oxygen became a constituent of the burnt and calcined bodies.

About ten years after the publication of the Micrographia, Dr. Mayow, of Oxford, published his Essays. In the first of which, De Sal-nitro et Spiritu Nitro-aëreo, he obviously adopts Dr. Hook's theory of combustion, and he applies it with great ingenuity to explain the nature of respiration. Dr. Mayow's book had been forgotten when the attention of men of science was attracted to it by Dr. Beddoes. Dr. Yeats, of Bedford, published a very interesting work on the merits of Mayow, in 1798. It will be admitted at once by every person who takes the trouble of perusing Mayow's tract, that he was not satisfied with mere theory; but proved by actual experiment that air was absorbed during combustion, and altered during respiration. He has given figures of his apparatus, and they are very much of the same nature with those afterwards made use of by Lavoisier. It would be wrong, therefore, to deprive Mayow of the reputation to which he is entitled for his ingeniously-contrived and well-executed experiments. It must be admitted that he proved both the absorption of air during combustion and respiration; but even this does not take much from the fair fame of Lavoisier. The analysis of air and the discovery of oxygen gas really diminish the analogy between the theories of Mayow and Lavoisier, or at any rate the full investigation of the subject and the generalization of it belong exclusively to Lavoisier.

Attempts were made by the other French chemists, about the beginning of the revolution, to associate themselves with Lavoisier, as equally entitled with himself to the merit of the antiphlogistic theory; but Lavoisier himself has disclaimed the partnership. Some years before his death, he had formed the plan of collecting together all his papers relating to the antiphlogistic theory and publishing them in one work; but his death interrupted the project. However, his widow afterwards published the first two volumes of the book, which were complete at the time of his death. In one of these volumes Lavoisier claims for himself the exclusive discovery of the cause of the augmentation of weight which bodies undergo during combustion and calcination. He informs us that a set of experiments, which he made in 1772, upon the different kinds of air which are disengaged in effervescence, and a great number of other chemical operations discovered to him demonstratively the cause of the augmentation of weight which metals experience when exposed to heat. " I was young," says he, " I had newly entered the lists of science, I was desirous of fame, and I thought it necessary to take some steps to secure to myself the property of my discovery. At that time there existed an habitual correspondence between the men of science of France and those of England. There was a kind of rivalry between the two nations, which gave importance to new experiments, and which sometimes was the cause that the writers of the one or the

other of the nations disputed the discovery with
the real author. Consequently, I thought it proper
to deposit on the 1st of November, 1772, the following note in the hands of the secretary of the
academy. This note was opened on the 1st of
May following, and mention of these circumstances
marked at the top of the note. It was in the
following terms :

" About eight days ago I discovered that sulphur
in burning, far from losing, augments in weight ;
that is to say, that from one pound of sulphur much
more than one pound of vitriolic acid is obtained,
without reckoning the humidity of the air. Phosphorus presents the same phenomenon. This augmentation of weight arises from a great quantity of
air, which becomes fixed during the combustion, and
which combines with the vapours.

" This discovery, which I confirmed by experiments which I regard as decisive, led me to think
that what is observed in the combustion of sulphur
and phosphorus, might likewise take place with
respect to all the bodies which augment in weight
by combustion and calcination ; and I was persuaded that the augmentation of weight in the
calces of metals proceeded from the same cause.
The experiment fully confirmed my conjectures. I
operated the reduction of litharge in close vessels
with Hales's apparatus, and I observed, that at the
moment of the passage of the calx into the metallic
state, there was a disengagement of air in considerable quantity, and that this air formed a volume
at least one thousand times greater than that of the
litharge employed. As this discovery appears to
me one of the most interesting which has been made
since Stahl, I thought it expedient to secure to myself the property, by depositing the present note in
the hands of the secretary of the academy, to re-

main secret till the period when I shall publish my experiments. " LAVOISIER.

" *Paris, November* 11, 1772."

This note leaves no doubt that Lavoisier had conceived his theory, and confirmed it by experiment, at least as early as November, 1772. But at that time the nature of air and the existence of oxygen were unknown. The theory, therefore, as he understood it at that time, was precisely the same as that of John Rey. It was not till the end of 1774 that his views became more precise, and that he was aware that oxygen is the portion of the air which unites with bodies during combustion, and calcination.

Nothing can be more evident from the whole history of the academy, and of the French chemists during this eventful period, for the progress of the science, that none of them participated in the views of Lavoisier, or had the least intention of giving up the phlogistic theory. It was not till 1785, after his experiments had been almost all published, and after all the difficulties had been removed by the two great discoveries of Mr. Cavendish, that Berthollet declared himself a convert to the Lavoisierian opinions. This was soon followed by others, and within a very few years almost all the chemists and men of science in France enlisted themselves on the same side. Lavoisier's objection, then, to the phrase *La Chimie Française*, is not without reason, the term *Lavoisierian Chemistry* should undoubtedly be substituted for it. This term, *La Chimie Française* was introduced by Fourcroy. Was Fourcroy anxious to clothe himself with the reputation of Lavoisier, and had this any connexion with the violent death of that illustrious man?

The first set of experiments which Lavoisier published on his peculiar views, was entitled, " A Me-

moir on the Calcination of Tin in close Vessels; and
on the Cause of the increase of Weight which the
Metal acquires during this Process." It appeared
in the Memoirs of the Academy, for 1774. In this
paper he gives an account of several experiments
which he had made on the calcination of tin in glass
retorts, hermetically sealed. He put a quantity of
tin (about half a pound) into a glass retort, sometimes of a larger and sometimes of a smaller size,
and then drew out the beak into a capillary tube.
The retort was now placed upon the sand-bath, and
heated till the tin just melted. The extremity of the
capillary beak of the retort was now fused so as to
seal it hermetically. The object of this heating was
to prevent the retort from bursting by the expansion
of the air during the process. The retort, with its
contents, was now carefully weighed, and the weight
noted. It was put again on the sand-bath, and
kept melted till the process of calcination refused to
advance any further. He observed, that if the retort was small, the calcination always stopped sooner
than it did if the retort was large. Or, in other
words, the quantity of tin calcined was always proportional to the size of the retort.

After the process was finished, the retort (still
hermetically sealed) was again weighed, and was
always found to have the same weight exactly as at
first. The beak of the retort was now broken off,
and a quantity of air entered with a hissing noise.
The increase of weight was now noted: it was obviously owing to the air that had rushed in. The
weight of air that had been at first driven out by the
fusion of the tin had been noted, and it was now
found that a considerably greater quantity had entered than had been driven out at first. In some experiments, as much as 10·06 grains, in others 9·87
grains, and in some less than this, when the size of

the retort was small. The tin in the retort was mostly unaltered, but a portion of it had been converted into a black powder, weighing in some cases above two ounces. Now it was found in all cases, that the weight of the tin had increased, and the increase of weight was always exactly equal to the diminution of weight which the air in the retort had undergone, measured by the quantity of new air which rushed in when the beak of the retort was broken, minus the air that had been driven out when the tin was originally melted before the retort was hermetically sealed.

Thus Lavoisier proved by these first experiments, that when tin is calcined in close vessels a portion of the air of the vessel disappears, and that the tin increases in weight just as much as is equivalent to the loss of weight which the air has sustained. He therefore inferred, that this portion of air had united with the tin, and that calx of tin is a compound of tin and air. In this first paper there is nothing said about oxygen, nor any allusion to lead to the suspicion that air is a compound of different elastic fluids. These, therefore, were probably the experiments to which Lavoisier alludes in the note which he lodged with the secretary of the academy in November, 1772.

He mentions towards the end of the Memoir that he had made similar experiments with lead; but he does not communicate any of the numerical results: probably because the results were not so striking as those with tin. The heat necessary to melt lead is so high that satisfactory experiments on its calcination could not easily be made in a glass retort.

Lavoisier's next Memoir appeared in the Memoirs of the Academy, for 1775, which were published in 1778. It is entitled, " On the Nature of the Principle which combines with the Metals during their Calcination, and which augments their Weight." He observes that when the metallic calces are reduced to the metallic state it is found necessary to heat them along with charcoal. In such cases a quantity of carbonic acid gas is driven off, which he assures us is the charcoal united to the elastic fluid contained in the calx. He tried to reduce the calx of iron by means of burning-glasses, while placed under large glass receivers standing over mercury; but as the gas thus evolved was mixed with a great deal of common air which was necessarily left in the receiver, he was unable to determine its nature. This induced him to have recourse to red oxide of mercury. He showed in the first place that this substance (*mercurius præcipitatus per se*) was a true calx, by mixing it with charcoal powder in a retort and heating it. The mercury was reduced and abundance of carbonic acid gas was collected in an inverted glass jar standing in a water-cistern into which the beak of the retort was plunged. On heating the red oxide of mercury by itself it was reduced to the metallic state, though not so easily, and at the same time a gas was evolved which possessed the following properties:

1. It did not combine with water by agitation.

2. It did not precipitate lime-water.

3. It did not unite with fixed or volatile alkalies.

4. It did not at all diminish their caustic quality.

5. It would serve again for the calcination of metals.

6. It was diminished like common air by addition of one-third of nitrous gas.

7. It had none of the properties of carbonic acid gas. Far from being fatal, like that gas, to animals, it seemed on the contrary more proper for the purposes of respiration. Candles and burning bodies were

not only not extinguished by it, but burned with an enlarged flame in a very remarkable manner. The light they gave was much greater and clearer than in common air.

He expresses his opinion that the same kind of air would be obtained by heating nitre without addition, and this opinion is founded on the fact that when nitre is detonated with charcoal it gives out abundance of carbonic acid gas.

Thus Lavoisier shows in this paper that the kind of air which unites with metals during their calcination is purer and fitter for combustion than common air. In short it is the gas which Dr. Priestley had discovered in 1774, and which is now known by the name of oxygen gas.

This Memoir deserves a few animadversions. Dr. Priestley discovered oxygen gas in August, 1774; and he informs us in his life, that in the autumn of that year he went to Paris and exhibited to Lavoisier, in his own laboratory the mode of obtaining oxygen gas by heating red oxide of mercury in a gun-barrel, and the properties by which this gas is distinguished—indeed the very properties which Lavoisier himself enumerates in his paper. There can, therefore, be no doubt that Lavoisier was acquainted with oxygen gas in 1774, and that he owed his knowledge of it to Dr. Priestley.

There is some uncertainty about the date of Lavoisier's paper. In the History of the Academy, for 1775, it is merely said about it, " Read at the resumption (*rentrée*) of the Academy, on the 26th of April, by M. Lavoisier," without naming the year. But it could not have been before 1775, because that is the year upon the volume of the Memoirs; and besides, we know from the Journal de Physique (v. 429), that 1775 was the year on which the paper of Lavoisier was read.

Yet in the whole of this paper the name of Dr. Priestley never occurs, nor is the least hint given that he had already obtained oxygen gas by heating red oxide of mercury. So far from it, that it is obviously the intention of the author of the paper to induce his readers to infer that he himself was the discoverer of oxygen gas. For after describing the process by which oxygen gas was obtained by him, he says nothing further remained but to determine its nature, and " I discovered with *much surprise* that it was not capable of combination with water by agitation," &c. Now why the expression of surprise in describing phenomena which had been already shown ? And why the omission of all mention of Dr. Priestley's name ? I confess that this seems to me capable of no other explanation than a wish to claim for himself the discovery of oxygen gas, though he knew well that that discovery had been previously made by another.

The next set of experiments made by Lavoisier to confirm or extend his theory, was " On the Combustion of Phosphorus, and the Nature of the Acid which results from that Combustion." It appeared in the Memoirs of the Academy, for 1777. The result of these experiments was very striking. When phosphorus is burnt in a given bulk of air in sufficient quantity, about four-fifths of the volume of the air disappears and unites itself with the phosphorus. The residual portion of the air is incapable of supporting combustion or maintaining animal life. Lavoisier gave it the name of *mouffette atmospherique,* and he describes several of its properties. The phosphorus by combining with the portion of air which has disappeared, is converted into phosphoric acid, which is deposited on the inside of the receiver in which the combustion is performed, in the state of fine white flakes. One grain by this process is converted into two and a half grains of phosphoric acid. These observations led to the conclusion that atmospheric air is a mixture or compound of two distinct gases, the one (*oxygen*) absorbed by burning phosphorus, the other (*azote*) not acted on by that principle, and not capable of uniting with or calcining metals. These conclusions had already been drawn by Scheele from similar experiments, but Lavoisier was ignorant of them.

In the second part of this paper, Lavoisier describes the properties of phosphoric acid, and gives an account of the salts which it forms with the different bases. The account of these salts is exceedingly imperfect, and it is remarkable that Lavoisier makes no distinction between phosphate of potash and phosphate of soda; though the different properties of these two salts are not a little striking. But these were not the investigations in which Lavoisier excelled.

The next paper in which the doctrines of the antiphlogistic theory were still further developed, was inserted in the Memoirs of the Academy, for 1777. It is entitled, " On the Combustion of Candles in atmospherical Air, and in Air eminently Respirable." This paper is remarkable, because in it he first notices Dr. Priestley's discovery of oxygen gas; but without any reference to the preceding paper, or any apology for not having alluded in it to the information which he had received from Dr. Priestley.

He begins by saying that it is necessary to distinguish four different kinds of air. 1. Atmospherical air in which we live, and which we breath. 2. Pure air *(oxygen)*, alone fit for breathing, constituting about the fourth of the volume of atmospherical air, and called by Dr. Priestley *dephlogisticated air*. 3. Azotic gas, which constitutes about three-fourths of the volume of atmospherical air, and whose properties are still unknown. 4. Fixed air, which he proposed to call (as Bucquet had done) *acide crayeux, acid of chalk.*

In this paper Lavoisier gives an account of a great many trials that he made by burning candles in given volumes of atmospherical air and oxygen gas enclosed in glass receivers, standing over mercury. The general conclusion which he deduces from these experiments are—that the azotic gas of the air contributes nothing to the burning of the candle; but the whole depends upon the oxygen gas of the air, constituting in his opinion one-fourth of its volume; that during the combustion of a candle in a given volume of air only two-fifths of the oxygen are converted into carbonic acid gas, while the remaining three-fifths remain unaltered; but when the combustion goes on in oxygen gas a much greater proportion (almost the whole) of this gas is converted into carbonic acid gas. Finally, that phosphorus, when burnt in air acts much more powerfully on the oxygen of the air than a lighted candle, absorbing four-fifths of the oxygen and converting it into phosphoric acid.

It is evident that at the time this paper was written, Lavoisier's theory was nearly complete. He considered air as a mixture of three volumes of azotic gas, and one volume of oxygen gas. The last alone was concerned in combustion and calcination. During these processes a portion of the oxygen united with the burning body, and the compound formed constituted the acid or the calx. Thus he was able to account for combustion and calcination without having recourse to phlogiston. It is true that several difficulties still lay in his way, which he was not yet able to obviate, and which prevented any other person from adopting his opinions. One of the greatest of these was the fact that hydrogen gas was evolved during the solution of several metals in dilute sulphuric or muriatic acid ; that by this solution these metals were converted into calces, and that calces, when heated in hydrogen gas, were reduced to the metallic state while the hydrogen disappeared. The simplest explanation of these phenomena was the one adopted by chemists at the time. Hydrogen was considered as phlogiston. By dissolving metals in acids, the phlogiston was driven off and the calx remained : by heating the calx in hydrogen, the phlogiston was again absorbed and the calx reduced to the metallic state.

This explanation was so simple and appeared so satisfactory, that it was universally adopted by chemists with the exception of Lavoisier himself. There was a circumstance, however, which satisfied him that this explanation, however plausible, was not correct. The calx was *heavier* than the metal from which it had been produced. And hydrogen, though a light body, was still possessed of weight. It was obviously impossible, then, that the metal could be a combination of the calx and hydrogen. Besides, he had ascertained by direct experiment, that the calces of mercury, tin, and lead are compounds of the respective metals and oxygen. And it was known that when the other calces were heated with charcoal, they were reduced to the metallic state, and at the same time carbonic acid gas is evolved. The very same evolution takes place when calces of mercury, tin, and lead, are heated with charcoal powder. Hence the inference was obvious that carbonic acid is a compound of charcoal and oxygen, and therefore that all calces are compounds of their respective metals and oxygen.

Thus, although Lavoisier was unable to account for the phenomena connected with the evolution and absorption of hydrogen gas, he had conclusive evidence that the orthodox explanation was not the true one. He wisely, therefore, left it to time to throw light upon those parts of the theory that were still obscure.

His next paper, which was likewise inserted in the Memoirs of the Academy, for 1777, had some tendency to throw light on this subject, or at least it elucidated the constitution of sulphuric acid, which bore directly upon the antiphlogistic theory. It was entitled, " On the Solution of Mercury in vitriolic Acid, and on the Resolution of that Acid into aeriform sulphurous Acid, and into Air eminently Respirable."

He had already proved that sulphuric acid is a compound of sulphur and oxygen; and had even shown how the oxygen which the acid contained might be again separated from it, and exhibited in a separate state. Dr. Priestley had by this time made known the method of procuring sulphurous acid gas, by heating a mixture of mercury and sulphuric acid in a phial. This was the process which Lavoisier analyzed in the present paper. He put into a retort a mixture of four ounces mercury and six ounces concentrated sulphuric acid. The beak of the retort was plunged into a mercurial cistern, to collect the sulphurous acid gas as it was evolved; and heat being applied to the belly of the retort, sulphurous acid gas passed over in abundance, and sulphate of mercury was formed. The process was continued till the whole liquid contents of the retort had disappeared : then a strong heat was applied to the salt. In the first place, a quantity of sulphurous acid gas passed over, and lastly a portion of oxygen gas. The quicksilver was reduced to the metallic state. Thus he resolved sulphuric acid into sulphurous acid and oxygen. Hence it followed as a consequence, that sulphurous acid differs from sulphuric merely by containing a smaller quantity of oxygen.

The object of his next paper, published at the same time, was to throw light upon the pyrophorus of Homberg, which was made by kneading alum into a cake, with flour, or some substance containing abundance of carbon, and then exposing the mixture to a strong heat in close vessels, till it ceased to give out smoke. It was known that a pyrophorus thus formed takes fire of its own accord, and burns when it comes in contact with common air. It will not be necessary to enter into a minute analysis of this paper, because, though the experiments were very carefully made, yet it was impossible, at the time when the paper was drawn, to elucidate the phenomena of this pyrophorus in a satisfactory manner. There can be little doubt that the pyrophorus owes its property of catching fire, when in contact with air or oxygen, to a little potassium, which has been reduced to the metallic state by the action of the charcoal and sulphur on the potash in the alum. This substance taking fire, heat enough is produced to set fire to the carbon and sulphur which the pyrophorus contains. Lavoisier ascertained that during its combustion a good deal of carbonic acid was generated.

There appeared likewise another paper by Lavoisier, in the same volume of the academy, which may be mentioned, as it served still further to demonstrate the truth of the antiphlogistic theory. It is entitled, " On the Vitriolization of Martial Pyrites." Iron pyrites is known to be a compound of *iron* and *sulphur*. Sometimes this mineral may be left exposed to the air without undergoing any alteration, while at other times it speedily splits, effloresces, swells, and is converted into sulphate of iron. There are two species of pyrites; the one composed of two atoms of sulphur and one atom of iron, the other of one atom of sulphur and one atom of iron. The first of these is called bisulphuret of iron; the second protosulphuret, or simply sulphuret of iron. The variety of pyrites which undergoes spontaneous decomposition in the air, is known to be a compound, or rather mixture of the two species of pyrites.

Lavoisier put a quantity of the decomposing pyrites under a glass jar, and found that the process went on just as well as in the open air. He found that the air was deprived of the whole of its oxygen by the process, and that nothing was left but azotic gas. Hence the nature of the change became evident. The sulphur, by uniting with oxygen, was converted into sulphuric acid, while the iron became oxide of iron, and both uniting, formed sulphate of iron. There are still some difficulties connected with this change that require to be elucidated.

We have still another paper by Lavoisier, bearing on the antiphlogistic theory, published in the same volume of the Memoirs of the Academy, for 1778, entitled, " On Combustion in general." He establishes that the only air capable of supporting combustion is oxygen gas : that during the burning of bodies in common air, a portion of the oxygen of the atmosphere disappears, and unites with the burning body, and that the new compound formed is either an acid or a metallic calx. When sulphur is burnt, sulphuric acid is formed; when phosphorus, phosphoric acid; and when charcoal, carbonic acid. The calcination of metals is a process analogous to combustion, differing chiefly by the slowness of the process : indeed when it takes place rapidly, actual combustion is produced. After establishing these general principles, which are deduced from his preceding papers, he proceeds to examine the Stahlian theory of phlogiston, and shows that no evidence of the existence of any such principle can be adduced, and that the phenomena can all be explained without having recourse to it. Powerful as these arguments were, they produced no immediate effects. Nobody chose to give up the phlogistic theory to which he had been so long accustomed.

The next two papers of Lavoisier require merely to be mentioned, as they do not bear immediately upon the antiphlogistic theory. They appeared in the Memoirs of the Academy, for 1780. These memoirs were,

1. Second Memoir on the different Combinations of Phosphoric Acid.

2. On a particular Process, by means of which Phosphorus may be converted into phosphoric Acid, without Combustion.

The process here described consisted in throwing phosphorus, by a few grains at a time, into warm nitric acid of the specific gravity 1·29895. It falls to the bottom like melted wax, and dissolves pretty rapidly with effervescence: then another portion is thrown in, and the process is continued till as much phosphorus has been employed as is wanted; then the phosphoric acid may be obtained pure by distilling off the remaining nitric acid with which it is still mixed.

Hitherto Lavoisier had been unable to explain the anomalies respecting hydrogen gas, or to answer the objections urged against his theory in consequence of these anomalies. He had made several attempts to discover what peculiar substance was formed during the combustion of hydrogen, but always without success: at last, in 1783, he resolved to make the experiment upon so large a scale, that whatever the product might be, it should not escape him; but Sir Charles Blagden, who had just gone to Paris, informed him that the experiment for which he was preparing had already been made by Mr. Cavendish, who had ascertained that the product of the combustion of hydrogen was *water*. Lavoisier saw at a glance the vast importance of this discovery for the establishment of the antiphlogistic theory, and with what ease it would enable him to answer all the plausible objections which had been brought forward against his opinions in consequence of the evolution of hydrogen, when metals were calcined by solution in acids, and the absorption of it when metals were reduced in an atmosphere of this gas. He therefore resolved to repeat the experiment of Cavendish with every possible care, and upon a scale sufficiently large to prevent ambiguity. The experiment was made on the 24th of June, 1783, by Lavoisier and Laplace, in the presence of M. Le Roi, M. Vandermonde, and Sir Charles Blagden, who was at that time secretary of the Royal Society. The quantity of water formed was considerable, and they found that water was a compound of

1 volume oxygen
1·91 volume hydrogen.

Not satisfied with this, he soon after made another experiment along with M. Meusnier to decompose water. For this purpose a porcelain tube, filled with iron wire, was heated red-hot by being passed through a furnace, and then the steam of water was made to traverse the red-hot wire. To the further extremity of the porcelain tube a glass tube was luted, which terminated in a water-trough under an inverted glass receiver placed to collect the gas. The steam was decomposed by the red-hot iron wire, its oxygen united to the wire, while the hydrogen passed on and was collected in the water-cistern.

Both of these experiments, though not made till

1783, and though the latter of them was not read to the academy till 1784, were published in the volume of the Memoirs for 1781.

It is easy to see how this important discovery enabled Lavoisier to obviate all the objections to his theory from hydrogen. He showed that it was evolved when zinc or iron was dissolved in dilute sulphuric acid, because the water underwent decomposition, its oxygen uniting to the zinc or iron, and converting it into an oxide, while its hydrogen made its escape in the state of gas. Oxide of iron was reduced when heated in contact with hydrogen gas, because the hydrogen united to the oxygen of the acid and formed water, and of course the iron was reduced to the state of a metal. I consider it unnecessary to enter into a minute detail of these experiments, because, in fact, they added very little to what had been already established by Cavendish. But it was this discovery that contributed more than any thing else to establish the antiphlogistic theory. Accordingly, the great object of Dr. Priestley, and other advocates of the phlogistic theory, was to disprove the fact that water is a compound of oxygen and hydrogen. Scheele admitted the fact that water is a compound of oxygen and hydrogen; and doubtless, had he lived, would have become a convert to the antiphlogistic theory, as Dr. Black actually did. In short, it was the discovery of the compound nature of water that gave the Lavoisierian theory the superiority over that of Stahl. Till the time of this discovery every body opposed the doctrine of Lavoisier; but within a very few years after it, hardly any supporters of phlogiston remained. Nothing could be more fortunate for Lavoisier than this discovery, or afford him greater reason for self-congratulation.

We see the effect of this discovery upon his next

paper, "On the Formation of Carbonic Acid," which appeared in the Memoirs of the Academy, for 1781. There, for the first time, he introduces new terms, showing, by that, that he considered his opinions as fully established. To the *dephlogisticated air* of Priestley, or his own *pure air*, he now gives the name of *oxygen*. The fixed air of Black he designates *carbonic acid*, because he considered it as a compound of *carbon* (the pure part of charcoal) and oxygen. The object of this paper is to determine the proportion of the constituents. He details a great many experiments, and deduces from them all, that carbonic acid gas is a compound of

Carbon . . . 0·75
Oxygen . . . 1·93

Now this is a tolerably near approximation to the truth. The true constituents, as determined by modern chemists, being

Carbon . . . 0·75
Oxygen . . . 2·00

The next paper of M. Lavoisier, which appeared in the Memoirs of the Academy, for 1782 (published in 1785), shows how well he appreciated the importance of the discovery of the composition of water. It is entitled, " General Considerations on the Solution of the Metals in Acids." He shows that when metals are dissolved in acids, they are converted into oxides, and that the acid does not combine with the metal, but only with its oxide. When nitric acid is the solvent the oxidizement takes place at the expense of the acid, which is resolved into nitrous gas and oxygen. The nitrous gas makes its escape, and may be collected; but the oxygen unites with the metal and renders it an oxide. He shows this with respect to the solution of mercury in nitric acid. He collected the nitrous gas given out during the solution of the metal in

the acid : then evaporated the solution to dryness,
and urged the fire till the mercury was converted
into red oxide. The fire being still further urged,
the red oxide was reduced, and the oxygen gas
given off was collected and measured. He showed
that the nitrous gas and the oxygen gas thus obtained, added together, formed just the quantity of
nitric acid which had disappeared during the process. A similar experiment was made by dissolving
iron in nitric acid, and then urging the fire till the
iron was freed from every foreign body, and obtained in the state of black oxide.

It is well known that many metals held in solution by acids may be precipitated in the metallic
state, by inserting into the solution a plate of some
other metal. A portion of that new metal dissolves,
and takes the place of the metal originally in solution. Suppose, for example, that we have a neutral
solution of copper in sulphuric acid, if we put into
the solution a plate of iron, the copper is thrown
down in the metallic state, while a certain portion
of the iron enters into the solution, combining with
the acid instead of the copper. But the copper,
while in solution, was in the state of an oxide, and
it is precipitated in the metallic state. The iron
was in the metallic state ; but it enters into the solution in the state of an oxide. It is clear from this
that the oxygen, during these precipitations, shifts
its place, leaving the copper, and entering into combination with the iron. If, therefore, in such a case
we determine the exact quantity of copper thrown
down, and the exact quantity of iron dissolved at
the same time, it is clear that we shall have the relative weight of each combined with the same weight
of oxygen. If, for example, 4 of copper be thrown
down by the solution of $3 \cdot 5$ of iron ; then it is clear
that $3 \cdot 5$ of iron requires just as much oxygen as 4

of copper, to turn both into the oxide that exists in the solution, which is the black oxide of each.

Bergman had made a set of experiments to determine the proportional quantities of phlogiston contained in the different metals, by the relative quantity of each necessary to precipitate a given weight of another from its acid solution. It was the opinion at that time, that metals were compounds of their respective calces and phlogiston. When a metal dissolved in an acid, it was known to be in the state of calx, and therefore had parted with its phlogiston: when another metal was put into this solution it became a calx, and the dissolved metal was precipitated in the metallic state. It had therefore united with the phlogiston of the precipitating metal. It is obvious, that by determining the quantities of the two metals precipitated and dissolved, the relative proportion of phlogiston in each could be determined. Lavoisier saw that these experiments of Bergman would serve equally to determine the relative quantity of oxygen in the different oxides. Accordingly, in a paper inserted in the Memoirs of the Academy, for 1782, he enters into an elaborate examination of Bergman's experiments, with a view to determine this point. But it is unnecessary to state the deductions which he drew, because Bergman's experiments were not sufficiently accurate for the object in view. Indeed, as the mutual precipitation of the metals is a galvanic phenomenon, and as the precipitated metal is seldom quite pure, but an alloy of the precipitating and precipitated metal; and as it is very difficult to dry the more oxidizable metals, as copper and tin, without their absorbing oxygen when they are in a state of very minute division; this mode of experimenting is not precise enough for the object which Lavoisier had in view. Accordingly the table of the

composition of the metallic oxides which Lavoisier
has drawn up is so very defective, that it is not worth
while to transcribe it.

The same remark applies to the table of the affinities of oxygen which Lavoisier drew up and inserted
in the Memoirs of the Academy, for the same year.
His data were too imperfect, and his knowledge too
limited, to put it in his power to draw up any such
table with any approach to accuracy. I shall have occasion to resume the subject in a subsequent chapter.

In the same volume of the Memoirs of the Academy, this indefatigable man inserted a paper in order
to determine the quantity of oxygen which combines
with iron. His method of proceeding was, to burn
a given weight of iron in oxygen gas. It is well
known that iron wire, under such circumstances,
burns with considerable splendour, and that the
oxide, by the heat, is fused into a black brittle matter, having somewhat of the metallic lustre. He
burnt 145·6 grains of iron in this way, and found
that, after combustion, the weight became 192
grains, and 97 French cubic inches of oxygen gas had
been absorbed. From this experiment it follows,
that the oxide of iron formed by burning iron in
oxygen gas is a compound of

Iron   3·5
Oxygen   1·11

This forms a tolerable approximation to the truth. It
is now known, that the quantity of oxygen in the
oxide of iron formed by the combustion of iron in
oxygen gas is not quite uniform in its composition;
sometimes it is a compound of

Iron $3\frac{1}{2}$
Oxygen $1\frac{1}{3}$

While at other times it consists very nearly of

Iron   3·5
Oxygen   1

and probably it may exist in all the intermediate

proportions between these two extremes. The last of these compounds constitutes what is now known by the name of *protoxide*, or *black oxide of iron*. The first is the composition of the ore of iron so abundant, which is distinguished by the name of *magnetic iron ore*.

Lavoisier was aware that iron combines with more oxygen than exists in the protoxide; indeed, his analysis of peroxide of iron forms a tolerable approximation to the truth; but there is no reason for believing that he was aware that iron is capable of forming only two oxides, and that all intermediate degrees of oxidation are impossible. This was first demonstrated by Proust.

I think it unnecessary to enter into any details respecting two papers of Lavoisier, that made their appearance in the Memoirs of the Academy, for 1783, as they add very little to what he had already done. The first of these describes the experiments which he made to determine the quantity of oxygen which unites with sulphur and phosphorus when they are burnt: it contains no fact which he had not stated in his former papers, unless we are to consider his remark, that the heat given out during the burning of these bodies has no sensible weight, as new.

The other paper is "On Phlogiston;" it is very elaborate, but contains nothing which had not been already advanced in his preceding memoirs. Chemists were so wedded to the phlogistic theory, their prejudices were so strong, and their understandings so fortified against every thing that was likely to change their opinions, that Lavoisier found it necessary to lay the same facts before them again and again, and to place them in every point of view. In this paper he gives a statement of his own theory of combustion, which he had previously done in several preceding papers. He examines the phlogistic theory of Stahl at great length, and refutes it.

In the Memoirs of the Academy, for 1784, Lavoisier published a very elaborate set of experiments on the combustion of alcohol, oil, and different combustible bodies, which gave a beginning to the analysis of vegetable substances, and served as a foundation upon which this most difficult part of chemistry might be reared. He showed that during the combustion of alcohol the oxygen of the air united to the vapour of the alcohol, which underwent decomposition, and was converted into water and carbonic acid. From these experiments he deduced as a consequence, that the constituents of alcohol are carbon, hydrogen, and oxygen, and nothing else; and he endeavoured from his experiments to determine the relative proportions of these different constituents. From these experiments he concluded, that the alcohol which he used in his experiments was a compound of

Carbon	. . .	2629·5 part.
Hydrogen	. .	725·5
Water	. . .	5861

It would serve no purpose to attempt to draw any consequences from these experiments; as Lavoisier does not mention the specific gravity of the alcohol, of course we cannot say how much of the water found was merely united with the alcohol, and how much entered into its composition. The proportion between the carbon and hydrogen, constitutes an approximation to the truth, though not a very near one.

Olive oil he showed to be a compound of hydrogen and carbon, and bees' wax to be a compound of the same constituents, though in a different proportion.

This subject was continued, and his views further extended, in a paper inserted in the Memoirs of the Academy, for 1786, entitled, "Reflections on the Decomposition of Water by Vegetable and Animal Substances." He begins by stating that when charcoal is exposed to a strong heat, it gives out a little carbonic acid gas and a little inflammable air, and after this nothing more can be driven off, however high the temperature be to which it is exposed; but if the charcoal be left for some time in contact with the atmosphere it will again give out a little carbonic acid .gas and inflammable gas when heated, and this process may be repeated till the whole charcoal disappears. This is owing to the presence of a little moisture which the charcoal imbibes from the air. The water is decomposed when the charcoal is heated and converted into carbonic acid and inflammable gas. When vegetable substances are heated in a retort, the water which they contain undergoes a similar decomposition, the carbon which forms one of their constituents combines with the oxygen and produces carbonic acid, while the hydrogen, the other constituent of the water, flies off in the state of gas combined with a certain quantity of carbon. Hence the substances obtained when vegetable or animal substances are distilled did not exist ready formed in the body operated on ; but proceeded from the double decompositions which took place by the mutual action of the constituents of the water, sugar, mucus, &c., which the vegetable body contains. The oil, the acid, &c., extracted by distilling vegetable bodies did not exist in them, but are formed during the mutual action of the constituents upon each other, promoted as their action is by the heat. These views were quite new and perfectly just, and threw a new light on the nature of vegetable substances and on the products obtained by distilling them. It showed the futility of all the pretended analyses of vegetable substances, which chemists had performed by simply subjecting them to distillation, and the error of drawing any conclu-

sions respecting the constituents of vegetable substances from the results of their distillation, except
indeed with respect to their elementary constituents.
Thus when by distilling a vegetable substance we
obtain water, oil, acetic acid, carbonic acid, and carburetted hydrogen, we must not conclude that these
principles existed in the substance, but merely that
it contained carbon, hydrogen, and oxygen, in such
proportions as to yield all these principles by decompositions.

As nitric acid acts upon metals in a very different
way from sulphuric and muriatic acids, and as it is
a much better solvent of metals in general than any
other, it was an object of great importance towards
completing the antiphlogistic theory to obtain an accurate knowledge of its constituents. Though Lavoisier did not succeed in this, yet he made at least a
certain progress, which enabled him to explain the
phenomena, at that time known, with considerable
clearness, and to answer all the objections to the antiphlogistic theory from the action of nitric acid on
metals. His first paper on the subject was published
in the Memoirs of the Academy, for 1776 He put
a quantity of nitric acid and mercury into a retort
with a long beak, which he plunged into the water-
trough. An effervescence took place and gas passed
over in abundance, and was collected in a glass jar;
the mercury being dissolved the retort was still further heated, till every thing liquid passed over into the
receiver, and a dry yellow salt remained. The beak of
the retort was now again plunged into the water-
trough, and the salt heated till all the nitric acid
which it contained was decomposed, and nothing remained in the retort but red oxide of mercury. During this last process much more gas was collected.
All the gas obtained during the solution of the mercury and the decomposition of the salt was nitrous

gas. The red oxide of mercury was now heated to redness, oxygen gas was emitted in abundance, and the mercury was reduced to the metallic state : its weight was found the very same as at first. It is clear, therefore, that the nitrous gas and the oxygen gas were derived, not from the mercury but from the nitric acid, and that the nitric acid had been decomposed into nitrous gas and oxygen: the nitrous gas had made its escape in the form of gas, and the oxygen had remained united to the metal.

From these experiments it follows clearly, that nitric acid is a compound of nitrous gas and oxygen. The nature of nitrous gas itself Lavoisier did not succeed in ascertaining. It passed with him for a simple substance; but what he did ascertain enabled him to explain the action of nitric acid on metals. When nitric acid is poured upon a metal which it is capable of dissolving, copper for example, or mercury, the oxygen of the acid unites to the metal, and converts into an oxide, while the nitrous gas, the other constituent of the acid, makes its escape in the gaseous form. The oxide combines with and is dissolved by another portion of the acid which escapes decomposition.

It was discovered by Dr. Priestley, that when nitrous gas and oxygen gas are mixed together in certain proportions, they instantly unite, and are converted into nitrous acid. If this mixture be made over water, the volume of the gases is instantly diminished, because the nitrous acid formed loses its elasticity, and is absorbed by the water. When nitrous gas is mixed with air containing oxygen gas, the diminution of volume after mixture is greater the more oxygen gas is present in the air. This induced Dr. Priestley to employ nitrous gas as a test of the purity of common air. He mixed together equal volumes of the nitrous gas and air to be examined, and he judged of the purity of the air by the degree of condensation : the greater the diminution of bulk, the greater did he consider the proportion of oxygen in the air under examination to be.   This method of proceeding was immediately adopted by chemists and physicians ; but there was a want of uniformity in the mode of proceeding, and a considerable diversity in the results.   M. Lavoisier endeavoured to improve the process, in a paper inserted in the Memoirs of the Academy, for 1782 ; but his method did not answer the purpose intended : it was Mr. Cavendish that first pointed out an accurate mode of testing air by means of nitrous gas, and who showed that the proportions of oxygen and azotic gas in common air are invariable.

Lavoisier, in the course of his investigations, had proved that carbonic acid is a compound of carbon and oxygen ; sulphuric acid, of sulphur and oxygen ; phosphoric acid, of phosphorus and oxygen ; and nitric acid, of nitrous gas and oxygen.   Neither the carbon, the sulphur, the phosphorus, nor the nitrous gas, possessed any acid properties when uncombined ; but they acquired these properties when they were united to oxygen.   He observed further, that all the acids known in his time which had been decomposed were found to contain oxygen, and when they were deprived of oxygen, they lost their acid properties.   These facts led him to conclude, that oxygen is an essential constituent in all acids, and that it is the principle which bestows acidity or the true acidifying principle.   This was the reason why he distinguished it by the name of oxygen.*   These views were fully developed by Lavoisier, in a paper inserted in the Memoirs of the Academy, for 1778,

---

* From ὀξύς, sour, and γίνομαι, which he defined the *producer of acids*, the *acidifying principle*.

entitled, " General Considerations on the Nature of Acids, and on the Principles of which they are composed." When this paper was published, Lavoisier's views were exceedingly plausible. They were gradually adopted by chemists in general, and for a number of years may be considered to have constituted a part of the generally-received doctrines. But the discovery of the nature of chlorine, and the subsequent facts brought to light respecting iodine, bromine, and cyanogen, have demonstrated that it is inaccurate; that many powerful acids exist which contain no oxygen, and that there is no one substance to which the name of acidifying principle can with justice be given. To this subject we shall again revert, when we come to treat of the more modern discoveries.

Long as the account is which we have given of the labours of Lavoisier, the subject is not yet exhausted. Two other papers of his remain to be noticed, which throw considerable light on some important functions of the living body : we allude to his experiments on *respiration* and *perspiration*.

It was known, that if an animal was confined beyond a certain limited time in a given volume of atmospherical air, it died of suffocation, in consequence of the air becoming unfit for breathing; and that if another animal was put into this air, thus rendered noxious by breathing, its life was destroyed almost in an instant. Dr. Priestley had thrown some light upon this subject by showing that air, in which an animal had breathed for some time, possessed the property of rendering lime-water turbid, and therefore contained carbonic acid gas. He considered the process of breathing as exactly analogous to the calcination of metals, or the combustion of burning bodies. Both, in his opinion acted by giving out phlogiston; which, uniting with the air of the atmosphere, converted it into phlo-gisticated air. Priestley found, that if plants were made to vegetate for some time in air that had been rendered unfit for supporting animal life by respira-tion, it lost the property of extinguishing a candle, and animals could breathe it again without injury. He concluded from this that animals, by breathing, phlogisticated air, but that plants, by vegetating, de-phlogisticated air : the former communicated phlo-giston to it, the latter took phlogiston from it.

After Lavoisier had satisfied himself that air is a mixture of oxygen and azote, and that oxygen alone is concerned in the processes of calcination and combustion, being absorbed and combined with the substances undergoing calcination and combustion, it was impossible for him to avoid drawing similar conclusions with respect to the breathing of animals. Accordingly, he made experiments on the subject, and the result was published in the Memoirs of the Academy, for 1777. From these experiments he drew the following conclusions :

1. The only portion of atmospherical air which is useful in breathing is the oxygen. The azote is drawn into the lungs along with the oxygen, but it is thrown out again unaltered.

2. The oxygen gas, on the contrary, is gradually, by breathing, converted into carbonic acid; and air becomes unfit for respiration when a certain portion of its oxygen is converted into carbonic acid gas.

3. Respiration is therefore exactly analogous to calcination. When air is rendered unfit for sup-porting life by respiration, if the carbonic acid gas formed be withdrawn by means of lime-water, or caustic alkali, the azote remaining is precisely the same, in its nature, as what remains after air is ex-hausted of its oxygen by being employed for cal-cining metals.

In this first paper Lavoisier went no further than establishing these general principles; but he afterwards made experiments to determine the exact amount of the changes which were produced in air by breathing, and endeavoured to establish an accurate theory of respiration. To this subject we shall have occasion to revert again, when we give an account of the attempts made to determine the phenomena of respiration by more modern experimenters.

Lavoisier's experiments on *perspiration* were made during the frenzy of the French revolution, when Robespierre had usurped the supreme power, and when it was the object of those at the head of affairs to destroy all the marks of civilization and science which remained in the country. His experiments were scarcely completed when he was thrown into prison, and though he requested a prolongation of his life for a short time, till he could have the means of drawing up a statement of their results, the request was barbarously refused. He has therefore left no account of them whatever behind him. But Seguin, who was associated with him in making these experiments, was fortunately overlooked, and escaped the dreadful times of the reign of terror: he afterwards drew up an account of the results, which has prevented them from being wholly lost to chemists and physiologists.

Seguin was usually the person experimented on. A varnished silk bag, perfectly air-tight, was procured, within which he was enclosed, except a slit over against the mouth, which was left open for breathing; and the edges of the bag were accurately cemented round the mouth, by means of a mixture of turpentine and pitch. Thus every thing emitted by the body was retained in the bag, except what made its escape from the lungs by respiration. By weighing himself in a delicate balance at the commencement of the experiment, and again after he had continued for some time in the bag, the quantity of matter carried off by respiration was determined. By weighing himself without this varnished covering, and repeating the operation after the same interval of time had elapsed, as in the former experiment, he determined the loss of weight occasioned by *perspiration* and *respiration* together. The loss of weight indicated by the first experiment being subtracted from that given by the second, the quantity of matter lost by *perspiration* through the pores of the skin was determined. The following facts were ascertained by these experiments:

1. The maximum of matter perspired in a minute amounted to 26·25 grains troy; the minimum to nine grains; which gives 17·63 grains, at a medium, in the minute, or 52·89 ounces in twenty-four hours.

2. The amount of perspiration is increased by drink, but not by solid food.

3. Perspiration is at its minimum immediately after a repast; it reaches its maximum during digestion.

Such is an epitome of the chemical labours of M. Lavoisier. When we consider that this prodigious number of experiments and memoirs were all performed and drawn up within the short period of twenty years, we shall be able to form some idea of the almost incredible activity of this extraordinary man: the steadiness with which he kept his own peculiar opinions in view, and the good temper which he knew how to maintain in all his publications, though his opinions were not only not supported, but actually opposed by the whole body of chemists in existence, does him infinite credit, and was undoubtedly the wisest line of conduct which he could possibly have adopted. The difficulties connected with the evolution and absorption of hydrogen, con-

stituted the stronghold of the phlogistians.   But
Mr. Cavendish's discovery, that water is a compound
of oxygen and hydrogen, was a death-blow to the
doctrine of Stahl.   Soon after this discovery was
fully established, or during the year 1785, M. Berthollet, a member of the academy, and fast rising
to the eminence which he afterwards acquired, declared himself a convert to the Lavoisierian theory.
His example was immediately followed by M. Fourcroy, also a member of the academy, who had succeeded Macquer as professor of chemistry in the
Jardin du Roi.

M. Fourcroy, who was perfectly aware of the
strong feeling of patriotism which, at that time,
actuated almost every man of science in France, hit
upon a most infallible way of giving currency to the
new opinions.   To the theory of Lavoisier he gave
the name of *La Chimie Française* (French Chemistry).   This name was not much relished by Lavoisier,
as, in his opinion, it deprived him of the credit which
was his due; but it certainly contributed, more than
any thing else, to give the new opinions currency, at
least, in France; they became at once a national
concern, and those who still adhered to the old
opinions, were hooted and stigmatized as enemies to
the glory of their country.   One of the most eminent
of those who still adhered to the phlogistic theory
was M. Guyton de Morveau, a nobleman of Burgundy, who had been educated as a lawyer, and
who filled a conspicuous situation in the Parliament
of Dijon: he had cultivated chemistry with great
zeal, and was at that time the editor of the chemical
part of the Encyclopédie Méthodique.   In the first
half-volume of the chemical part of this dictionary,
which had just appeared, Morveau had supported
the doctrine of phlogiston, and opposed the opinions
of Lavoisier with much zeal and considerable skill:

on this account, it became an object of considerable consequence to satisfy Morveau that his opinions were inaccurate, and to make him a convert to the antiphlogistic theory ; for the whole matter was managed as if it had been a political intrigue, rather than a philosophical inquiry.

Morveau was accordingly invited to Paris, and Lavoisier succeeded without difficulty in bringing him over to his own opinions. We are ignorant of the means which he took ; no doubt friendly discussion and the repetition of the requisite experiments, would be sufficient to satisfy a man so well acquainted with the subject, and whose mode of thinking was so liberal as Morveau. Into the middle of the second half-volume of the chemical part of the Encyclopédie Méthodique he introduced a long advertisement, announcing this change in his opinions, and assigning his reasons for it.

The chemical nomenclature at that time in use had originated with the medical chemists, and contained a multiplicity of unwieldy and unmeaning, and even absurd terms. It had answered the purposes of chemists tolerably well while the science was in its infancy ; but the number of new substances brought into view had of late years become so great, that the old names could not be applied to them without the utmost straining : and the chemical terms in use were so little systematic that it required a considerable stretch of memory to retain them. These evils were generally acknowledged and lamented, and various attempts had been made to correct them. Bergman, for instance, had contrived a new nomenclature, confined chiefly to the salts and adapted to the Latin language. Dr. Black had done the same thing : his nomenclature possessed both elegance and neatness, and was, in several respects, superior to the terms ultimately

K 2

adopted; but with his usual indolence and disregard of reputation, he satisfied himself merely with drawing it up in the form of a table and exhibiting it to his class. Morveau contrived a new nomenclature of the salts, and published it in 1783; and it appears to have been seen and approved of by Bergman.

The old chemical phraseology as far as it had any meaning was entirely conformable to the phlogistic theory. This was so much the case that it was with difficulty that Lavoisier was able to render his opinions intelligible by means of it. Indeed it would have been out of his power to have conveyed his meaning to his readers, had he not invented and employed a certain number of new terms. Lavoisier, aware of the defects of the chemical nomenclature, and sensible of the advantage which his own doctrine would acquire when dressed up in a language exactly suited to his views, was easily prevailed upon by Morveau to join with him in forming a new nomenclature to be henceforth employed exclusively by the antiphlogistians, as they called themselves. For this purpose they associated with themselves Berthollet, and Fourcroy. We do not know what part each took in this important undertaking; but, if we are to judge from appearances, the new nomenclature was almost exclusively the work of Lavoisier and Morveau. Lavoisier undoubtedly contrived the general phrases, and the names applied to the simple substances, so far as they were new, because he had employed the greater number of them in his writings before the new nomenclature was concocted. That the mode of naming the salts originated with Morveau is obvious; for it differs but little from the nomenclature of the salts published by him four years before.

The new nomenclature was published by Lavoisier and his associates in 1787, and it was ever after employed by them in all their writings. Aware of the importance of having a periodical work in which they could register and make known their opinions, they established the *Annales de Chimie*, as a sort of counterpoise to the *Journal de Physique*, the editor of which, M. Delametherie, continued a zealous votary of phlogiston to the end of his life. This new nomenclature very soon made its way into every part of Europe, and became the common language of chemists, in spite of the prejudices entertained against it, and the opposition which it every where met with. In the year 1796, or nine years after the appearance of the new nomenclature, when I attended the chemistry-class in the College of Edinburgh, it was not only in common use among the students, but was employed by Dr. Black, the professor of chemistry, himself; and I have no doubt that he had introduced it into his lectures several years before. This extraordinary rapidity with which the new chemical language came into use, was doubtless owing to two circumstances. First, the very defective, vague, and barbarous state of the old chemical nomenclature : for although, in consequence of the prodigious progress which the science of chemistry has made since the time of Lavoisier, his nomenclature is now nearly as inadequate to express our ideas as that of Stahl was to express his ; yet, at the time of its appearance, its superiority over the old nomenclature was so great, that it was immediately felt and acknowledged by all those who were acquiring the science, who are the most likely to be free from prejudices, and who, in the course of a few years, must constitute the great body of those who are interested in the science. 2. The second circumstance, to which the rapid triumph of the new nomenclature was owing, is the superiority of

Lavoisier's theory over that of Stahl. The subsequent progress of the science has betrayed many weak points in Lavoisier's opinions; yet its superiority over that of Stahl was so obvious, and the mode of interrogating nature introduced by him was so good, and so well calculated to advance the science, that no unprejudiced person, who was at sufficient pains to examine both, could hesitate about preferring that of Lavoisier. It was therefore generally embraced by all the young chemists in every country; and they became, at the same time, partial to the new nomenclature, by which only that theory could be explained in an intelligible manner.

When the new nomenclature was published, there were only three nations in Europe who could be considered as holding a distinguished place as cultivators of chemistry : France, Germany, and Great Britain. For Sweden had just lost her two great chemists, Bergman and Scheele, and had been obliged, in consequence, to descend from the high chemical rank which she had formerly occupied. In France the fashion, and of course almost the whole nation, were on the side of the new chemistry. Macquer, who had been a stanch phlogistian to the last, was just dead. Monnet was closing his laborious career. Baumé continued to adhere to the old opinions; but he was old, and his chemical skill, which had never been *accurate*, was totally eclipsed by the more elaborate researches of Lavoisier and his friends. Delametherie was a keen phlogistian, a man of some abilities, of remarkable honesty and integrity, and editor of the Journal de Physique, at that time a popular and widely-circulating scientific journal. But his habits, disposition, and conduct, were by no means suited to the taste of his countrymen, or conformable to the practice of his contemporaries. The consequence was, that he was shut out of all the scientific coteries of Paris; and that his opinions, however strongly, or rather violently expressed, failed to produce the intended effect. Indeed, as his views were generally inaccurate, and expressed without any regard to the rules of good manners, they in all probability rather served to promote than to injure the cause of his opponents. Lavoisier and his friends appear to have considered the subject in this light: they never answered any of his attacks, or indeed took any notice of them. France, then, from the date of the publication of the new nomenclature, might be considered as enlisted on the side of the antiphlogistic theory.

The case was very different in Germany. The national prejudices of the Germans were naturally enlisted on the side of Stahl, who was their countryman, and whose reputation would be materially injured by the refutation of his theory. The cause of phlogiston, accordingly, was taken up by several German chemists, and supported with a good deal of vigour; and a controversy was carried on for some years in Germany between the old chemists who adhered to the doctrine of Stahl, and the young chemists who had embraced the theory of Lavoisier. Gren, who was at that time the editor of a chemical journal, deservedly held in high estimation, and whose reputation as a chemist stood rather high in Germany, finding it impossible to defend the Stahlian theory as it had been originally laid down, introduced a new modification of phlogiston, and attempted to maintain it against the antiphlogistians. The death of Gren and of Wiegleb, who were the great champions of phlogiston, left the field open to the antiphlogistians, who soon took possession of all the universities and scientific journals in Germany. The most eminent chemist in Germany, or perhaps in Europe at that time, was Martin Henry

Klaproth, professor of chemistry at Berlin, to whom analytical chemistry lies under the greatest obligations. In the year 1792 he proposed to the Academy of Sciences of Berlin of which he was a member, to repeat all the requisite experiments before them, that the members of the academy might be able to determine for themselves which of the two theories deserved the preference. This proposal was acceded to. All the fundamental experiments were repeated by Klaproth with the most scrupulous attention to accuracy : the result was a full conviction, on the part of Klaproth and the academy, that the Lavoisierian theory was the true one. Thus the Berlin Academy became antiphlogistians in 1792 : and as Berlin has always been the focus of chemistry in Germany, the determination of such a learned body must have had a powerful effect in accelerating the propagation of the new theory through that vast country.

In Great Britain the investigation of gaseous bodies, to which the new doctrines were owing, had originated. Dr. Black had begun the inquiry—Mr. Cavendish had prosecuted it with unparalleled accuracy—and Dr. Priestley had made known a great number of new gaseous bodies, which had hitherto escaped the attention of chemists. As the British chemists had contributed more than those of any other nation to the production of the new facts on which Lavoisier's theory was founded, it was natural to expect that they would have embraced that theory more readily than the chemists of any other nation : but the matter of fact was somewhat different. Dr. Black, indeed, with his characteristic candour, speedily embraced the opinions, and even adopted the new nomenclature : but Mr. Cavendish new modelled the phlogistic theory, and published a defence of phlogiston, which it was impossible at that time to refute. The French chemists had the good sense not to attempt to overturn it. Mr. Cavendish after this laid aside the cultivation of chemistry altogether, and never acknowledged himself a convert to the new doctrines.

Dr. Priestley continued a zealous advocate for phlogiston till the very last, and published what he called a refutation of the antiphlogistic theory about the beginning of the present century: but Dr. Priestley, notwithstanding his merit as a discoverer and a man of genius, was never, strictly speaking, entitled to the name of chemist; as he was never able to make a chemical analysis. In his famous experiments, for example, on the composition of water, he was obliged to procure the assistance of Mr. Keir to determine the nature of the blue-coloured liquid which he had obtained, and which Mr. Keir showed to be nitrate of copper. Besides, Dr. Priestley, though perfectly honest and candid, was so hasty in his decisions, and so apt to form his opinions without duly considering the subject, that his chemical theories are almost all erroneous and sometimes quite absurd.

Mr. Kirwan, who had acquired a high reputation, partly by his *mineralogy*, and partly by his experiments on the composition of the salts, undertook the task of refuting the antiphlogistic theory, and with that view published a work to which he gave the name of " An Essay on Phlogiston and the Composition of Acids." In that book he maintained an opinion which seems to have been pretty generally adopted by the most eminent chemists of the time; namely, that phlogiston is the same thing with what is at present called *hydrogen*, and which, when Kirwan wrote, was called light *inflammable air*. Of course Mr. Kirwan undertook to prove that every combustible substance and every

metal contains hydrogen as a constituent, and that
hydrogen escapes in every case of combustion and
calcination. On the other hand, when calces are reduced to the metallic state hydrogen is absorbed.
The book was divided into thirteen sections. In the
first the specific gravity of the gases was stated according to the best data then existing. The second
section treats of the composition of acids, and the
composition and decomposition of water. The
third section treats of sulphuric acid; the fourth, of
nitric acid; the fifth, of muriatic acid; the sixth, of
aqua regia; the seventh, of phosphoric acid; the eighth,
of oxalic acid; the ninth, of the calcination and reduction of metals and the formation of fixed air; the tenth,
of the dissolution of metals; the eleventh, of the precipitation of metals by each other; the twelfth, of the
properties of iron and steel; while the thirteenth sums
up the whole argument by way of conclusion.

In this work Mr. Kirwan admitted the truth of
M. Lavoisier's theory, that during combustion and
calcination, oxygen united with the burning and
calcining body. He admitted also that water is a
compound of oxygen and hydrogen. Now these
admissions, which, however, it was scarcely possible
for a man of candour to refuse, rendered the whole
of his arguments in favour of the identity of hydrogen and phlogiston, and of the existence of hydrogen
in all combustible bodies, exceedingly inconclusive.
Kirwan's book was laid hold of by the French
chemists, as affording them an excellent opportunity
of showing the superiority of the new opinions over
the old. Kirwan's view of the subject was that which
had been taken by Bergman and Scheele, and indeed by every chemist of eminence who still adhered
to the phlogistic system. A satisfactory refutation
of it, therefore, would be a death-blow to phlogiston
and would place the antiphlogistic theory upon a

basis so secure that it would be henceforth impossible to shake it.

Kirwan's work on phlogiston was accordingly translated into French, and published in Paris. At the end of each section was placed an examination and refutation of the argument contained in it by some one of the French chemists, who had now associated themselves in order to support the antiphlogistic theory. The introduction, together with the second, third, and eleventh sections were examined and refuted by M. Lavoisier; the fourth, the fifth, and sixth sections fell to the share of M. Berthollet; the seventh and thirteenth sections were undertaken by M. de Morveau; the eighth, ninth, and tenth, by M. De Fourcroy; while the twelfth section, on iron and steel was animadverted on by M. Monge. These refutations were conducted with so much urbanity of manner, and were at the same time so complete, that they produced all the effects expected from them. Mr. Kirwan, with a degree of candour and liberality of which, unfortunately, very few examples can be produced, renounced his old opinions, abandoned phlogiston, and adopted the antiphlogistic doctrines of his opponents. But his advanced age, and a different mode of experimenting from what he had been accustomed to, induced him to withdraw himself entirely from experimental science and to devote the evening of his life to metaphysical and logical and moral investigations.

Thus, soon after the year 1790, a kind of interregnum took place in British chemistry. Almost all the old British chemists had relinquished the science, or been driven out of the field by the superior prowess of their antagonists. Dr. Austin and Dr. Pearson will, perhaps, be pointed out as exceptions. They undoubtedly contributed somewhat to the progress of the science. But they were arranged on the side of the antiphlogistians. Dr. Crawford, who had done so much for the theory of heat, was about this time ruined in his circumstances by the bankruptcy of a house to which he had intrusted his property. This circumstance preyed upon a mind which had a natural tendency to morbid sensibility, and induced this amiable and excellent man to put an end to his existence. Dr. Higgins had acquired some celebrity as an experimenter and teacher ; but his disputes with Dr. Priestley, and his laying claim to discoveries which certainly did not belong to him, had injured his reputation, and led him to desert the field of science. Dr. Black was an invalid, Mr. Cavendish had renounced the cultivation of chemistry, and Dr. Priestley had been obliged to escape from the iron hand of theological and political bigotry, by leaving the country. He did little as an experimenter after he went to America; and, perhaps, had he remained in England, his reputation would rather have diminished than increased. He was an admirable pioneer, and as such, contributed more than any one to the revolution which chemistry underwent; though he was himself utterly unable to rear a permanent structure capable, like the Newtonian theory, of withstanding all manner of attacks, and becoming only the firmer and stronger the more it is examined. Mr. Keir, of Birmingham, was a man of great eloquence, and possessed of all the chemical knowledge which characterized the votaries of phlogiston. In the year 1789 he attempted to stem the current of the new opinions by publishing a dictionary of chemistry, in which all the controversial points were to be fully discussed, and the antiphlogistic theory examined and refuted. Of this dictionary only one part appeared, constituting a very thin volume of two hundred and eight quarto pages, and treating almost entirely of *acids*.

Finding that the sale of this work did not answer his expectations, and probably feeling, as he proceeded, that the task of refuting the antiphlogistic opinions was much more difficult, and much more hopeless than he expected, he renounced the undertaking, and abandoned altogether the pursuit of chemistry.

It will be proper in this place to introduce some account of the most eminent of those French chemists who embraced the theory of Lavoisier, and assisted him in establishing his opinions.

Claude-Louis Berthollet was born at Talloire, near Annecy, in Savoy, on the 9th of December, 1748. He finished his school education at Chambéry, and afterwards studied at the College of Turin, a celebrated establishment, where many men of great scientific celebrity have been educated. Here he attached himself to medicine, and after obtaining a degree he repaired to Paris, which was destined to be the future theatre of his speculations and pursuits.

In Paris he had not a single acquaintance, nor did he bring with him a single introductory letter; but understanding that M. Tronchin, at that time a distinguished medical practitioner in Paris, was a native of Geneva, he thought he might consider him as in some measure a countryman. On this slender ground he waited on M. Tronchin, and what is rather surprising, and reflects great credit on both, this acquaintance, begun in so uncommon a way, soon ripened into friendship. Tronchin interested himself for his young *protégée*, and soon got him into the situation of physician in ordinary to the Duke of Orleans, father of him who cut so conspicuous a figure in the French revolution, under the name of M. Égalité. In this situation he devoted himself to the study of chemistry, and soon made himself known by his publications on the subject.

In 1781 he was elected a member of the Academy of Sciences of Paris : one of his competitors was M. Fourcroy. No doubt Berthollet owed his election to the influence of the Duke of Orleans. In the year 1784 he was again a competitor with M. de Fourcroy for the chemical chair at the Jardin du Roi, left vacant by the death of Macquer. The chair was in the gift of M. Buffon, whose vanity is said to have been piqued because the Duke of Orleans, who supported Berthollet's interest, did not pay him sufficient court. This induced him to give the chair to Fourcroy; and the choice was a fortunate one, as his uncommon vivacity and rapid elocution particularly fitted him for addressing a Parisian audience. The chemistry-class at the Jardin du Roi immediately became celebrated, and attracted immense crowds of admiring auditors.

But the influence of the Duke of Orleans was sufficient to procure for Berthollet another situation which Macquer had held. This was government commissary and superintendent of the dyeing processes. It was this situation which naturally turned his attention to the phenomena of dyeing, and occasioned afterwards his book on dyeing; which at the time of its publication was excellent, and exhibited a much better theory of dyeing, and a better account of the practical part of the art than any work which had previously appeared. The arts of dyeing and calico-printing have been very much improved since the time that Berthollet's book was written; yet if we except Bancroft's work on the permanent colours, nothing very important has been published on the subject since that period. We are at present almost as much in want of a good work on dyeing as we were when Berthollet's book appeared.

In the year 1785 Berthollet, at a meeting of the Academy of Sciences, informed that learned body that he had become a convert to the antiphlogistic doctrines of Lavoisier. There was one point, however, upon which he entertained a different opinion from Lavoisier, and this difference of opinion continued to the last. Berthollet did not consider oxygen as the acidifying principle. On the contrary, he was of opinion that acids existed which contained no oxygen whatever. As an example, he mentioned sulphuretted hydrogen, which possessed the properties of an acid, reddening vegetable blues, and combining with and neutralizing bases, and yet it was a compound of sulphur and hydrogen, and contained no oxygen whatever. It is now admitted that Berthollet was accurate in his opinion, and that oxygen is not of itself an acidifying principle.

Berthollet continued in the uninterrupted prosecution of his studies, and had raised himself a very high reputation when the French revolution burst upon the world in all its magnificence. It is not our business here to enter into any historical details, but merely to remind the reader that all the great powers of Europe combined to attack France, assisted by a formidable army of French emigrants assembled at Coblentz. The Austrian and Prussian armies hemmed her in by land, while the British fleets surrounded her by sea, and thus shut her out from all communication with other nations. Thus France was thrown at once upon her own resources. She had been in the habit of importing her saltpetre, and her iron, and many other necessary implements of war: these supplies were suddenly withdrawn; and it was expected that France, thus deprived of all her resources, would be obliged to submit to any terms imposed upon her by her adversaries. At this time she summoned her men of science to her assistance, and the call was speedily answered. Berthollet and Monge were particularly active, and saved the French nation from destruction by their activity, intelligence, and zeal. Berthollet traversed France from one extremity to the other; pointed out the mode of extracting saltpetre from the soil, and of purifying it. Saltpetre-works were instantly established in every part of France, and gunpowder made of it in prodigious quantity, and with incredible activity. Berthollet even attempted to manufacture a new species of gunpowder still more powerful than the old, by substituting chlorate of potash for saltpetre : but it was found too formidable a substance to be made with safety.

The demand for cannon, muskets, sabres, &c., was equally urgent and equally difficult to be supplied. A committee of men of science, of which Berthollet and Monge were the leading members, was established, and by them the mode of smelting iron, and of converting it into steel, was instantly communicated, and numerous manufactories of these indispensable articles rose like magic in every part of France.

This was the most important period of the life of of Berthollet. It was in all probability his zeal, activity, sagacity, and honesty, which saved France from being overrun by foreign troops. But perhaps the moral conduct of Berthollet was not less conspicuous than his other qualities. During the reign of terror, a short time before the 9th Thermidor, when it was the system to raise up pretended plots, to give pretexts for putting to death those that were obnoxious to Robespierre and his friends, a hasty notice was given at a sitting of the Committee of Public Safety, that a conspiracy had just been discovered to destroy the soldiers, by poisoning the brandy which was just going to be served out to them previous to an engagement. It was said that the sick in the hospitals who had tasted this brandy, all perished in consequence of it. Immediate orders were issued to arrest those previously marked for execution. A quantity of the brandy was sent to Berthollet to be examined. He was informed, at the same time, that Robespierre wanted a conspiracy to be established, and all knew that opposition to his will was certain destruction. Having finished his analysis, Berthollet drew up his results in a Report, which he accompanied with a written explanation of his views; and he there stated, in the plainest language, that nothing poisonous was mixed with the brandy, but that it had been diluted with water holding small particles of slate in suspension, an ingredient which filtration would remove. This report deranged the plans of the Committee of Public Safety. They sent for the author, to convince him of the inaccuracy of his analysis, and to persuade him to alter its results. Finding that he remained unshaken in his opinion, Robespierre exclaimed, " What, Sir ! darest thou affirm that the muddy brandy is free from poison?" Berthollet immediately filtered a glass of it in his presence, and drank it off. " Thou art daring, Sir, to drink that liquor," exclaimed the ferocious president of the committee. " I dared much more," replied Berthollet, " when I signed my name to that Report." There can be no doubt that he would have paid the penalty of this undaunted honesty with his life, but that fortunately the Committee of Public Safety could not at that time dispense with his services.

In the year 1792 Berthollet was named one of the commissioners of the Mint, into the processes of which he introduced considerable improvements. In 1794 he was appointed a member of the Commission of Agriculture and the Arts : and in the course of the same year he was chosen professor of chemistry at the Polytechnic School and also in the Normal School. But his turn of mind did not fit him for a public teacher. He expected too much information to be possessed by his hearers, and did not, therefore, dwell sufficiently upon the elementary details. His pupils were not able to follow his metaphysical disquisitions on subjects totally new to them ; hence, instead of inspiring them with a love for chemistry, he filled them with langour and disgust.

In 1795, at the organization of the Institute, which was intended to include all men of talent or celebrity in France, we find Berthollet taking a most active lead ; and the records of the Institute afford abundant evidence of the perseverance and assiduity with which he laboured for its interests. Of the committees to which all original memoirs are in the first place referred, we find Berthollet, oftener than any other person, a member, and his signature to the report of each work stands generally first.

In the year 1796, after the subjugation of Italy by Bonaparte, Berthollet and Monge were selected by the Directory to proceed to that country, in order to select those works of science and art with which the Louvre was to be filled and adorned. While engaged in the prosecution of that duty, they became acquainted with the victorious general. He easily saw the importance of their friendship, and therefore cultivated it with care ; and was happy afterwards to possess them, along with nearly a hundred other philosophers, as his companions in his celebrated expedition to Egypt, expecting no doubt an eclat from such a halo of surrounding science, as might favour the development of his schemes of future greatness. On this expedition, which promised so favourably for the French nation, and which was intended to inflict a mortal stab upon the commercial greatness of Great Britain, Bonaparte set out in the year 1798, accompanied by a crowd of the most eminent men of science that France could boast of. That they might co-operate more effectually in the cause of knowledge, these gentlemen formed themselves into a society, named " The Institute of Egypt," which was constituted on the same plan as the National Institute of France. Their first meeting was on the 6th Fructidor (24th of August, 1798; and after that they continued to assemble, at stated intervals. At these meetings papers were read, by the respective members, on the climate, the inhabitants, and the natural and artificial productions of the country to which they had gone. These memoirs were published in 1800, in Paris, in a single volume entitled, " Memoirs of the Institute of Egypt."

The history of the Institute of Egypt, as related by Cuvier, is not a little singular, and deserves to be stated. Bonaparte, during his occasional intercourse with Berthollet in Italy, was delighted with the simplicity of his manners, joined to a force and depth of thinking which he soon perceived to characterize our chemist. When he returned to Paris, where he enjoyed some months of comparative leisure, he resolved to employ his spare time in studying chemistry under Berthollet. It was at this period that his illustrious pupil imparted to our philosopher his intended expedition to Egypt, of which no whisper was to be spread abroad till the blow was ready to fall; and he begged of him not merely to accompany the army himself, but to choose such men of talent and experience as he conceived fitted to find there an employment worthy of the country which they visited, and of that which sent them forth. To invite men to a hazardous expedition, the nature and destination of which he was not permitted to disclose, was rather a delicate task; yet Berthollet undertook it. He could simply inform them that he would himself accompany them; yet such was the universal esteem in which he was held, such was the confidence universally placed in his honesty and integrity, that all the men of science agreed at once, and without hesitation, to embark on an unknown expedition, the dangers of which he was to share along with them. Had it not been for the link which Berthollet supplied between the commander-in-chief and the men of science, it would have been impossible to have united, as was done on this occasion, the advancement of knowledge with the progress of the French arms.

During the whole of this expedition, Berthollet and Monge distinguished themselves by their firm friendship, and by their mutually braving every danger to which any of the common soldiers could be exposed. Indeed, so intimate was their association that many of the army conceived Berthollet and Monge to be one individual; and it is no small proof of the intimacy of these philosophers with Bonaparte, that the soldiers had a dislike at this double personage, from a persuasion that it had been at his suggestion that they were led into a country which they detested. It happened on one occasion that a boat, in which Berthollet and some others were conveyed up the Nile, was assailed by a troop of Mamelukes, who poured their small shot into it from the banks. In the midst of this perilous voyage, M. Berthollet began very coolly to pick up stones and stuff his pockets with them. When his motive for this conduct was asked, "I am desirous," said he,

" that in case of my being shot, my body may sink at once to the bottom of this river, and may escape the insults of these barbarians."

In a conjuncture where a courage of a rarer kind was required, Berthollet was not found wanting. The plague broke out in the French army, and this, added to the many fatigues they had previously endured, the diseases under which they were already labouring, would, it was feared, lead to insurrection on the one hand, or totally sink the spirits of the men on the other. Acre had been besieged for many weeks in vain. Bonaparte and his army had been able to accomplish nothing against it : he was anxious to conceal from his army this disastrous intelligence. When the opinion of Berthollet was asked in council, he spoke at once the plain, though unwelcome truth. He was instantly assailed by the most violent reproaches. " In a week," said he, " my opinion will be unfortunately but too well vindicated." It was as he foretold : and when nothing but a hasty retreat could save the wretched remains of the army of Egypt, the carriage of Berthollet was seized for the convenience of some wounded officers. On this, he travelled on foot, and without the smallest discomposure, across twenty leagues of the desert.

When Napoleon abandoned the army of Egypt, and traversed half the Mediterranean in a single vessel, Berthollet was his companion. After he had put himself at the head of the French government, and had acquired an extent of power, which no modern European potentate had ever before realized, he never forgot his associate. He was in the habit of placing all chemical discoveries to his account, to the frequent annoyance of our chemist; and when an unsatisfactory answer was given him upon any scientific subject, he was in the habit of saying,

" Well; I shall ask this of Berthollet." But he did
not limit his affection to these proofs of regard.
Having been informed that Berthollet's earnest pursuits of science had led him into expenses which
had considerably deranged his fortune, he sent for
him, and said, in a tone of affectionate reproach,
" M. Berthollet, I have always one hundred thousand crowns at the service of my friends." And,
in fact, this sum was immediately presented to him.

Upon his return from Egypt, Berthollet was nominated a senator by the first consul; and afterwards received the distinction of grand officer of the
Legion of Honour; grand cross of the Order of
Reunion; titulary of the Senatory of Montpellier;
and, under the emperor, he was created a peer of
France, receiving the title of Count. The advancement
to these offices produced no change in the manners
of Berthollet. Of this he gave a striking proof, by
adopting, as his armorial bearing (at the time that
others eagerly blazoned some exploit), the plain
unadorned figure of his faithful and affectionate
dog. He was no courtier before he received these
honours, and he remained equally simple and un-
assuming, and not less devoted to science after they
were conferred.

As we advance towards the latter period of his
life, we find the same ardent zeal in the cause of
science which had glowed in his early youth, accompanied by the same generous warmth of heart
that he ever possessed, and which displayed itself
in his many intimate friendships still subsisting,
though mellowed by the hand of time. At this
period La Place lived at Arcueil, a small village
about three miles from Paris. Between him and
Berthollet there had long subsisted a warm affection, founded on mutual esteem. To be near this
illustrious man Berthollet purchased a country-seat

in the village : there he established a very complete laboratory, fit for conducting all kinds of experiments in every branch of natural philosophy. Here he collected round him a number of distinguished young men, who knew that in his house their ardour would at once receive fresh impulse and direction from the example of Berthollet. These youthful philosophers were organized by him into a society, to which the name of Société d'Arcueil was given. M. Berthollet was himself the president, and the other members were La Place, Biot, Gay-Lussac, Thenard, Collet-Descotils, Decandolle, Humboldt, and A. B. Berthollet. This society published three volumes of very valuable memoirs. The energy of this society was unfortunately paralyzed by an untoward event, which imbittered the latter days of this amiable man. His only son, M. A. B. Berthollet, in whom his happiness was wrapped up, was unfortunately afflicted with a lowness of spirits which rendered his life wholly insupportable to him. Retiring to a small room, he locked the door, closed up every chink and crevice which might admit the air, carried writing materials to a table, on which he placed a second-watch, and then seated himself before it. He now marked precisely the hour, and lighted a brasier of charcoal beside him. He continued to note down the series of sensations he then experienced in succession, detailing the approach and rapid progress of delirium; until, as time went on, the writing became confused and illegible, and the young victim dropped dead upon the floor.

After this event the spirits of the old man never again rose. Occasionally some discovery, extending the limits of his favourite science, engrossed his interest and attention for a short time : but such intervals were rare, and shortlived. The restoration of the Bourbons, and the downfal of his friend and patron Napoleon, added to his sufferings by diminishing his income, and reducing him from a state of affluence to comparative embarrassment. But he was now old, and the end of his life was approaching. In 1822 he was attacked by a slight fever, which left behind it a number of boils: these were soon followed by a gangrenous ulcer of uncommon size. Under this he suffered for several months with surprising fortitude. He himself, as a physician, knew the extent of his danger, felt the inevitable progress of the malady, and calmly regarded the slow approach of death. At length, after a tedious period of suffering, in which his equanimity had never once been shaken, he died on the 6th of November, when he had nearly completed the seventy-fourth year of his age.

His papers are exceedingly numerous, and of a very miscellaneous nature, amounting to more than eighty. The earlier were chiefly inserted into the various volumes of the Memoirs of the Academy. He furnished many papers to the Annales de Chimie and the Journal de Physique, and was also a frequent contributor to the Society of Arcueil, in the different volumes of whose transactions several memoirs of his are to be found. He was the author likewise of two separate works, comprising each two octavo volumes. These were his Elements of the Art of Dyeing, first published in 1791, in a single volume: but the new and enlarged edition of 1814 was in two volumes; and his Essay on Chemical Statics, published about the beginning of the present century. I shall notice his most important papers.

His earlier memoirs on sulphurous acid, on volatile alkali, and on the decomposition of nitre, were encumbered by the phlogistic theory, which at that time he defended with great zeal, though he afterwards retracted these his first opinions upon all these subjects. Except his paper on soaps, in which he shows that they are chemical compounds of an oil (acting the part of an acid) and an alkaline base, and his proof that phosphoric acid exists ready formed in the body (a fact long before demonstrated by Gahn and Scheele), his papers published before he became an antiphlogistian are of inferior merit.

In 1785 he demonstrated the nature and proportion of the constituents of ammonia, or volatile alkali. This substance had been collected in the gaseous form by the indefatigable Priestley, who had shown also that when electric sparks are made to pass for some time through a given volume of this gas, its bulk is nearly doubled. Berthollet merely repeated this experiment of Priestley, and analyzed the new gases evolved by the action of electricity. This gas he found a mixture of three volumes hydrogen and one volume azotic gas: hence it was evident that ammoniacal gas is a compound of three volumes of hydrogen and one volume of azotic gas united together, and condensed into two volumes. The same discovery was made about the same time by Dr. Austin, and published in the Philosophical Transactions. Both sets of experiments were made without any knowledge of what was done by the other: but it is admitted, on all hands, that Berthollet had the priority in point of time.

It was about this time, likewise, that he published his first paper on chlorine. He observed, that when water, impregnated with chlorine, is exposed to the light of the sun, the water loses its colour, while, at the same time, a quantity of oxygen gas is given out. If we now examine the water, we find that it contains no chlorine, but merely a little muriatic acid. This fact, which is undoubted, led him to conclude that chlorine is decomposed by the action of solar light, and that its two elements are muriatic acid and oxygen. This led to the notion that the basis of muriatic acid is capable of combining with various doses of oxygen, and of forming various acids, one of which is chlorine : on that account it was called *oxygenized muriatic acid* by the French chemists, which unwieldy appellation was afterwards shortened by Kirwan into *oxymuriatic acid.*

Berthollet observed that when a current of chlorine gas is passed through a solution of carbonate of potash an effervescence takes place owing to the disengagement of carbonic acid gas. By-and-by crystals are deposited in fine silky scales, which possess the property of detonating with combustible bodies still more violently than saltpetre. Berthollet examined these crystals and showed that they were compounds of potash with an acid containing much more oxygen than oxymuriatic acid. He considered its basis as muriatic acid, and distinguished it by the name of hyper-oxymuriatic acid.

It was not till the year 1810, that the inaccuracy of these opinions was established. Gay-Lussac and Thenard attempted in vain to extract oxygen from chlorine. They showed that not a trace of that principle could be detected. Next year Davy took up the subject and concluded from his experiments that *chlorine* is a simple substance, that muriatic acid is a compound of chlorine and hydrogen, and hyper-oxymuriatic acid of chlorine and oxygen. Gay-Lussac obtained this acid in a separate state, and gave it the name of *chloric acid,* by which it is now known

Scheele, in his original experiments on chlorine, had noticed the property which it has of destroying vegetable colours. Berthollet examined this pro-

perty with care, and found it so remarkable that he
proposed it as a substitute for exposure to the sun
in bleaching.   This suggestion alone would have
immortalized Berthollet had he done nothing else;
since its effect upon some of the most important of
the manufactures of Great Britain has been scarcely
inferior to that of the steam-engine itself.   Mr. Watt
happened to be in Paris when the idea suggested
itself to Berthollet.   He not only communicated
it to Mr. Watt, but showed him the process in all
its simplicity.   It consisted in nothing else than in
steeping the cloth to be bleached in water impregnated with chlorine gas.   Mr. Watt, on his return
to Great Britain, prepared a quantity of this liquor,
and sent it to his father-in-law, Mr. Macgregor,
who was a bleacher in the neighbourhood of Glasgow.   He employed it successfully, and thus was
the first individual who tried the new process of
bleaching in Great Britain.   For a number of years
the bleachers in Lancashire and the neighbourhood
of Glasgow were occupied in bringing the process to
perfection.   The disagreeable smell of the chlorine
was a a great annoyance.   This was attempted to
be got rid of by dissolving potash in the water to be
impregnated with chlorine ; but it was found to
injure considerably the bleaching powers of the
gas.   The next method tried was to mix the water
with quicklime, and then to pass a current of
chlorine through it.   The quicklime was dissolved,
and the liquor thus constituted was found to answer
very well.   The last improvement was to combine
the chlorine with dry lime.   At first two atoms of
lime were united to one atom of chlorine ; but of
late years it is a compound of one atom of lime,
and one of chlorine.   This chloride is simply dissolved
in water, and the cloth to be bleached is steeped in
it.   For all these improvements, which have brought
the method of bleaching by means of chlorine to

great simplicity and perfection, the bleachers are indebted to Knox, Tennant, and Mackintosh, of Glasgow; by whose indefatigable exertions the mode of manufacturing chloride of lime has been brought to a state of perfection.

Berthollet's experiments on prussic acid and the prussiates deserve also to be mentioned, as having a tendency to rectify some of the ideas at that time entertained by chemists, and to advance their knowledge of one of the most difficult departments of chemical investigation. In consequence of his experiments on the nature and constituents of sulphuretted hydrogen, he had already concluded that it was an acid, and that it was destitute of oxygen : this had induced him to refuse his assent to the hypothesis of Lavoisier, that *oxygen* is the *acidifying principle*. Scheele, in his celebrated experiments on prussic acid, had succeeded in ascertaining that its constituents were carbon and azote ; but he had not been able to make a rigid analysis of that acid, and consequently to demonstrate that oxygen did not enter into it as a constituent. Berthollet took up the subject, and though his analysis was also incomplete, he satisfied himself, and rendered it exceedingly probable, that the only constituents of this acid were, carbon, azote, and hydrogen, and that oxygen did not enter into it as a constituent. This was another reason for rejecting the notion of *oxygen* as an acidifying principle. Here were two acids capable of neutralizing bases, namely, sulphuretted hydrogen and prussic acid, and yet neither of them contained oxygen. He found that when prussic acid was treated with chlorine, its properties were altered ; it acquired a different smell and taste, and no longer precipitated iron blue, but green. From his opinion respecting the nature of chlorine, that it was a compound of muriatic acid and oxygen, he naturally concluded that by this process he had formed a new prussic acid by adding oxygen to the old constituents. He therefore called this new substance *oxyprussic acid*. It has been proved by the more recent experiments of Gay-Lussac, that the new acid of Berthollet is a compound of *cyanogen* (the prussic acid deprived of hydrogen) and *chlorine:* it is now called *chloro-cyanic acid*, and is known to possess the characters assigned it by Berthollet: it constitutes, therefore, a new example of an acid destitute of oxygen. Berthollet was the first person who obtained prussiate of potash in regular crystals; the salt was known long before, but had been always used in a state of solution.

Berthollet's discovery of fulminating silver, and his method of obtaining pure hydrated potash and soda, by means of alcohol, deserve to be mentioned. This last process was of considerable importance to analytical chemistry. Before he published his process, these substances in a state of purity were not known.

I think it unnecessary to enter into any details respecting his experiments on sulphuretted hydrogen, and the hydrosulphurets and sulphurets. They contributed essentially to elucidate that obscure part of chemistry. But his success was not perfect; nor did we understand completely the nature of these compounds, till the nature of the alkaline bases had been explained by the discoveries of Davy.

The only other work of Berthollet, which I think it necessary to notice here, is his book entitled " Chemical Statics," which he published in 1803. He had previously drawn up some interesting papers on the subject, which were published in the Memoirs of the Institute. Though chemical affinity constitutes confessedly the basis of the science, it had been almost completely overlooked by Lavoisier, who had done nothing more on the subject than drawn up some tables of affinity, founded on very imperfect data. Morveau had attempted a more profound investigation of the subject in the article *Affinité*, inserted in the chemical part of the Encyclopédie Méthodique. His object was, in imitation of Buffon, who had preceded him in the same investigation, to prove that chemical affinity is merely a case of the *attraction of gravitation*. But it is beyond our reach, in the present state of our knowledge, to determine the amount of attraction which the atoms of bodies exert with respect to each other. This was seen by Newton, and also by Bergman, who satisfied themselves with considering it as an attraction, without attempting to determine its amount; though Newton, with his usual sagacity, was inclined, from the phenomena of light, to consider the attraction of affinity as much stronger than that of gravitation, or at least as increasing much more rapidly, as the distances between the attracting particles diminished.

Bergman, who had paid great attention to the subject, considered affinity as a certain determinate attraction, which the atoms of different bodies exerted towards each other. This attraction varies in intensity between every two bodies, though it is constant between each pair. The consequence is, that these intensities may be denoted by numbers. Thus, suppose a body m, and the atoms of six other bodies, a, b, c, d, e, f, to have an affinity for m, the forces by which they are attracted towards each other may be represented by the numbers x, x+1, x+2, x+3, x+4, x+5. And the attractions may be represented thus:

Attraction between m & a = x

$\qquad$ m & b = x+1

$\qquad$ m & c = x+2

$\qquad$ m & d = x+3

$\qquad$ m & e = x+4

$\qquad$ m & f = x+5

Suppose we have the compound $m\,a$, if we present b,

it will unite with m and displace a, because the attraction between m and a is only x, while that between m & b is x+1: c will displace b; d will displace c, and so on, for the same reason. On this account Bergman considered affinity as an *elective attraction*, and in his opinion the intensity may always be estimated by decomposition. That substance which displaces another from a third, has a greater affinity than the body which is displaced. If b displace a from the compound $a\,m$, then b has a greater affinity for m than a has.

The object of Berthollet in his Chemical Statics, was to combat this opinion of Bergman, which had been embraced without examination by chemists in general. If affinity be an attraction, Berthollet considered it as evident that it never could occasion decomposition. Suppose a to have an affinity for m, and b to have an affinity for the same substances. Let the affinity between b and m be greater than that between $a\,m$. Let b be mixed with a solution of the compound $a\,m$, then in that case b would unite with $a\,m$, and form the triple compound $a\,m\,b$. Both a and b would at once unite with m. No reason can be assigned why a should separate from m, and b take its place. Berthollet admitted that in fact such decompositions often happened; but he accounted for them from other causes, and not from the superior affinity of one body over another. Suppose we have a solution of *sulphate of soda* in water. This salt is a compound of *sulphuric acid* and *soda*; two substances between which a strong affinity subsists, and which therefore always unites whenever they come in contact. Suppose we have dissolved in another portion of water, a quantity of barytes, just sufficient to saturate the sulphuric acid in the sulphate of soda. If we mix these two solutions together. The barytes will combine with the sulphuric acid and the compound (*sulphate of barytes*) will fall to the bottom, leaving a pure solution of soda in the water. In this case the barytes has seized all the sulphuric acid, and displaced the soda. The reason of this, according to Berthollet, is not that barytes has a stronger affinity for sulphuric acid than soda has; but because sulphate of barytes is insoluble in water. It therefore falls down, and of course the sulphuric acid is withdrawn from the soda. But if we add to a solution of sulphate of soda as much potash as will saturate all the sulphuric acid, no such decomposition will take place; at least, we have no evidence that it does. Both the alkalies, in this case, will unite to the acid and form a triple compound, consisting of potash, sulphuric acid, and soda. Let us now concentrate the solution by evaporation, and crystals of sulphate of potash will fall down. The reason is, that sulphate of potash is not nearly so soluble in water as sulphate of soda. Hence it separates; not because sulphuric acid has a greater affinity for potash than for soda, but because sulphate of potash is a much less soluble salt than sulphate of soda.

This mode of reasoning of Berthollet is plausible, but not convincing: it is merely an *argumentum ad ignorantiam.* We can only prove the decomposition by separating the salts from each other, and this can only be done by their difference of solubility. But cases occur in which we can judge that decomposition has taken place from some other phenomena than precipitation. For example, *nitrate of copper* is a *blue* salt, while *muriate of copper* is *green*. If into a solution of nitrate of copper we pour muriatic acid, no precipitation appears, but the colour changes from blue to green. Is not this an evidence that the muriatic acid has displaced the nitric, and that the salt held in solution is not nitrate of copper, as it was at first, but muriate of copper?

Berthollet accounts for all decompositions which take place when a third body is added, either by insolubility or by *elasticity*: as, for example, when sulphuric acid is poured into a solution of carbonate of ammonia, the carbonic acid all flies off, in consequence of its elasticity, and the sulphuric acid combines with the ammonia in its place. I confess that this explanation, of the reason why the carbonic acid flies off, appears to me very defective. The ammonia and carbonic acid are united by a force quite sufficient to overcome the elasticity of the carbonic acid. Accordingly, it exhibits no tendency to escape. Now, why should the elasticity of the acid cause it to escape when sulphuric acid is added? It certainly could not do so, unless it has weakened the affinity by which it is kept united to the ammonia. Now this is the very point for which Bergman contends. The subject will claim our attention afterwards, when we come to the electro-chemical discoveries, which distinguished the first ten years of the present century.

Another opinion supported by Berthollet in his Chemical Statics is, that quantity may be made to overcome force; or, in other words, that if we mix a great quantity of a substance which has a weaker affinity with a small quantity of a substance which has a stronger affinity, the body having the weaker affinity will be able to overcome the other, and combine with a third body in place of it. He gave a number of instances of this; particularly, he showed that a large quantity of potash, when mixed with a small quantity of sulphate of barytes, is able to deprive the barytes of a portion of its sulphuric acid. In this way he accounted for the decomposition of the common salt, by carbonate of lime in the soda lakes in Egypt; and the decomposition of the same salt by iron, as noticed by Scheele.

I must acknowledge myself not quite satisfied with Berthollet's reasoning on this subject. No doubt if two atoms of a body having a weaker affinity, and one atom of a body having a stronger affinity, were placed at equal distances from an atom of a third body, the force of the two atoms might overcome that of the one atom. And it is possible that such cases may occasionally occur : but such a balance of distances must be rare and accidental. I cannot but think that all the cases adduced by Berthollet are of a complicated nature, and admit of an explanation independent of the efficacy of mass. And at any rate, abundance of instances might be stated, in which mass appears to have no preponderating effect whatever. Chemical decomposition is a phenomenon of so complicated a nature, that it is more than doubtful whether we are yet in possession of data sufficient to enable us to analyze the process with accuracy.

Another opinion brought forward by Berthollet in his work was of a startling nature, and occasioned a controversy between him and Proust which was carried on for some years with great spirit, but with perfect decorum and good manners on both sides. Berthollet affirmed that bodies were capable of uniting with each other in all possible proportions, and that there is no such thing as a definite compound, unless it has been produced by some accidental circumstances, as insolubility, volatility, &c. Thus every metal is capable of uniting with all possible doses of oxygen. So that instead of one or two oxides of every metal, an infinite number of oxides of each metal exist. Proust affirmed that all compounds are definite. Iron, says he, unites with oxygen only in two proportions; we have either a compound of 3·5 iron and 1 oxygen, or of 3·5 iron and 1·5 oxygen. The first constitutes the *black*, and the second the *red* oxide of iron ; and beside these there is no other. Every one is now satisfied that Proust's view of the subject was correct, and Berthollet's erroneous. But a better opportunity will occur hereafter to explain this subject, or at least to give the information respecting it which we at present possess.

Berthollet in this book points out the quantity of each base necessary to neutralize a given weight of acid, and he considers the strength of affinity as inversely that quantity. Now of all the bases known when Berthollet wrote, ammonia is capable of saturating the greatest quantity of acid. Hence he considered its affinity for acids as stronger than that of any other base. Barytes, on the contrary, saturates the smallest quantity of acid ; therefore its affinity for acids is smallest. Now ammonia is separated from acids by all the other bases ; while there is not one capable of separating barytes. It is surprising that the notoriety of this fact did not induce him to hesitate, before he came to so problematical a conclusion. Mr. Kirwan had already considered the force of affinity as directly proportional to the quantity of base necessary to saturate a given weight of acid. When we consider the subject metaphysically, Berthollet's opinion is most plausible ; for it is surely natural to consider that body as the strongest which produces the greatest effect. Now when we deprive an acid of its properties, or neutralize it by adding a base, one would be disposed to consider that base as acting with most energy, which with the smallest quantity of matter is capable of producing a given effect. This was the way that Berthollet reasoned. But if we attend to the power which one base has of displacing another, we shall find it very nearly proportional to the weight of it necessary to saturate a given weight of acid; or, at least those bases act most powerfully in displacing others of which the greatest quantity is necessary to saturate a given weight of acid. Kirwan's opinion, therefore, was more conformable to the order of decomposition. These two opposite views of the subject show clearly that neither Kirwan nor Berthollet had the smallest conception of the atomic theory; and, consequently, that the allegation of Mr. Higgens, that he had explained the atomic theory in his book on phlogiston, published in the year 1789, was not well founded. Whether Berthollet had read that book I do not know, but there can be no doubt that it was perused by Kirwan; who, however, did not receive from it the smallest notions respecting the atomic theory. Had he imbibed any such notions, he never would have considered chemical affinity as capable of being measured by the weight of base capable of neutralizing a given weight of acid.

Berthollet was not only a man of great energy of character, but of the most liberal feelings and benevolence. The only exception to this is his treatment of M. Clement. This gentleman, in company with M. Desormes, had examined the carbonic oxide of Priestley, and had shown as Cruikshanks had done before them, that it is a compound of carbon and oxygen, and that it contains no hydrogen whatever. Berthollet examined the same gas, and he published a paper to prove that it was a triple compound of oxygen, carbon, and hydrogen. This occasioned a controversy, which chemists have finally determined in favour of the opinion of Clement and Desormes. Berthollet, during this discussion, did not on every occasion treat his opponents with his accustomed temper and liberality; and ever after he opposed all attempts on the part of Clement to be admitted a member of the Institute. Whether there was any other reason for this conduct on the part of Berthollet, besides difference of opinion respecting the composition of carbonic oxide, I do not know : nor would it be right to condemn him without a more exact knowledge of all the circumstances han I can pretend to.

Antoine François de Fourcroy, was born at Paris on the 15th of June, 1755. His family had long resided in the capital, and several of his ancestors had distinguished themselves at the bar. But the branch from which he sprung had gradually sunk into poverty. His father exercised in Paris the trade of an apothecary, in consequence of a charge which he held in the house of the Duke of Orleans. The corporation of apothecaries having obtained the general suppression of all such charges, M. de Fourcroy, the father, was obliged to renounce his mode of livelihood ; and his son grew up in the midst of the poverty produced by the monopoly of the privileged bodies in Paris. He felt this situation the more keenly, because he possessed from nature an extreme sensibility of temper. When he lost his mother, at the age of seven years, he attempted to throw himself into her grave. The care of an elder sister preserved him with difficulty till he reached the age at which it was usual to be sent to college. There he was unlucky enough to meet with a brutal master, who conceived an aversion for him and treated him with cruelty : the consequence was, a dislike to study ; and he quitted the college at the age of fourteen, somewhat less informed than when he went to it.

His poverty now was such that he was obliged to endeavour to support himself by becoming writing-master. He had even some thoughts of going on

the stage; but was prevented by the hisses bestowed on a friend of his who had unadvisedly
entered upon that perilous career, and was treated
in consequence ·without mercy by the audience.
While uncertain what plan to follow, the advice
of Viq. d'Azyr induced him to commence the study
of medicine.

This great anatomist was an acquaintance of M.
de Fourcroy, the father. Struck with the appearance of his son, and the courage with which he
struggled with his bad fortune, he conceived an
affection for him, and promised to direct his studies,
and even to assist him during their progress. The
study of medicine to a man in his situation was by
no means an easy task. He was obliged to lodge
in a garret, so low in the roof that he could only
stand upright in the middle of the room. Beside him
lodged a water-carrier with twelve children. Fourcroy acted as physician.to this numerous family, and
in recompence was always supplied with abundance
of water. He contrived to support himself by giving
lessons to other students, by facilitating the researches
of richer writers, and by some translations which he
sold to a bookseller. For these he was only half
paid ; but the conscientious bookseller offered thirty
years afterwards to make up the deficiency, when
his creditor was become director-general of public
instruction.

Fourcroy studied with so much zeal and ardour
that he soon became well acquainted with the subject of medicine. But this was not sufficient. It
was necessary to get a doctor's degree, and all the
expenses at that time amounted to 250*l*. An old
physician, Dr. Diest, had left funds to the faculty
to give a gratuitous degree and licence, once every
two years, to the poor student who should best deserve them. Fourcroy was the most conspicuous

student at that time in Paris. He would therefore have reaped the benefit of this benevolent institution had it not been for the unlucky situation in which he was placed. There happened to exist a quarrel between the faculty charged with the education of medical men and the granting of degrees, and a society recently formed by government for the improvement of the medical art. This dispute had been carried to a great length, and had attracted the attention of all the frivolous and idle inhabitants of Paris. Viq. d'Azyr was secretary to the society, and of course one of its most active champions ; and was, in consequence, particularly obnoxious to the faculty of medicine at Paris. Fourcroy was unluckily the acknowledged *protégée* of this eminent anatomist. This was sufficient to induce the faculty of medicine to refuse him a gratuitous degree. He would have been excluded in consequence of this from entering on the career of a practitioner, had not the society, enraged at this treatment, and influenced by a violent party spirit, formed a subscription, and contributed the necessary expenses.

It was no longer possible to refuse M. de Fourcroy the degree of doctor, when he was thus enabled to pay for it. But above the simple degree of doctor there was another, entitled *docteur regent*, which depended entirely on the votes of the faculty. It was unanimously refused to M. de Fourcroy. This refusal put it out of his power afterwards to commence teacher in the medical school, and gave the medical faculty the melancholy satisfaction of not being able to enroll among their number the most celebrated professor in Paris. This violent and unjust conduct of the faculty of medicine made a deep impression on the mind of Fourcroy, and contributed not a little to the subsequent downfal of that powerful body.

Fourcroy being thus entitled to practise in Paris, his success depended entirely on the reputation which he could contrive to establish. For this purpose he devoted himself to the sciences connected with medicine, as the shortest and most certain road by which he could reach his object. His first writings showed no predilection for any particular branch of science. He wrote upon *chemistry*, *anatomy*, and *natural history*. He published an Abridgment of the History of Insects, and a Description of the Bursæ Mucosæ of the Tendons. This last piece seems to have given him the greatest celebrity; for in 1785 he was admitted, in consequence of it, into the academy as an anatomist. But the reputation of Bucquet, at that time very high, gradually drew his particular attention to chemistry, and he retained this predilection during the rest of his life.

Bucquet was at that time professor of chemistry in the Medical School of Paris, and was greatly celebrated and followed on account of his eloquence, and the elegance of his language. Fourcroy became in the first place his pupil, and afterwards his particular friend. One day, when a sudden attack of disease prevented him from lecturing as usual, he entreated Fourcroy to supply his place. Our young chemist at first declined, and alleged his ignorance of the method of addressing a public audience. But, overcome by the persuasions of Bucquet, he at last consented : and in this, his first essay, he spoke two hours without disorder or hesitation, and acquitted himself to the satisfaction of his whole audience. Bucquet soon after substituted him in his place, and it was in his laboratory and in his class-room that he first made himself acquainted with chemistry. He was enabled at the death of Bucquet, in consequence of an advantageous marriage that he had made, to purchase the apparatus and cabinet of his master ; and although the faculty of medicine would not allow him to succeed to the chair of Bucquet, they could not prevent him from succeeding to his reputation.

There was a kind of college which had been established in the Jardin du Roi, which at that time was under the superintendence of Buffon, and Macquer was the professor of chemistry in this institution. On the death of this chemist, in 1784, both Berthollet and Fourcroy offered themselves as candidates for the vacant chair. The voice of the public was so loud in favour of Fourcroy, that he was appointed to the situation in spite of the high character of his antagonist and the political influence which was exerted in his favour. He filled this chair for twenty-five years, with a reputation for eloquence continually on the increase. Such were the crowds, both of men and women, who flocked to hear him, that it was twice necessary to enlarge the size of the lecture room.

After the revolution had made some progress, he was named a member of the National Convention in the autumn of the memorable year 1793. It was during the reign of terror, when the Convention itself, and with it all France, was under the absolute dominion of one of the most sanguinary monsters that ever existed : it was almost equally dangerous for the members of the Convention to remain silent, or to take an active part in the business of that assembly. Fourcroy never opened his mouth in the Convention till after the death of Robespierre; at this period he had influence enough to save the lives of some men of merit : among others, of Darcet, who did not know the obligation under which he lay to him till long after; at last his own life was threatened, and his influence, of course, completely annihilated.

It was during this unfortunate and disgraceful period, that many eminent men lost their lives; among others, Lavoisier; and Fourcroy is accused of having contributed to the death of this illustrious chemist: but Cuvier entirely acquits him of this atrocious charge, and assures us that it was urged against him merely out of envy at his subsequent elevation. "If in the rigorous researches which we have made," says Cuvier in his Eloge of Fourcroy, "we had found the smallest proof of an atrocity so horrible, no human power could have induced us to sully our mouths with his Eloge, or to have pronounced it within the walls of this temple, which ought to be no less sacred to honour than to genius."

Fourcroy began to acquire influence only after the 9th Thermidor, when the nation was wearied with destruction, and when efforts were making to restore those monuments of science, and those public institutions for education, which during the wantonness and folly of the revolution had been overturned and destroyed. Fourcroy was particularly active in this renovation, and it was to him, chiefly, that the schools established in France for the education of youth are to be ascribed. The Convention had destroyed all the colleges, universities, and academies throughout France. The effects of this absurd abolition soon became visible; the army stood in need of surgeons and physicians, and there were none educated to supply the vacant places: three new schools were founded for educating medical men; they were nobly endowed. The term *schools of medicine* was proscribed as too aristocratical; they were distinguished by the ridiculous appellation of *schools of health*. The *Polytechnic School* was next instituted, as a kind of preparation for the exercise of the military profession, where young men could be instructed in mathematics and natural philosophy, to make them fit for entering the schools of the artillery, of engineers, and of the marine. The *Central Schools* was another institution for which France was indebted to the efforts of Fourcroy. The idea was good, though it was very imperfectly executed. It was to establish a kind of university in every department, for which the young men were to be prepared by a sufficient number of inferior schools scattered through the department. But unfortunately these inferior schools were never properly established or endowed; and even the central schools themselves were never supplied with proper masters. Indeed, it was found impossible to furnish such a number of masters at once. On that account, an institution was established in Paris, called the *Normal School*, for the express purpose of educating a sufficient number of masters to supply the different central schools.

Fourcroy, either as a member of the Convention or of the *Council of the Ancients*, took an active part in all these institutions, as far as regarded the plan and the establishment. He was equally concerned in the establishment of the Institute and of the *Musée d'Histoire Naturelle*. This last was endowed with the utmost liberality, and Fourcroy was one of the first professors; as he was also in the School of Medicine and the Polytechnic School. He was equally concerned in the restoration of the university, which constituted one of the most useful parts of Bonaparte's reign.

The violent exertions which he made in the numerous situations which he filled, and the prodigious activity which he displayed, gradually undermined his constitution. He himself was sensible of his approaching death, and announced it to his friends as an event which would speedily take place. On the 16th of December, 1809, after signing some despatches, he suddenly cried out, *Je suis mort (I am dead)*, and dropped lifeless on the ground.

He was twice married : first to Mademoiselle Bettinger, by whom he had two children, a son and a daughter, who survived him. He was married for the second time to Madame Belleville, the widow of Vailly, by whom he had no family. He left but little fortune behind him ; and two maiden sisters, who lived with him, depended afterwards for their support on his friend M. Vauquelin.

Notwithstanding the vast quantity of papers which he published, it will be admitted, without dispute, that the prodigious reputation which he enjoyed during his lifetime was more owing to his eloquence than to his eminence as a chemist—though even as a chemist he was far above mediocrity. He must have possessed an uncommon facility of writing. Five successive editions of his System of Chemistry appeared, each of them gradually increasing in size and value : the first being in two volumes and the last in ten. This last edition he wrote in sixteen months : it contains much valuable information, and doubtless contributed considerably to the general diffusion of chemical knowledge. Its style is perhaps too diffuse, and the spirit of generalizing from particular, and often ill-authenticated facts, is carried to a vicious length. Perhaps the best of all his productions is his Philosophy of Chemistry. It is remarkable for its conciseness, its perspicuity, and the neatness of its arrangement.

Besides these works, and the periodical publication entitled " Le Médecin éclairé," of which he was the editor, there are above one hundred and sixty papers on chemical subjects, with his name attached to them, which appeared in the Memoirs of the Academy and of the Institute ; in the Annales de Chimie, or the Annales de Musée d'Histoire Naturelle ; of which last work he was the original projector. Many of these papers contained analyses both animal, vegetable, and mineral, of very considerable value. In most of them, the name of Vauquelin is associated with his own as the author; and the general opinion is, that the experiments were all made by Vauquelin; but that the papers themselves were drawn up by Fourcroy.

It would serve little purpose to go over this long list of papers; because, though they contributed essentially to the progress of chemistry, yet they exhibit but few of those striking discoveries, which at once alter the face of the science, by throwing a flood of light on every thing around them. I shall merely notice a few of what I consider as his best papers.

1. He ascertained that the most common biliary calculi are composed of a substance similar to spermaceti. This substance, in consequence of a subsequent discovery which he made during the removal of the dead bodies from the burial-ground of the Innocents at Paris; namely, that these bodies are converted into a fatty matter, he called *adipocire*. It has since been distinguished by the name of *cholestine;* and has been shown to possess properties different from those of adipocire and spermaceti.

2. It is to him that we are indebted for the first knowledge of the fact, that the salts of magnesia and ammonia have the property of uniting together, and forming double salts.

3. His dissertation on the sulphate of mercury contains some good observations. The same remark applies to his paper on the action of ammonia on the sulphate, nitrate, and muriate of mercury. He first described the double salts which are formed.

4. The analysis of urine would have been valuable had not almost all the facts contained in it been

anticipated by a paper of Dr. Wollaston, published in the Philosophical Transactions. It is to him that we are indebted for almost all the additions to our knowledge of calculi since the publication of Scheele's original paper on the subject.

5. I may mention the process of Fourcroy and Vauquelin for obtaining pure barytes, by exposing nitrate of barytes to a red heat, as a good one. They discovered the existence of phosphate of magnesia in bones, of phosphorus in the brain and in the milts of fishes, and of a considerable quantity of saccharine matter in the bulb of the common onion ; which, by undergoing a kind of spontaneous fermentation was converted into *manna*.

In these, and many other similar discoveries, which I think it unnecessary to notice, we do not know what fell to the share of Fourcroy and what to Vauquelin; but there is one merit at least to which Fourcroy is certainly entitled, and it is no small one : he formed and brought forward Vauquelin, and proved to him, ever after, a most steady and indefatigable friend. This is bestowing no small panegyric on his character; for it would have been impossible to have retained such a friend through all the horrors of the French revolution, if his own qualities had not been such as to merit so steady an attachment.

Louis Bernard Guyton de Morveau was born at Dijon on the 4th of January, 1737. His father, Anthony Guyton, was professor of civil law in the University of Dijon, and descended from an ancient and respectable family. At the age of seven he showed an uncommon mechanical turn : being with his father at a small village near Dijon, he there happened to meet a public officer returning from a sale, whence he had brought back a clock that had remained unsold on account of its very bad condition. Morveau supplicated his father to buy it. The purchase was made for six francs. Young Morveau took it to pieces and cleaned it, supplied some parts that were wanting, and put it up again without any assistance. In 1799 this very clock was resold at a higher price, together with the estate and house in which it had been originally placed; having during the whole of that time continued to go in the most satisfactory manner. When only eight years of age, he took his mother's watch to pieces, cleaned it, and put it up again to the satisfaction of all parties.

After finishing his preliminary studies in his father's house, he went to college, and terminated his attendance on it at the age of sixteen. About this time he was instructed in botany by M. Michault, a friend of his father, and a naturalist of some eminence. He now commenced law student in the University of Dijon; and, after three years of intense application, he went to Paris to acquire a knowledge of the practice of the law.

While in Paris, he not only attended to law, but cultivated at the same time several branches of polite literature. In 1756 he paid a visit to Voltaire, at Ferney. This seems to have inspired him with a love of poetry, particularly of the descriptive and satiric kind. About a year afterwards, when only twenty, he published a poem called " Le Rat Iconoclaste, ou le Jesuite croquée." It was intended to throw ridicule on a well-known anecdote of the day, and to assist in blowing the fire that already threatened destruction to the obnoxious order of Jesuits. The adventure alluded to was this: Some nuns, who felt a strong predilection for a Jesuit, their spiritual director, were engaged in their accustomed Christmas occupation of modelling a representation of a religious mystery, decorated with several small statues representing the holy personages connected with the subject, and among them that of the ghostly father; but, to mark their favourite, his statue was made of loaf sugar. The following day was destined for the triumph of the Jesuit: but, meanwhile, a rat had devoured the valuable puppet. The poem is written after the agreeable manner of the celebrated poem, "Ververt."

At the age of twenty-four he had already pleaded several important causes at the bar, when the office of advocate-general, at the parliament of Dijon, was advertised for sale. At that time all public situations, however important, were sold to the best bidder. His father having ascertained that this place would be acceptable to his son, purchased it for forty thousand francs. The reputation of the young advocate, and his engaging manners, facilitated the bargain.

In 1764 he was admitted an honorary member of the Academy of Sciences, Arts, and Belles Lettres, of Dijon. Two months after, he presented to the assembled chamber of the parliament of Burgundy, a memoir on public instruction, with a plan for a college, on the principles detailed in his work. The encomiums which every public journal of the time passed on this production, and the flattering letters which he received, were unequivocal proofs of its value. In this memoir he endeavoured to prove that man is *bad* or *good*, according to the education which he has received. This doctrine was contrary to the creed of Diderot, who affirmed, in his Essay on the Life of Seneca, that nature makes wicked persons, and that the best institutions cannot render them good. But this mischievous opinion was successfully refuted by Morveau, in a letter to an anonymous friend.

The exact sciences were so ill taught, and lamely

cultivated at Dijon, during the time of his university education, that after his admission into the academy his notions on mechanics and natural philosophy were scanty and inaccurate. Dr. Chardenon was in the habit of reading memoirs on chemical subjects; and on one occasion Morveau thought it necessary to hazard some remarks which were ill received by the doctor, who sneeringly tōld him that having obtained such success in literature, he had better rest satisfied with the reputation so justly acquired, and leave chemistry to those who knew more of the matter.

Provoked at this violent remark, he resolved upon taking an honourable revenge. He therefore applied himself to the study of Macquer's Theoretical and Practical Chemistry, and of the Manual of Chemistry which Beaumé had just published. To the last chemist he also sent an extensive order for chemical preparations and utensils, with a view of forming a small laboratory near his office. He began by repeating many of Beaumé's experiments, and then trying his inexperienced hand at original researches. He soon found himself strong enough to attack the doctor. The latter had just been reading a memoir on the analysis of different kinds of oil; and Morveau combated some of his opinions with so much skill and sagacity, as astonished every one present. After the meeting, Dr. Chardenon addressed him thus : " You are born to be an honour to chemistry. So much knowledge could only have been gained by genius united with perseverance. Follow your new pursuit, and confer with me in your difficulties."

But this new pursuit did not prevent Morveau from continuing to cultivate literature with success. He wrote an *Eloge* of Charles V. of France, surnamed *the Wise*, which had been given out as the

subject of a prize, by the academy. A few months
afterwards, at the opening of the session of parliament, he delivered a discourse on the actual state of
jurisprudence; on which subject, three years after, he
composed a more extensive and complete work.   No
code of laws demanded reform more urgently than
those of France, and none saw more clearly the
necessity of such a reformation.

About this time a young gentleman of Dijon had
taken into his house an adept, who offered, upon
being furnished with the requisite materials, to produce gold in abundance; but, after six months of
expensive and tedious operations (during which period the roguish pretender had secretly distilled
many oils, &c., which he disposed of for his own
profit), the gentleman beginning to doubt the sin-
cerity of his instructer, dismissed him from his service and sold the whole of his apparatus and materials
to Morveau for a trifling sum.

Soon after he repaired to Paris, to visit the
scientific establishments of that metropolis, and to
purchase preparations and apparatus which he still
wanted to enable him to pursue with effect his favourite study.   For this purpose he applied to
Beaumé, then one of the most conspicuous of the
French chemists.   Pleased with his ardour, Beaumé
inquired what courses of chemistry he had attended.
" None," was the answer.—" How then could you
have learned to make experiments, and above all,
how could you have acquired the requisite dexterity?"—" Practice," replied the young chemist,
" has been my master; melted crucibles and broken
retorts my tutors."—" In that case," said Beaumé,
" you have not learned, you have invented."

About this time Dr. Chardenon read a paper before
the Dijon Academy on the causes of the augmentation of weight which metals experience when cal-

cined. He combated the different explanations which had been already advanced, and then proceeded to show that it might be accounted for in a satisfactory manner by the *abstraction* of phlogiston. This drew the attention of Morveau to the subject : he made a set of experiments a few months afterwards, and read a paper on the *phénomena of the air during combustion.* It was soon after that he made a set of experiments on the time taken by different substances to absorb or emit a given quantity of heat. These experiments, if properly followed out, would have led to the discovery of *specific heat;* but in his hands they seem to have been unproductive.

In the year 1772 he published a collection of scientific essays under the title of " Digressions Académiques." The memoirs on *phlogiston, crystallization,* and *solution,* found in this book deserve particular attention, and show the superiority of Morveau over most of the chemists of the time.

About this time an event happened which deserves to be stated. It had been customary in one of the churches of Dijon to bury considerable numbers of dead bodies. From these an infectious exhalation had proceeded, which had brought on a malignant disorder, and threatened the inhabitants of Dijon with something like the plague. All attempts to put an end to this infectious matter had failed, when Morveau tried the following method with complete success : A mixture of common salt and sulphuric acid in a wide-mouthed vessel was put upon chafing-dishes in various parts of the church. The doors and windows were closed and left in this state for twenty-four hours. They were then thrown open, and the chafing-dishes with the mixtures removed. Every remains of the bad smell was gone, and the church was rendered quite clean and free from infection. The same process was tried soon after in the prisons of Dijon, and with the same success. Afterwards chlorine gas was substituted for muriatic acid gas, and found still more efficacious. The present practice is to employ chloride of lime, or chloride of soda, for the purpose of fumigating infected apartments, and the process is found still more effectual than the muriatic acid gas, as originally employed by Morveau. The nitric acid fumes, proposed by Dr. Carmichael Smith, are also efficacious, but the application of them is much more troublesome and more expensive than of chloride of lime, which costs very little.

In the year 1774 it occurred to Morveau, that a course of lectures on chemistry, delivered in his native city, might be useful. Application being made to the proper authorities, the permission was obtained, and the necessary funds for supplying a laboratory granted. These lectures were begun on the 29th of April, 1776, and seem to have been of the very best kind. Every thing was stated with great clearness, and illustrated by a sufficient number of experiments. His fame now began to extend, and his name to be known to men of science in every part of Europe; and, in consequence, he began to experience the fate of almost all eminent men—to be exposed to the attacks of the malignant and the envious. The experiments which he exhibited to determine the properties of *carbonic acid gas* drew upon him the animadversions of several medical men, who affirmed that this gas was nothing else than a peculiar state of sulphuric acid. Morveau answered these animadversions in two pamphlets, and completely refuted them.

About this time he got metallic conductors erected on the house of the Academy at Dijon. On this account he was attacked violently for his presumption in disarming the hand of the Supreme Being. A multitude of fanatics assembled to pull down the conductors, and they would probably have done much mischief, had it not been for the address of M. Maret, the secretary, who assured them that the astonishing virtue of the apparatus resided in the gilded point, which had purposely been sent from Rome by the holy father! Will it excite any surprise, that within less than twenty years after this the mass of the French people not only renounced the Christian religion, and the spiritual dominion of the pope, but declared themselves atheists!

In 1777 Morveau published the first volume of a course of chemistry, which was afterwards followed by three other volumes, and is known by the name of " Elémens de Chimie de l'Académie de Dijon." This book was received with universal approbation, and must have contributed very much to increase the value of his lectures. Indeed, a text-book is essential towards a successful course of lectures : it puts it in the power of the students to understand the lecture if they be at the requisite pains ; and gives them a means of clearing up their difficulties, when any such occur. I do not hesitate to say, that a course of chemical lectures is twice as valuable when the students are furnished with a good text-book, as when they are left to interpret the lectures by their own unassisted exertions.

Soon after he undertook the establishment of a manufacture of saltpetre upon a large scale. For this he received the thanks of M. Necker, who was at that time minister of finance, in the name of the King of France. This manufactory he afterwards gave up to M. Courtois, whose son still carries it on, and is advantageously known to the public as the discoverer of *iodine*.

His next object was to make a collection of minerals, and to make himself acquainted with the science of mineralogy. All this was soon accomplished. In 1777 he was charged to examine the slate-quarries and the coal-mines of Burgundy, for which purpose he performed a mineralogical tour through the province. In 1779 he discovered a lead-mine in that country, and a few years afterwards, when the attention of chemists had been drawn to sulphate of barytes and its base, by the Swedish chemists, he sought for it in Burgundy, and found it in considerable quantity at Thôte. This enabled him to draw up a description of the mineral, and to determine the characters of the base, to which he gave the name of *burote;* afterwards altered to that of barytes. This paper was published in the third volume of the Memoirs of the Dijon Academy. In this paper he describes his method of decomposing sulphate of barytes, by heating it with charcoal—a method now very frequently followed.

In the year 1779 he was applied to by Pankouke, who meditated the great project of the *Encyclopédie Méthodique,* to undertake the chemical articles in that immense dictionary, and the demand was supported by a letter from Buffon, whose request he did not think that he could with propriety refuse. The engagement was signed between them in September, 1780. The first half-volume of the chemical part of this Encyclopédie did not appear till 1786, and Morveau must have been employed during the interval in the necessary study and researches. Indeed, it is obvious, from many of the articles, that he had spent a good deal of time in experiments of research.

The state of the chemical nomenclature was at that period peculiarly barbarous and defective. He found himself stopped at every corner for want of words to express his meaning. This state of things he resolved to correct, and accordingly in 1782 pub-

lished his first essay on a new chemical nomenclature. No sooner did this essay appear than it was
attacked by almost all the chemists of Paris, and
by none more zealously than by the chemical members of the academy. Undismayed by the violence
of his antagonists, and satisfied with the rectitude of
his views, and the necessity of the reform, he went
directly to Paris to answer the objections in person.
He not only succeeded in convincing his antagonists
of the necessity of reform ; but a few years afterwards prevailed upon the most eminent chemical
members of the academy, Lavoisier, Berthollet,
and Fourcroy, to unite with him in rendering the reform still more complete and successful. He drew
up a memoir, exhibiting a plan of a methodical chemical nomenclature, which was read at a meeting of
the Academy of Sciences, in 1787. Morveau, then,
was in reality the author of the new chemical nomenclature, if we except a few terms, which had been
already employed by Lavoisier. Had he done nothing
more for the science than this, it would deservedly
have immortalized his name. For every one must
be sensible how much the new nomenclature contributed to the subsequent rapid extension of chemical
science.

It was during the repeated conferences held with
Lavoisier and the other two associates that Morveau
became satisfied of the truth of Lavoisier's new doctrine, and that he was induced to abandon the phlogistic theory. We do not know the methods employed to convert him. Doubtless both reasoning
and experiment were made use of for the purpose.

It was during this period that Morveau published
a French translation of the Opuscula of Bergman.
A society of friends, under his encouragement, translated the chemical memoirs of Scheele and many
other foreign books of importance, which by their

means were made known to the men of science in France.

In 1783, in consequence of a favourable report by Macquer, Morveau obtained permission to establish a manufactory of carbonate of soda, the first of the kind ever attempted in France. It was during the same year that he published his collection of pleadings at the bar, among which [we find his Discours sur la Bonhomie, delivered at the opening of the sessions at Dijon, with which he took leave of his fellow-magistrates, surrendering the insignia of office, as he had determined to quit the profession of the law.

On the 25th of April, 1784, Morveau, accompanied by President Virly, ascended from Dijon in a balloon, which he had himself constructed, and repeated the ascent on the 12th of June following, with a view of ascertaining the possibility of directing these aerostatic machines, by an apparatus of his own contrivance. The capacity of the balloon was 10,498,074 French cubic feet. The effect produced by this bold undertaking by two of the most distinguished characters in the town was beyond description. Such ascents were then quite new, and looked upon with a kind of reverential awe. Though Morveau failed in his attempts to direct these aerial vessels, yet his method was ingenious and exceedingly plausible.

In 1786 Dr. Maret, secretary to the Dijon Academy, having fallen a victim to an epidemic disease, which he had in vain attempted to arrest, Morveau was appointed perpetual secretary and chancellor of the institution. Soon after this the first half-volume of the chemical part of the Encyclopédie Méthodique made its appearance, and drew the attention of every person interested in the science of chemistry. No chemical treatise had hitherto appeared worthy of being compared to it. The article *Acid*, which occupies a considerable part, is truely admirable; and whether we consider the historical details, the completeness of the accounts, the accuracy of the description of the experiments, or the elegance of the style, constitutes a complete model of what such a work should be. I may, perhaps, be partial, as it was from this book that I imbibed my own first notions in chemistry, but I never perused any book with more delight, and when I compared it with the best chemical books of the time, whether German, French, or English, its superiority became still more striking.

In the article *Acier*, Morveau had come to the very same conclusions, with respect to the nature of *steel*, as had been come to by Berthollet, Monge, and Vandermonde, in their celebrated paper on the subject, just published in the Memoirs of the Academy. His own article had been printed, though not published, before the appearance of the Memoir of the Academicians. This induced him to send an explanation to Berthollet, which was speedily published in the Journal de Physique.

In September, 1787, he received a visit from Lavoisier, Berthollet, Fourcroy, Monge, and Vandermonde. Dr. Beddoes, who was travelling through France at the time, and happened to be in Dijon, joined the party. The object of the meeting was to discuss several experiments explanatory of the new doctrine. In 1789 an attempt was made to get him admitted as a member of the Academy of Sciences; but it failed, notwithstanding the strenuous exertions of Berthollet and his other chemical friends.

The French revolution had now broken out, occasioned by the wants of the state on the one hand, and the resolute determination of the clergy and the nobility on the other, not to submit to bear any share in the public burdens. During the early part of this revolution Morveau took no part whatever in politics. In 1790, when France was divided into departments, he was named one of a commission by the National Assembly for the formation of the department of the Côte d'Or. On the 25th of August, 1791, he received from the Academy of Sciences the annual prize of 2000 francs, for the most useful work published in the course of the year. This was decreed him for his Dictionary of Chemistry, in the Encyclopédie Méthodique. Aware of the pressing necessities of the state, Morveau seized the opportunity of showing his desire of contributing towards its relief, by making a patriotic offering of the whole amount of his prize.

When the election of the second Constitutional Assembly took place, he was nominated a member by the electoral college of his department. A few months before, his name had appeared among the list of members proposed by the assembly, for the election of a governor to the heir-apparent. All this, together with the dignity of solicitor-general of the department to which he had recently been raised, not permitting him to continue his chemical lectures at Dijon, of which he had already delivered fifteen gratuitous courses, he resigned his chair in favour of Dr. Chaussier, afterwards a distinguished professor at the Faculty of Medicine of Paris; and, bidding adieu to his native city, proceeded to Paris.

On the ever memorable 16th of January, 1793, he voted with the majority of deputies. He was therefore, in consequence of this vote, a regicide. During the same year he resigned, in favour of the republic, his pension of two thousand francs, together with the arrears of that pension.

In 1794 he received from government different

commissions to act with the French armies in the Low Countries. Charged with the direction of a great aerostatic machine for warlike purposes, he superintended that one in which the chief of the staff of General Jourdan and himself ascended during the battle of Fleurus, and which so materially contributed to the success of the French arms on that day. On his return from his various missions, he received from the three committees of the executive government an invitation to co-operate with several learned men in the instruction of the *central schools*, and was named professor of chemistry at the *Ecole Centrale des Travaux publiques*, since better known under the name of the *Polytechnic School*.

In 1795 he was re-elected member of the Council of Five Hundred, by the electoral assemblies of Sarthe and Ile et Vilaine. The executive government, at this time, decreed the formation of the National Institute, and named him one of the forty-eight members chosen by government to form the nucleus of that scientific body.

In 1797 he resigned all his public situations, and once more attached himself exclusively to science and to the establishments for public instruction. In 1798 he was appointed a provisional director of the Polytechnic School, to supply the place of Monge, who was then in Egypt. He continued to exercise its duties during eighteen months, to the complete satisfaction of every person connected with that establishment. With much delicacy and disinterestedness, he declined accepting the salary of 2000 francs attached to this situation, which he thought belonged to the proper director, though absent from his duties.

In 1799 Bonaparte appointed him one of the administrators-general of the Mint; and the year following he was made director of the Polytechnic

School. In 1803 he received the cross of the Legion of Honour, then recently instituted; and in 1805 was made an officer of the same order. These honours were intended as a reward for the advantage which had accrued from the mineral acid fumigations which he had first suggested. In 1811 he was created a baron of the French empire.

After having taught in the *Ecole Polytechnique* for sixteen years, he obtained leave, on applying to the proper authorities, to withdraw into the retired station of private life, crowned with years and reputation, and followed with the blessings of the numerous pupils whom he had brought up in the career of science. In this situation he continued about three years, during which he witnessed the downfal of Bonaparte, and the restoration of the Bourbons. On the 21st of December, 1815, he was seized with a total exhaustion of strength; and, after an illness of three days only, expired in the arms of his disconsolate wife, and a few trusty friends, having nearly completed the eightieth year of his age. On the 3d of January, 1816, his remains were followed to the grave by the members of the Institute, and many other distinguished men: and Berthollet, one of his colleagues, pronounced a short but impressive funeral oration on his departed friend.

Morveau had married Madame Picardet, the widow of a Dijon academician, who had distinguished himself by numerous scientific translations from the Swedish, German, and English languages. The marriage took place after they were both advanced in life, and he left no children behind him. His publications on chemical subjects were exceedingly numerous, and he contributed as much as any of his contemporaries to the extension of the science; but as he was not the author of any striking chemical discoveries, it would be tedious to give a catalogue of his numerous productions which were scattered through the Dijon Memoirs, the Journal de Physique, and the Annales de Chimie.

CHAPTER IV.

PROGRESS OF ANALYTICAL CHEMISTRY.

ANALYSIS, or the art of determining the constituents of which every compound is composed, constitutes the essence of chemistry : it was therefore attempted as soon as the science put on any thing like a systematic form.   At first, with very little success ; but as knowledge became more and more general, chemists became more expert, and something like regular analysis began to appeàr. Thus, Brandt showed that *white vitriol* is a compound of sulphuric acid and oxide of zinc ; and Margraaf, that *selenite* or *gypsum* is a compound of sulphuric acid and lime.   Dr. Black made analyses of several of the salts of magnesia, so far at least as to determine the nature of the constituents.   For hardly any attempt was made in that early period of the art to determine the weight of the respective constituents.   The first person who attempted to lay down rules for the regular analysis of minerals, and to reduce these rules to practice, was Bergman.   This he did in his papers " De Docimasia Minerarum Humida," " De Terra Gemmarum," and " De Terra Tourmalini," published between the years 1777 and 1780.

To analyze a mineral, or to separate it into its constituent parts, it is necessary in the first place, to be able to dissolve it in an acid.   Bergman showed that most minerals become soluble in muriatic acid if they be reduced to a very fine powder, and then heated to redness, or fused with an alkaline carbonate. After obtaining a solution in this way he pointed out methods by which the different constituents may be separated one after another, and their relative quantities determined. The fusion with an alkaline carbonate required a strong red heat. An earthenware crucible could not be employed, because at a fusing temperature it would be corroded by the alkaline carbonate, and thus the mineral under analysis would be contaminated by the addition of a quantity of foreign matter. Bergman employed an iron crucible. This effectually prevented the addition of any earthy matter. But at a red heat the iron crucible itself is apt to be corroded by the action of the alkali, and thus the mineral under analysis becomes contaminated with a quantity of that metal. This iron might easily be separated again by known methods, and would therefore be of comparatively small consequence, provided we were sure that the mineral under examination contained no iron ; but when that happens (and it is a very frequent occurrence), an error is occasioned which we cannot obviate. Klaproth made a vast improvement in the art of analysis, by substituting crucibles of fine silver for the iron crucibles of Bergman. The only difficulty attending their use was, that they were apt to melt unless great caution was used in heating them. Dr. Wollaston introduced crucibles of platinum about the beginning of the present century. It is from that period that we may date the commencement of accurate analyzing.

Bergman's processes, as might have been expected, were rude and imperfect. It was Klaproth who first systematized chemical analysis and brought the art to such a state, that the processes followed could be imitated by others with nearly the same results, thus offering a guarantee for the accuracy of the process.

Martin Henry Klaproth, to whom chemistry lies under so many and such deep obligations, was born at Wernigerode, on the 1st of December, 1743. His father had the misfortune to lose his whole goods by a great fire, on the 30th of June, 1751, so that he was able to do little or nothing for the education of his children. Martin was the second of three brothers, the eldest of whom became a clergyman, and the youngest private secretary at war, and keeper of the archives of the cabinet of Berlin. Martin survived both his brothers. He procured such meagre instruction in the Latin language as the school of Wernigerode afforded, and he was obliged to procure his small school-fees by singing as one of the church choir. It was at first his intention to study theology; but the unmerited hard treatment which he met with at school so disinclined him to study, that he determined, in his sixteenth year, to learn the trade of an apothecary. Five years which he was forced to spend as an apprentice, and two as an assistant in the public laboratory in Quedlinburg, furnished him with but little scientific information, and gave him little else than a certain mechanical adroitness in the most common pharmaceutical preparations.

He always regarded as the epoch of his scientific instruction, the two years which he spent in the public laboratory at Hanover, from Easter 1766, till the same time in 1768. It was there that he first met with some chemical books of merit, especially those of Spielman, and Cartheuser, in which a higher scientific spirit already breathed. He was now anxious to go to Berlin, of which he had formed a high idea from the works of Pott, Henkel, Rose, and Margraaf. An opportunity presenting itself about Easter, 1768, he was placed as assistant in the laboratory of Wendland, at the sign of the Golden Angel, in the Street of the Moors. Here he employed all the time which a conscientious discharge of the duties of his station left him, in completing his own scientific education. And as he considered a profounder acquaintance with the ancient languages, than he had been able to pick up at the school of Wernigerode, indispensable for a complete scientific education, he applied himself with great zeal to the study of the Greek and Latin languages, and was assisted in his studies by Mr. Poppelbourn, at that time a preacher.

About Michaelmas, 1770, he went to Dantzig, as assistant in the public laboratory: but in March of the following year he returned to Berlin, as assistant in the office of the elder Valentine Rose, who was one of the most distinguished chemists of his day. But this connexion did not continue long; for Rose died in 1771. On his deathbed he requested Klaproth to undertake the superintendence of his office. Klaproth not only superintended this office for nine years with the most exemplary fidelity and conscientiousness, but undertook the education of the two sons of Rose, as if he had been their father. The younger died before reaching the age of manhood: the elder became his intimate friend, and the associate of all his scientific researches. For several years before the death of Rose (which happened in 1808) they wrought together, and Klaproth was seldom satisfied with the results of his experiments till they had been repeated by Rose.

In the year 1780 Klaproth went through his trials for the office of apothecary with distinguished applause. His thesis, " On Phosphorus and distilled Waters," was printed in the Berlin Miscellanies for

1782. Soon after this, Klaproth bought what had formerly been the Flemming laboratory in Spandau-street: and he married Sophia Christiana Lekman, with whom he lived till 1803 (when she died) in a happy state. They had three daughters and a son, who survived their parents. He continued in possession of this laboratory, in which he had arranged a small work-room of his own, till the year 1800, when he purchased the room of the Academical Chemists, in which he was enabled, at the expense of the academy, to furnish a better and more spacious apartment for his labours, for his mineralogical and chemical collection, and for his lectures.

As soon as he had brought the first arrangements of his office to perfection—an office which, under his inspection and management, became the model of a laboratory, conducted upon the most excellent principles, and governed with the most conscientious integrity, he published in the various periodical works of Germany, such as " Crell's Chemical Annals," the " Writings of the Society for the promotion of Natural Knowledge," " Selle's Contributions to the Science of Nature and of Medicine," " Köhler's Journal," &c.; a multitude of papers which soon drew the attention of chemists; for example, his Essay on Copal—on the Elastic Stone—on Proust's Sel perlée—on the Green Lead Spar of Tschoppau—on the best Method of preparing Ammonia—on the Carbonate of Barytes—on the Wolfram of Cornwall—on Wood Tin—on the Violet Schorl—on the celebrated Aerial Gold—on Apatite, &c. All these papers, which secured him a high reputation as a chemist, appeared before 1788, when he was chosen an ordinary member of the physical class of the Royal Berlin Academy of Sciences. The Royal Academy of Arts had elected him a member a year earlier. From this time, every volume of the

Memoirs of the Academy, and many other periodical works besides, contained numerous papers by this accomplished chemist; and there is not one of them which does not furnish us with a more exact knowledge of some one of the productions of nature or art. He has either corrected false representations, or extended views that were before partially known, or has revealed the composition and mixture of the parts of bodies, and has made us acquainted with a variety of new elementary substances. Amidst all these labours, it is difficult to say whether we should most admire the fortunate genius, which, in all cases, readily and easily divined the point where any thing of importance lay concealed; or the acuteness which enabled him to find the best means of accomplishing his object; or the unceasing labour and incomparable exactness with which he developed it; or the pure scientific feeling under which he acted, and which was removed at the utmost possible distance from every selfish, every avaricious, and every contentious purpose.

In the year 1795 he began to collect his chemical works which lay scattered among so many periodical publications, and gave them to the world under the title of " Beitrage zur Chemischen Kenntniss der Mineralkörper" (Contributions to the Chemical Knowledge of Mineral Bodies). Of this work, which consists of six volumes, the last was published in 1815, about a year before the author's death. It contains no fewer than two hundred and seven treatises, the most valuable part of all that Klaproth had done for chemistry and mineralogy. It is a pity that the sale of this work did not permit the publication of a seventh volume, which would have included the rest of his papers, which he had not collected, and given us a good index to the whole work, which would have been of great importance to the practical chemist. There is, indeed, an index to the first five volumes ; but it is meagre and defective, containing little else than the names of the substances on which his experiments were made.

Besides his own works, the interest which he took in the labours of others deserves to be noticed. He superintended a new edition of Gren's Manual of Chemistry, remarkable not so much for what he added as for what he took away and corrected. The part which he took in Wolff's Chemical Dictionary was of great importance. The composition of every particular treatise was by Professor Wolff ; but Klaproth read over every important article before it was printed, and assisted the editor on all occasions with the treasures of his experience and knowledge. Nor was he less useful to Fischer in his translation of Berthollet on Affinity and on Chemical Statics.

These meritorious services, and the lustre which his character and discoveries conferred on his country were duly appreciated by his sovereign. In 1782 he had been made assessor in the Supreme College of Medicine and of Health, which then existed. At a more recent period he enjoyed the same rank in the Supreme Council of Medicine and of Health ; and when this college was subverted, in 1810, he became a member of the medical deputation attached to the ministry of the interior. He was also a member of the perpetual court commission for medicines. His lectures, too, procured for him several municipal situations. As soon as the public became acquainted with his great chemical acquirements he was permitted to give yearly two private courses of lectures on chemistry ; one for the officers of the royal artillery corps, the other for officers not connected with the army, who wished to accomplish themselves for some practical employment. Both of these lectures assumed afterwards a municipal character. The former led to his appointment as professsor of the Artillery Academy instituted at Tempelhoff; and, after its dissolution, to his situation as professor in the Royal War School. The other lecture procured for him the professorship of chemistry in the Royal Mining Institutute. On the establishment of the university, Klaproth's lectures became those of the university, and he himself was appointed ordinary professor of chemistry, and member of the academical senate. From 1797 to 1810 he was an active member of a small scientific society, which met yearly during a few weeks for the purpose of discussing the more recondite mysteries of the science. In the year 1811, the King of Prussia added to all his other honours the order of the Red Eagle of the third class.

Klaproth spent the whole of a long life in the most active and conscientious discharge of all the duties of his station, and in an uninterrupted course of experimental investigations. He died at Berlin on the 1st of January, 1817, in the 70th year of his age.

Among the remarkable traits in his character was his incorruptible regard for every thing that he believed to be true, honourable, and good; his pure love of science, with no reference whatever to any selfish, ambitious, and avaricious feeling; his rare modesty, undebased by the slightest vainglory or boasting. He was benevolently disposed towards all men, and never did a slighting or contemptuous word respecting any person fall from him. When forced to blame, he did it briefly, and without bitterness, for his blame always applied to actions, not to persons. His friendship was never the result of selfish calculation, but was founded on his opinion of the personal worth of the individual. Amidst all the unpleasant accidents of his life, which were far from few, he evinced the greatest firmness of mind. In his common behaviour he was pleasant and composed, and was indeed rather inclined to a joke. To all this may be added a true religious feeling, so uncommon among men of science of his day. His religion consisted not in words and forms, not in positive doctrines, nor in ecclesiastical observances, which, however, he believed to be necessary and honourable; but in a zealous and conscientious discharge of all his duties, not only of those which are imposed by the laws of men, but of those holy duties of love and charity, which no human law, but only that of God can command, and without which the most enlightened of men is but " as sounding brass, or a tinkling cymbal." He early showed this religious feeling by the honourable care which he bestowed on the education of the children of Valentine Rose. Nor did he show less care at an after-period towards his assistants and apprentices, to whom he refused no instruction, and in whose success he took the most active concern. He took a pleasure in every thing that was good and excellent, and felt a lively interest in every undertaking which he believed to be of general utility. He was equally removed from the superstition and infidelity of his age, and carried the principles of religion, not on his lips, but in the inmost feelings of his heart, from whence they emanated in actions which pervaded and ennobled his whole being and conduct.

When we take a view of the benefits which Klaproth conferred upon chemistry, we must not look so much at the new elementary substances which he discovered, though they must not be forgotten, as at the new analytical methods which he introduced, the precision, and neatness, and order, and regularity with which his analyses were conducted, and the scrupulous fidelity with which every thing was faithfully stated as he found it.

1. When a mineral is subjected to analysis, whatever care we take to collect all the constituents, and to weigh them without losing any portion whatever, it is generally found that the sum of the constituents obtained fall a little short of the weight of the mineral employed in the analysis. Thus, if we take 100 grains of any mineral, and analyze it, the weights of all the constituents obtained added together will rarely make up 100 grains, but generally somewhat less; perhaps only 99, or even 98 grains. But some cases occur, when the analysis of 100 grains of a mineral gives us constituents that weigh, when added together, more than 100 grains; perhaps 105, or, in some rare cases, as much as 110. It was the custom with Bergman, and other analysts of his time, to consider this deficiency or surplus as owing to errors in the analysis, and therefore to slur it over in the statement of the analysis, by bringing the weight of the constituents, by calculation, to amount exactly to 100 grains. Klaproth introduced the method of stating the results exactly as he got them. He gives the weight of mineral employed in all his analyses, and the weight of each constituent extracted. These weights, added together, generally show a loss, varying from two per cent. to a half per cent. This improvement may appear at first sight trifling; yet I am persuaded that to it we are indebted for most of the subsequent improvements introduced into analytical chemistry. If the loss sustained was too great, it was obvious either that the analysis had been badly performed, or that the mineral contains some constituent which had been overlooked, and not obtained. This laid him under the necessity of repeating the analysis; and if the loss continued, he naturally looked out for

some constituent which his analysis had not enabled him to obtain. It was in this way that he discovered the presence of potash in minerals; and Dr. Kennedy afterwards, by following out his processes, discovered soda as a constituent. It was in this way that water, phosphoric acid, arsenic acid, fluoric acid, boracic acid, &c., were also found to exist as constituents in various mineral bodies, which, but for the accurate mode of notation introduced by Klaproth, would have been overlooked and neglected.

2. When Klaproth first began to analyze mineral bodies, he found it extremely difficult to bring them into a state capable of being dissolved in acids, without which an accurate analysis was impossible. Accordingly corundum, adamantine spar, and the zircon, or hyacinth, baffled his attempts for a considerable time, and induced him to consider the earth of corundum as of a peculiar nature. He obviated this difficulty by reducing the mineral to an extremely fine powder, and, after digesting it in caustic potash ley till all the water was dissipated, raising the temperature, and bringing the whole into a state of fusion. This fusion must be performed in a silver crucible. Corundum, and every other mineral which had remained insoluble after fusion with an alkaline carbonate, was found to yield to this new process. This was an improvement of considerable importance. All those stony minerals which contain a notable proportion of silica, in general become soluble after having been kept for some time in a state of ignition with twice their weight of carbonate of soda. At that temperature the silica of the mineral unites with the soda, and the carbonic acid is expelled. But when the quantity of silica is small, or when it is totally absent, heating with carbonate of soda does not answer so well. With such minerals, caustic potash or soda may be substituted with ad-

vantage; and there are some of them that cannot be analyzed without having recourse to that agent. I have succeeded in analyzing corundum and chrysoberyl, neither of which, when pure, contain any silica, by simply heating them in carbonate of soda; but the process does not succeed unless the minerals be reduced to an exceedingly minute powder.

3. When Klaproth discovered potash in the idocrase, and in some other minerals, it became obvious that the old mode of rendering minerals soluble in acids by heating them with caustic potash, or an alkaline carbonate, could answer only for determining the quantity of silica, and of earths or oxides, which the mineral contained; but that it could not be used when the object was to determine its potash. This led him to substitute *carbonate of barytes* instead of potash or soda, or their carbonates. After having ascertained the quantity of silica, and of earths, and metallic oxides, which the mineral contained, his last process to determine the potash in it was conducted in this way : A portion of the mineral reduced to a fine powder was mixed with four or five times its weight of carbonate of barytes, and kept for some time (in a platinum crucible) in a red heat. By this process, the whole becomes soluble in muriatic acid. The muriatic acid solution is freed from silica, and afterwards from barytes, and all the earths and oxides which it contains, by means of carbonate of ammonia. The liquid, thus freed from every thing but the alkali, which is held in solution by the muriatic acid, and the ammonia, used as a precipitant, is evaporated to dryness, and the dry mass, cautiously heated in a platinum crucible till the ammoniacal salts are driven off. Nothing now remains but the potash, or soda, in combination with muriatic acid. The addition of muriate of platinum enables us to determine whether the alkali

be potash or soda: if it be potash, it occasions a yellow precipitate; but nothing falls if the alkali be soda.

This method of analyzing minerals containing potash or soda is commonly ascribed to Rose. Fescher, in his Eloge of Klaproth, informs us that Klaproth said to him, more than once, that he was not quite sure whether he himself, or Rose, had the greatest share in bringing this method to a state of perfection. From this, I think it not unlikely that the original suggestion might have been owing to Rose, but that it was Klaproth who first put it to the test of experiment.

The objection to this mode of analyzing is the high price of the carbonate of barytes. This is partly obviated by recovering the barytes in the state of carbonate; and this, in general, may be done, without much loss. Berthier has proposed to substitute oxide of lead for carbonate of barytes. It answers very well, is sufficiently cheap, and does not injure the crucible, provided the oxide of lead be mixed previously with a little nitrate of lead, to oxidize any fragments of metallic lead which it may happen to contain. Berthier's mode, therefore, in point of cheapness, is preferable to that of Klaproth. It is equally efficacious and equally accurate. There are some other processes which I myself prefer to either of these, because I find them equally easy, and still less expensive than either carbonate of barytes or oxide of lead. Davy's method with boracic acid is exceptionable, on account of the difficulty of separating the boracic acid completely again.

4. The mode of separating iron and manganese from each other employed by Bergman was so defective, that no confidence whatever can be placed in his results. Even the methods suggested by Vauquelin, though better, are still defective. But

the process followed by Klaproth is susceptible of very great precision. He has (we shall suppose) the mixture of iron and manganese to be separated from each other, in solution, in muriatic acid. The first step of the process is to convert the protoxide of iron (should it be in that state) into peroxide. For this purpose, a little nitric acid is added to the solution, and the whole heated for some time. The liquid is now to be rendered as neutral as possible ; first, by driving off as much of the excess of acid as possible, by concentrating the liquid; and then by completing the neutralization, by adding very dilute ammonia, till no more can be added without occasioning a permanent precipitation. Into the liquid thus neutralized, succinate or benzoate of ammonia is dropped, as long as any precipitate appears. By this means, the whole peroxide of iron is thrown down in combination with succinic, or benzoic acid, while the whole manganese remains in solution. The liquid being filtered, to separate the benzoate of iron, the manganese may now (if nothing else be in the liquid) be thrown down by an alkaline carbonate; or, if the liquid contain magnesia, or any other earthy matter, by hydrosulphuret of ammonia, or chloride of lime.

This process was the contrivance of Gehlen ; but it was made known to the public by Klaproth, who ever after employed it in his analyses. Gehlen employed succinate of ammonia; but Hisinger afterwards showed that benzoate of ammonia might be substituted without any diminution of the accuracy of the separation. This last salt, being much cheaper than succinate of ammonia, answers better in this country. In Germany, the succinic acid is the cheaper of the two, and therefore the best.

5. But it was not by new processes alone that Klaproth improved the mode of analysis, though

they were numerous and important ; the improvements in the apparatus contributed not less essentially to the success of his experiments. When he had to do with very hard minerals, he employed a mortar of flint, or rather of agate. This mortar he, in the first place, analyzed, to determine exactly the nature of the constituents. He then weighed it. When a very hard body is pounded in such a mortar, a portion of the mortar is rubbed off, and mixed with the pounded mineral. What the quantity thus abraded was, he determined by weighing the mortar at the end of the process. The loss of weight gave the portion of the mortar abraded ; and this portion must be mixed with the pounded mineral.

When a hard stone is pounded in an agate mortar it is scarcely possible to avoid losing a little of it. The best method of proceeding is to mix the matter to be pounded (previously reduced to a coarse powder in a diamond mortar) with a little water. This both facilitates the trituration, and prevents any of the dust from flying away; and not more than a couple of grains of the mineral should be pounded at once. Still, owing to very obvious causes, a little of the mineral is sure to be lost during the pounding. When the process is finished, the whole powder is to be exposed to a red heat in a platinum crucible, and weighed. Supposing no loss, the weight should be equal to the quantity of the mineral pounded together with the portion abraded from the mortar. But almost always the weight will be found less than this. Suppose the original weight of the mineral before pounding was a, and the quantity abraded from the mortar 1; then, if nothing were lost, the weight should be $a + 1$; but we actually find it only b, a quantity less than $a + 1$. To determine the weight of matter abraded from the mortar contained in this powder, we say $a + 1 : b :: 1 : x$, the

quantity from the mortar in our powder, and $x = \dfrac{b}{a+1}$.
In performing the analysis, Klaproth attended to
this quantity, which was silica, and subtracted it.
Such minute attention may appear, at first sight
superfluous; but it is not so. In analyzing sapphire, chrysoberyl, and some other very hard minerals, the quantity of silica abraded from the mortar
sometimes amounts to five per cent. of the weight of
the mineral; and if we were not to attend to the
way in which this silica has been introduced into the
powder, we should give an erroneous view of the constitution of the mineral under analysis. All the
analyses of chrysoberyl hitherto published, give a
considerable quantity of silica as a constituent of it.
This silica, if really found by the analysts, must
have been introduced from the mortar, for pure
chrysoberyl contains no silica whatever, but is a definite compound of glucina, alumina, and oxide of iron.

When Klaproth operated with fire, he always selected his vessels, whether of earthenware, glass,
plumbago, iron, silver, or platinum, upon fixed
principles; and showed more distinctly than chemists had previously been aware of, what an effect
the vessel frequently has upon the result. He also
prepared his reagents with great care, to ensure
their purity; for obtaining several of which in their
most perfect state, he invented several efficient
methods. It is to the extreme care with which he
selected his minerals for analysis, and to the purity
of his reagents, and the fitness of his vessels for the
objects in view, that the great accuracy of his analyses is to be, in a great measure, ascribed. He
must also have possessed considerable dexterity in
operating, for when he had in view to determine any
particular point with accuracy, his results came,
in general, exceedingly near the truth. I may no-

tice, as an example of this, his analysis of sulphate of barytes, which was within about one-and-a-half per cent. of absolute correctness. When we consider the looseness of the data which chemists were then obliged to use, we cannot but be surprised at the smallness of the error. Berzelius, in possession of better data, and possessed of much dexterity, and a good apparatus, when he analyzed this salt many years afterwards, committed an error of a half per cent.

Klaproth, during a very laborious life, wholly devoted to analytical chemistry, entirely altered the face of mineralogy. When he began his labours, chemists were not acquainted with the true composition of a single mineral. He analyzed above 200 species, and the greater number of them with so much accuracy, that his successors have, in most cases, confirmed the results which he obtained. The analyses least to be depended on, are of those minerals which contain both lime and magnesia ; for his process for separating lime and magnesia from each other was not a good one ; nor am I sure that he always succeeded completely in separating silica and magnesia from each other. This branch of analysis was first properly elucidated by Mr. Chenevix.

6. Analytical chemistry was, in fact, systematized by Klaproth ; and it is by studying his numerous and varied analyses, that modern chemists have learned this very essential, but somewhat difficult art ; and have been able, by means of still more accurate data than he possessed, to bring it to a still greater degree of perfection. But it must not be forgotten, that Klaproth was in reality the creator of this art, and that on that account the greatest part of the credit due to the progress that has been made in it belongs to him.

It would be invidious to point out the particular

analyses which are least exact; perhaps they ought rather to be ascribed to an unfortunate selection of specimens, than to any want of care or skill in the operator. But, during his analytical processes, he discovered a variety of new elementary substances which it may be proper to enumerate.

In 1789 he examined a mineral called *pechblende*, and found in it the oxide of a new metal, to which he gave the name of *uranium*. He determined its characters, reduced it to the metallic state, and described its properties. It was afterwards examined by Richter, Bucholz, Arfvedson, and Berzelius.

It was in the same year, 1789, that he published his analysis of the zircon; he showed it to be a compound of silica and a new earth, to which he gave the name of zirconia. He determined the properties of this new earth, and showed how it might be separated from other bodies and obtained in a state of purity. It has been since ascertained, that it is a metallic oxide, and the metallic basis of it is now distinguished by the name of *zirconium*. In 1795 he showed that the *hyacinth* is composed of the same ingredients as the zircon; and that both, in fact, constitute only one species. This last analysis was repeated by Morveau, and has been often confirmed by modern analytical chemists.

It was in 1795 that he analyzed what was at that time called *red schorl*, and now *titanite*. He showed that it was the oxide of a new metallic body, to which he gave the name of *titanium*. He described the properties of this new body, and pointed out its distinctive characters. It must not be omitted, however, that he did not succeed in obtaining oxide of titanium, or *titanic acid*, as it is now called, in a state of purity. He was not able to separate a quantity of oxide of iron, with which it was united, and which gave it a reddish colour. It was first obtained pure by H. Rose, the son of his friend and pupil, who took so considerable a part in his scientfic investigations.

Titanium, in the metallic state, was some years ago discovered by Dr. Wollaston, in the slag at the bottom of the iron furnace, at Merthyr Tydvil, in Wales. It is a yellow-coloured, brittle, but very hard metal, possessed of considerable beauty; but not yet applied to any useful purpose.

In 1797 he examined the menachanite, a black sand from Cornwall, which had been subjected to a chemical analysis by Gregor, in 1791, who had extracted from it a new metallic substance, which Kirwan distinguished by the name of *menachine*. Klaproth ascertained that the new metal of Gregor was the very same as his own titanium, and that menachanite is a compound of titanic acid and oxide of iron. Thus Mr. Gregor had anticipated him in the discovery of titanium, though he was not aware of the circumstance till two years after his own experiments had been published.

In the year 1793 he published a comparative set of experiments on the nature of carbonates of barytes and strontian; showing that their bases are two different earths, and not the same, as had been hitherto supposed in Germany. This was the first publication on strontian which appeared on the continent; and Klaproth seems to have been ignorant of what had been already done on it in Great Britain; at least, he takes no notice of it in his paper, and it was not his character to slur over the labours of other chemists, when they were known to him. Strontian was first mentioned as a peculiar earth by Dr. Crawford, in his paper on the medicinal properties of the muriate of barytes, published in 1790. The experiments on which he founded his opinions were made, he informs us, by Mr. Cruikshanks. A paper on the same subject, by Dr. Hope, was read to the Royal Society of Edinburgh, in 1793 ; but they had been begun in 1791. In this paper Dr. Hope establishes the peculiar characters of strontian, and describes its salts with much precision.

Klaproth had been again anticipated in his experiments on strontian ; but he could not have become aware of this till afterwards. For his own experiments were given to the public before those of Dr. Hope.

On the 25th of January, 1798, his paper on the gold ores of Transylvania was read at a meeting of the Academy of Sciences at Berlin. During his analysis of these ores, he detected a new white metal, to which he gave the name of *tellurium*. Of this metal he describes the properties, and points out its distinguishing characters.

These ores had been examined by Muller, of Reichenstein, in the year 1782 ; and he had extracted from them a metal which he considered as differing from every other. Not putting full confidence in his own skill, he sent a specimen of his new metal to Bergman, requesting him 'to examine it and give his opinion respecting its nature. All that Bergman did was to show that the metallic body which he had got was not antimony, to which alone, of all known metals, it bore any resemblance. It might be inferred from this, that Muller's metal was new. But the subject was lost sight of, till the publication of Klaproth's experiments, in 1802, recalled it to the recollection of chemists. Indeed, Klaproth relates all that Muller had done, with the most perfect fairness.

In the year 1804 he published the analysis of a red-coloured mineral, from Bastnäs in Sweden, which had been at one time confounded with tungsten ; but which the Elhuyarts had shown to contain none of that metal.   Klaproth showed that it contained a new substance, as one of its constituents, which he considered as a new earth, and which he called *ochroita*, because it forms coloured salts with acids. Two years after, another analysis of the same mineral was published by Berzelius and Hisinger.   They considered the new substance which the mineral contained as a metallic oxide, and to the unknown metallic base they gave the name of *cerium*, which has been adopted by chemists in preference to Klaproth's name.   The characters of oxide of cerium given by Berzelius and Hisinger, agree with those given by Klaproth to ochroita, in all the essential circumstances.   Of course Klaproth must be considered as the discoverer of this new body.   The distinction between *earth* and *metallic oxide* is now known to be an imaginary one.   All the substances formerly called earths are, in fact, metallic oxides.

Besides these new substances, which he detected by his own labours, he repeated the analyses of others, and confirmed and extended the discoveries they had made.   Thus, when Vauquelin discovered the new earth *glucina*, in the emerald and beryl, he repeated the analysis of these minerals, confirmed the discovery of Vauquelin, and gave a detailed account of the characters and properties of glucina. Gadolin had discovered another new earth in the mineral called gadolinite.   This discovery was confirmed by the analysis of Ekeberg, who distinguished the new earth by the name of yttria.   Klaproth immediately repeated the analysis of the gadolinite, confirmed the results of Ekeberg's analysis, and examined and described the properties of *yttria*.

When Dr. Kennedy discovered soda in basalt, Klaproth repeated the analysis of this mineral, and confirmed the results obtained by the Edinburgh analyst.

But it would occupy too much room, if I were to enumerate every example of such conduct. Whoever will take the trouble to examine the different volumes of the Beitrage, will find several others not less striking or less useful.

The service which Klaproth performed for mineralogy, in Germany, was performed equally in France by the important labours of M. Vauquelin. It was in France, in consequence of the exertions of Romé de Lisle, and the mathematical investigations of the Abbé Hauy, respecting the structure of crystals, which were gradually extended over the whole mineral kingdom, that the reform in mineralogy, which has now become in some measure general, originated. Hauy laid it down as a first principle, that every mineral species is composed of the same constituents united in the same proportion. He therefore considered it as an object of great importance, to procure an exact chemical analysis of every mineral species. Hitherto no exact analysis of minerals had been performed by French chemists ; for Sage, who was the chemical mineralogist connected with the academy, satisfied himself with ascertaining the nature of the constituents of minerals, without determining their proportions. But Vauquelin soon displayed a knowledge of the mode of analysis, and a dexterity in the use of the apparatus which he employed, little less remarkable than that of Klaproth himself.

Of Vauquelin's history I can give but a very imperfect account, as I have not yet had an opportunity of seeing any particulars of his life. He was a peasant-boy of Normandy, with whom Fourcroy accidentally met. He was pleased with his quickness and parts, and delighted with the honesty and integrity of his character. He took him with him to Paris, and gave him the superintendence of his laboratory. His chemical knowledge speedily became great, and his practice in experimenting gave him skill and dexterity: he seems to have performed all the analytical experiments which Fourcroy was in the habit of publishing. He speedily became known by his publications and discoveries. When the scientific institutions were restored or established, after the death of Robespierre, Vauquelin became a member of the Institute and chemist to the School of Mines. He was made also assay-master of the Mint. He was a professor of chemistry in Paris, and delivered, likewise, private lectures, and took in practical pupils into his laboratory. His laboratory was of considerable size, and he was in the habit of preparing both medicines and chemical reagents for sale. It was he chiefly that supplied the French chemists with phosphorus, &c., which cannot be conveniently prepared in a laboratory fitted up solely for scientific purposes.

Vauquelin was by far the most industrious of all the French chemists, and has published more papers, consisting of mineral, vegetable, and animal analyses, than any other chemist without exception. When he had the charge of the laboratory of the School of Mines, Hauy was in the habit of giving him specimens of all the different minerals which he wished analyzed. The analyses were conducted with consummate skill, and we owe to him a great number of improvements in the methods of analysis. He is not entitled to the same credit as Klaproth, because he had the advantage of many analyses of Klaproth to serve him as a guide. But he had no model before him in France; and both the apparatus used by him, and the reagents which he employed, were of his own contrivance and preparation. I have sometimes suspected that his reagents were not always very pure; but I believe the true reason of the unsatisfactory nature of many of his analyses, is the bad choice made of the specimens selected for analysis. It is obvious from his papers, that Vauquelin was not a mineralogist; for he never attempts a description of the mineral which he subjects to analysis, satisfying himself with the specimen put into his hands by Hauy. Where that specimen was pure, as was the case with emerald and beryl, his analysis is very good ; but when the specimen was impure or ill-chosen, then the result obtained could not convey a just notion of the constituents of the mineral. That Hauy would not be very difficult to please in his selection of specimens, I think myself entitled to infer from the specimens of minerals contained in his own cabinet, many of which were by no means well selected. I think, therefore, that the numerous analyses published by Vauquelin, in which the constituents assigned by him are not those, or, at least, not in the same proportions, as have been found by succeeding analysts, are to be ascribed, not to errors in the analysis, which, on the contrary, he always performed carefully, and with the requisite attention to precision, but to the bad selection of specimens put into his hand by Hauy, or those other indviduals who furnished him with the specimens which he employed in his analyses. This circumstance is very much to be deplored; because it puts it out of our power to confide in an analysis of Vauquelin, till it has been repeated and confirmed by somebody else.

Vauquelin not only improved the analytical methods, and reduced the art to a greater degree of simplicity and precision, but he discovered, likewise, new elementary bodies.

The red lead ore of Siberia had early drawn the attention of chemists, on account of its beauty ; and various attempts had been made to analyze it.

Among others, Vauquelin tried his skill upon it, in 1789, in concert with M. Macquart, who had brought specimens of it from Siberia; but at that time he did not succeed in determining the nature of the acid with which the oxide of lead was combined in it. He examined it again in 1797, and now succeeded in separating an acid to which, from the beautiful coloured salts which it forms, he gave the name of *chromic.* He determined the properties of this acid, and showed that its basis was a new metal to which he gave the name of *chromium.* He succeeded in obtaining this metal in a separate state, and showed that its protoxide is an exceedingly beautiful green powder. This discovery has been of very great importance to different branches of manufacture in this country. The green oxide is used pretty extensively in painting green on porcelain. It constitutes an exceedingly beautiful green pigment, very permanent, and easily applied. The chromic acid, when combined with oxide of lead, forms either a yellow or an orange colour upon cotton cloth, both very fixed and exceedingly beautiful colours. In that way it is extensively used by the calico-printers; and the bichromate of potash is prepared, in a crystalline form, to a very considerable amount, both in Glasgow and Lancashire, and doubtless in other places.

Vauquelin was requested by Hauy to analyze the *beryl*, a beautiful light-green mineral, crystallized in six-sided prisms, which occurs not unfrequently in granite rocks, especially in Siberia. He found it to consist chiefly of silica, united to alumina, and to another earthy body, very like alumina in many of its properties, but differing in others. To this new earth he gave the name of *glucina*, on account of the sweet taste of its salts; a name not very appropriate, as alumina, yttria, lead, protoxide of chromium, and even protoxide of iron, form salts which are distinguished by a sweet taste likewise. This discovery of glucina confers honour on Vauquelin, as it shows the care with which his analyses must have been conducted. A careless experimenter might easily have confounded *glucina* with *alumina*. Vauquelin's mode of distinguishing them was, to add sulphate of potash to their solution in sulphuric acid If the earth in solution was alumina, crystals of alum would form in the course of a short time ; but if the earth was glucina, no such crystals would make their appearance, alumina being the basis of alum, and not glucina. He showed, too, that glucina is easily dissolved in a solution of carbonate of ammonia, while alumina is not sensibly taken up by that solution.

Vauquelin died in 1829, after having reached a good old age. His character was of the very best kind, and his conduct had always been most exemplary. He never interfered with politics, and steered his way through the bloody period of the revolution, uncontaminated by the vices or violence of any party, and respected and esteemed by every person.

Mr. Chenevix deserves also to be mentioned as an improver of analytical chemistry. He was an Irish gentleman, who happened to be in Paris during the reign of terror, and was thrown into prison and put into the same apartment with several French chemists, whose whole conversation turned upon chemical subjects. He caught the infection, and, after getting out of prison, began to study the subject with much energy and success, and soon distinguished himself as an analytical chemist.

His analysis of corundum and sapphire, and his observations on the affinity between magnesia and silica, are valuable, and led to considerable improvements in the method of analysis. His analyses of the arseniates of copper, though he demonstrated that several different species exist, are not so much to be depended on; because his method of separating and estimating the quantity of arsenic acid is not good. This difficult branch of analysis was not fully understood till afterwards.

Chenevix was for several years a most laborious and meritorious chemical experimenter. It is much to be regretted that he should have been induced, in consequence of the mistake into which he fell respecting palladium, to abandon chemistry altogether. Palladium was originally made known to the public by an anonymous handbill which was circulated in London, announcing that *palladium*, or new silver, was on sale at Mrs. Forster's, and describing its properties. Chenevix, in consequence of the unusual way in which the discovery was announced, naturally considered it as an imposition on the public. He went to Mrs. Forster's, and purchased the whole palladium in her possession, and set about examining it, prepossessed with the idea that it was an alloy of some two known metals. After a laborious set of experiments, he considered that he had ascertained it to be a compound of platinum and mercury, or an amalgam of platinum made in a peculiar way, which he describes. This paper was read at a meeting of the Royal Society by Dr. Wollaston, who was secretary, and afterwards published in their Transactions. Soon after this publication, another anonymous handbill was circulated, offering a considerable price for every grain of palladium *made* by Mr. Chenevix's process, or by any other process whatever. No person appearing to claim the money thus offered, Dr. Wollaston, about a year after, in a paper read to the Royal Society, acknowledged himself to have been the discoverer of palladium, and related the process by which he had obtained it.

from the solution of crude platina in aqua regia. There could be no doubt after this, that palladium was a peculiar metal, and that Chenevix, in his experiments, had fallen into some mistake, probably by inadvertently employing a solution of palladium, instead of a solution of his amalgam of platinum; and thus giving the properties of the one solution to the other. It is very much to be regretted, that Dr. Wollaston allowed Mr. Chenevix's paper to be printed, without informing him, in the first place, of the true history of palladium : and I think that if he had been aware of the bad consequences that were to follow, and that it would ultimately occasion the loss of Mr. Chenevix to the science, he would have acted in a different manner. I have more than once conversed with Dr. Wollaston on the subject, and he assured me that he did every thing that he could do, short of betraying his secret, to prevent Mr. Chenevix from publishing his paper; that he had called upon, and assured him, that he himself had attempted his process without being able to succeed, and that he was satisfied that he had fallen into some mistake. As Mr. Chenevix still persisted in his conviction of the accuracy of his own experiments after repeated warnings, perhaps it is not very surprising that Dr. Wollaston allowed him to publish his paper, though ; had he been aware of the consequences to their full extent, I am persuaded that he would not have done so. It comes to be a question whether, had Dr. Wollaston informed him of the whole secret, Mr. Chenevix would have been convinced.

Another chemist, to whom the art of analyzing minerals lies under great obligations, is Dr. Frederick Stromeyer, professor of chemistry and pharmacy, in the University of Gottingen. He was originally a botanist, and only turned his attention to chemistry when he had the offer of the chemical chair at Got-

tingen.   He then went to Paris, and studied practical chemistry for some years in Vauquelin's laboratory.   He has devoted most of his attention to the
analysis of minerals ; and in the year 1821 published
a volume of analyses under the title of "Untersuchungen über die Mischung der Mineralkörper und
anderer damit verwandten Substanzen." It contains
thirty analyses, which constitute perfect models of
analytical sagacity and accuracy.   After Klaproth's
Beitrage, no book can be named more highly deserving the study of the analytical chemist than
Stromeyer's Untersuchungen.

The first paper in this work contains the analysis
of arragonite.   Chemists had not been able to discover any difference in the chemical constitution of
arragonite and calcareous spar, both being compounds of

$$\text{Lime} \quad . \ . \ . \ . \ . \ . \ 3{\cdot}5$$
$$\text{Carbonic acid} \quad . \ 2{\cdot}75$$

Yet the minerals differ from each other in their hardness, specific gravity, and in the shape of their crystals.   Many attempts had been made to account for
this difference in characters between these two minerals, but in vain.   Mr. Holme showed that arragonite contained about one per cent. of water, which
is wanting in calcareous spar ; and that when arragonite is heated, it crumbles into powder, which is
not the case with calcareous spar.   But it is not easy
to conceive how the addition of one per cent. of water should increase the specific gravity and the hardness, and quite alter the shape of the crystals of
calcareous spar.   Stromeyer made a vast number of
experiments upon arragonite, with very great care,
and the result was, that the arragonite from Bastenes,
near Dax, in the department of Landes, and likewise
that from Molina, in Arragon, was a compound of

96 carbonate of lime
4 carbonate of strontian.

This amounts to about thirty-five atoms of carbonate of lime, and one atom of carbonate of strontian. Now as the hardness and specific gravity of carbonate of strontian is greater than that of carbonate of lime, we can see a reason why arragonite should be heavier and harder than calcareous spar. More late researches upon different varieties of arragonite enabled him to ascertain that this mineral exists with different proportions of carbonate of strontian. Some varieties contain only 2 per cent., some only 1 per cent., and some only 0·75, or even 0·5 per cent.; but he found no specimen among the great number which he analyzed totally destitute of carbonate of strontian. It is true that Vauquelin afterwards examined several varieties in which he could detect no strontian whatever; but as Vauquelin's mineralogical knowledge was very deficient, it comes to be a question, whether the minerals analyzed by him were really arragonites, or only varieties of calcareous spar.

To Professor Stromeyer we are likewise indebted for the discovery of the new metal called *cadmium;* and the discovery does great credit to his sagacity and analytical skill. He is inspector-general of the apothecaries for the kingdom of Hanover. While discharging the duties of his office at Hildesheim, in the year 1817, he found that the carbonate of zinc had been substituted for the oxide of zinc, ordered in the Hanoverian Pharmacopœia. This carbonate of zinc was manufactured at Salzgitter. On inquiry he learned from Mr. Jost, who managed that manufactory, that they had been obliged to substitute the carbonate for the oxide of zinc, because the oxide had a yellow colour which rendered it unsaleable. On examining this oxide, Stromeyer found that it owed its yellow colour to the presence of a small quantity of the oxide of a new metal, which he separated, reduced, and examined, and to which he gave the name of *cadmium*, because it occurs usually associated with zinc. The quantity of cadmium which he was able to obtain from this oxide of zinc was but small. A fortunate circumstance, however, supplied him with an additional quantity, and enabled him to carry his examination of cadmium to a still greater length. During the apothecaries' visitation in the state of Magdeburg, there was found, in the possession of several apothecaries, a preparation of zinc from Silesia, made in Hermann's laboratory at Schönebeck, which was confiscated on the supposition that it contained arsenic, because its solution gave a yellow precipitate with sulphuretted hydrogen, which was considered as orpiment. This statement could not be indifferent to Mr. Hermann, as it affected the credit of his manufactory; especially as the medicinal counsellor, Roloff, who had assisted at the visitation, had drawn up a statement of the circumstances which occasioned the confiscation, and caused it to be published in Hofeland's Medical Journal. He subjected the suspected oxide to a careful examination; but he could not succeed in detecting any arsenic in it. He then requested Roloff to repeat his experiments. This he did; and now perceived that the precipitate, which he had taken for orpiment, was not so in reality, but owed its existence to the presence of another metallic oxide, different from arsenic and probably new. Specimens of this oxide of zinc, and of the yellow precipitate, were sent to Stromeyer for examination, who readily recognised the presence of cadmium, and was able to extract from it a considerable quantity of that metal.

It is now nine years since the first volume of the

Untersuchungen was published. All those who are interested in analytical chemistry are anxious for the continuance of that admirable work. By this time he must have collected ample materials for an additional volume; and it could not but add considerably to a reputation already deservedly high.

There is no living chemist, to whom analytical chemistry lies under greater obligations than to Berzelius, whether we consider the number or the exactness of the analyses which he has made.

Jacob Berzelius was educated at Upsala, when Professor Afzelius, a nephew of Bergman, filled the chemical chair, and Ekeberg was *magister docens* in chemistry. Afzelius began his chemical career with considerable *éclat*, his paper on sulphate of barytes being possessed of very considerable merit. But he is said to have soon lost his health, and to have sunk, in consequence, into listless inactivity.

Andrew Gustavus Ekeberg was born in Stockholm, on the 16th of January, 1767. His father was a captain in the Swedish navy. He was educated at Calmar; and in 1784 went to Upsala, where he devoted himself chiefly to the study of mathematics. He took his degree in 1788, when he wrote a thesis " De Oleis Seminum expressis." In 1789 he went to Berlin; and on his return, in 1790, he gave a specimen of his poetical talents, by publishing a poem entitled "Tal öfver Freden emellan Sverige och Ryssland" (Discourse about the Peace between Sweden and Russia). After this he turned his attention to chemistry; and in 1794 was made *chemiæ docens*. In this situation he continued till 1813, when he died on the 11th of February. He had been in such bad health for some time before his death, as to be quite unable to discharge the duties of his situation. He published but little, and that little consisted almost entirely of chemical analyses.

His first attempt was on phosphate of lime; then he wrote a paper on the analysis of the topaz, the object of which was to explain Klaproth's method of dissolving hard stony bodies.

He made an analysis of gadolinite, and determined the chemical properties of yttria. During these experiments he discovered the new metal to which he gave the name of *tantalum*, and which Dr. Wollaston afterwards showed to be the same with the *columbium* of Mr. Hatchett. He also published an analysis of the automalite, of an ore of titanium, and of the mineral water of Medevi. In this last analysis he was assisted by Berzelius, who was then quite unknown to the chemical world.

Berzelius has been much more industrious than his chemical contemporaries at Upsala. His first publication was a work in two volumes on animal chemistry, chiefly a compilation, with the exception of his experiments on the analysis of blood, which constitute an introduction to the second volume. This book was published in 1806 and 1808. In the year 1806 he and Hisinger began a periodical work, entitled "Afhandlingar i Fysik, Kemi och Mineralogi," of which six volumes in all were published, the last in 1818. In this work there occur forty-seven papers by Berzelius, some of them of great length and importance, which will be noticed afterwards; but by far the greatest part of them consist of mineral analyses. We have the analysis of cerium by Hisinger and Berzelius, together with an account of the chemical characters of the two oxides of cerium. In the fourth volume he gives us a new chemical arrangement of minerals, founded on the supposition that they are all chemical compounds in definite proportions. Mr. Smithson had thrown out the opinion that *silica* is an acid: which opinion was taken up by Berzelius, who showed, by decisive experiments, that it enters into definite combinations with most of the bases. This happy idea enabled him to show, that most of the stony minerals are definite compounds of silica, with certain earths or metallic oxides. This system has undergone several modifications since he first gave it to the world; and I think it more than doubtful whether his last correction of it, published in the Memoirs of the Stockholm Academy, for 1824, be quite as good as the first, which he published in 1815. The first arrangement was founded on the bases, the last upon the acids with which these bases are united. He was induced to alter his arrangement, in consequence of Mitcherlich's doctrine of isomorphism But I conceive that the alterations which exist in the constitution of pyroxene, amphibole, garnet, and a few other minerals, might be explained in a very simple way, without admitting this doctrine of isomorphism; which if it do not, like Berthollet's hypothesis of indefinite combinations, overturn the whole principles of chemistry, seems scarcely consistent with what we know respecting chemical combination.

In the same volume we have a set of experiments on columbium, and its characters when reduced to the metallic state; together with an analysis of all the minerals containing columbicum that were known in the year 1815.

We have also a new examination of the properties of yttria, together with the analysis of a number of minerals, containing both cerium and yttria, and the mode of separating these two substances from each other by means of sulphate of potash.

In the sixth volume we have his discovery of selenium, with an account of selenic acid, and the different compounds which it forms.

Since the year 1818 his papers have been all published in the Memoirs of the Stockholm Academy;

but he has taken care to have translations of them inserted into Poggensdorf's Annalen, and the Annales de Chimie et de Physique.

In the Stockholm Memoirs, for 1819, we have his analysis of wavellite, showing that this mineral is a hydrous phosphate of alumina. The same analysis and discovery had been made by Fuchs, who published his results in 1818 ; but probably Berzelius had not seen the paper ; at least he takes no notice of it. We have also in the same volume his analysis of euclase, of silicate of zinc, and his paper on the prussiates.

In the Memoirs for 1820 we have, besides three others, his paper on the mode of analyzing the ores of nickel. In the Memoirs for 1821 we have his paper on the alkaline sulphurets, and his analysis of achmite. The specimen selected for this analysis was probably impure ; for two successive analyses of it, made in my laboratory by Captain Lehunt, gave a considerable difference in the proportion of the constituents, and a different formula for the composition than that resulting from the constituents found by Berzelius.

In the Memoirs for 1822 we have his analysis of the mineral waters of Carlsbad. In 1823 he published his experiments on uranium, which were meant as a confirmation and extension of the examination of this substance previously made by Arfvedson. In the same year appeared his experiments on fluoric acid and its combinations, constituting one of the most curious and important of all the numerous additions which he has made to analytical chemistry. In 1824 we have his analysis of phosphate of yttria, a mineral found in Norway ; of polymignite, a mineral from the neighbourhood of Christiania, where it occurs in the zircon sienite, and remarkable for the great number of bases which it contains united to titanic acid; namely, zirconia, oxide of iron, lime, oxide of manganese, oxide of cerium, and yttria. We have also his analysis of arseniate of iron, from Brazil and from Cornwall; and of chabasite from Ferro. In this last analysis he mentions chabasites from Scotland, containing soda instead of lime. The only chabasites in Scotland, that I know of, occur in the neighbourhood of Glasgow; and in none of these have I found any soda. But I have found soda instead of lime in chabasites from the north of Ireland, always crystallized in the form to which Hauy has given the name of *trirhomboidale*. I think, therefore, that the chabasites analyzed by Arfvedson, to which Berzelius refers, must have been from Ireland, and not from Scotland; and I think it may be a question whether this form of crystal, if it should always be found to contain soda instead of lime, ought not to constitute a peculiar species.

In 1826 we have his very elaborate and valuable paper on sulphur salts. In this paper he shows that sulphur is capable of combining with bodies, in the same way as oxygen, and of converting the acidifiable bases into acids, and the alkalifiable bases into alkalies. These sulphur acids and alkalies unite with each other, and form a new class of saline bodies, which may be distinguished by the name of *sulphur salts*. This subject has been since carried a good deal further by M. H. Rose, who has by means of it thrown much light on some mineral species hitherto quite inexplicable. Thus, what is called *nickel glance*, is a sulphur salt of nickel. The acid is a compound of sulphur and arsenic, the base a compound of sulphur and nickel. Its composition may be represented thus :

1 atom disulphide of arsenic
1 atom disulphide of nickel.

In like manner glance cobalt is

1 atom disulphide of arsenic
1 atom disulphide of nickel.

Zinkenite is composed of

3 atoms sulphide of antimony
1 atom sulphide of lead ;

and jamesonite of

$2\frac{1}{2}$ atoms sulphide of antimony
1 atom sulphide of lead.

Feather ore of antimony, hitherto confounded with sulphuret of antimony, is a compound of

5 atoms sulphide of antimony
3 atoms sulphide of lead.

Gray copper ore, which has hitherto appeared so difficult to be reduced to any thing like regularity, is composed of

1 atom sulphide of antimony or arsenic
2 atoms sulphide of copper or silver.

Dark red silver ore is composed of

1 atom sulphide of antimony
1 atom sulphide of silver ;

and light red silver ore of

2 atoms sesquisulphide of arsenic
3 atoms sulphide of silver.

These specimens show how much light the doctrine of sulphur salts has thrown on the mineral kingdom.

In 1828 he published his experimental investigation of the characters and compounds of palladium, rhodium, osmium, and iridium; and upon the mode of analyzing the different ores of platinum.

One of the greatest improvements which Berzelius has introduced into analytical chemistry, is his mode of separating those bodies which become acid when united to oxygen, as sulphur, selenium, arsenic, &c., from those that become alkaline, as copper, lead, silver, &c. His method is to put the alloy or ore to be analyzed into a glass tube, and to pass over it a current of dry chlorine gas, while the powder in the tube is heated by a lamp. The acidifiable bodies are volatile, and pass over along the tube into a vessel of water placed to receive them, while the alkalifiable bodies remain fixed in the tube. This mode of analysis has been considerably improved by Rose, who availed himself of it in his analysis of gray copper ore, and other similar compounds.

Analytical chemistry lies under obligations to Berzelius, not merely for what he has done himself, but for what has been done by those pupils who were educated in his laboratory. Bonsdorf, Nordenskiöld, C. G. Gmelin, Rose, Wöhler, Arfvedson, have given us some of the finest examples of analytical investigations with which the science is furnished.

P. A. Von Bonsdorf was a professor of Abo, and after that university was burnt down, he moved to the new locality in which it was planted by the Russian government. His analysis of the minerals which crystallize in the form of the amphibole, constitutes a model for the young analysts to study, whether we consider the precision of the analyses, or the methods by which the different constituents were separated and estimated. His analysis of red silver ore first demonstrated that the metals in it were not in the state of oxides. The nature of the combination was first completely explained by Rose, after Berzelius's paper on the sulphur salts had made its appearance. His paper on the acid properties of several of the chlorides, has served considerably to extend and to rectify the views first proposed by Berzelius respecting the different classes of salts.

Nils Nordenskiöld is superintendent of the mines in Finland: his " Bidrag till närmare kännedom af Finland's Mineralier och Geognosie" was published in 1820. It contains a description and analysis of fourteen species of Lapland minerals, several of them new, and all of them interesting. The analyses were conducted in Berzelius's laboratory, and are excellent. In 1827 he published a tabular view of the mineral species, arranged chemically, in which he gives the crystalline form, hardness, and specific gravity, together with the chemical formulas for the composition.

C. G. Gmelin is professor of chemistry at Tubingen; he has devoted the whole of his attention to chemical analysis, and has published a great number of excellent ones, particularly in Schweigger's Journal. His analysis of helvine, and of the tourmalin, may be specified as particularly valuable. In this last mineral, he demonstrated the presence of boracic acid. Leopold Gmelin, professor of chemistry at Heidelberg, has also distinguished himself as an analytical chemist. His System of Chemistry, which is at present publishing, promises to be the best and most perfect which Germany has produced.

Henry Rose, of Berlin, is the son of that M. Rose who was educated by Klaproth, and afterwards became the intimate friend and fellow-labourer of that illustrious chemist. He has devoted himself to analytical chemistry with indefatigable zeal, and has favoured us with a prodigious number of new and admirably-conducted analyses. His analyses of pyroxenes, of the ores of titanium, of gray copper ore, of silver glance, of red silver ore, miargyrite, polybasite, &c., may be mentioned as examples. In 1829 he published a volume on analytical chemistry, which is by far the most complete and valuable work of the kind that has hitherto appeared; and ought to be carefully studied by all those who wish to make themselves masters of the difficult, but necessary art of analyzing compound bodies.*

* An excellent English translation of this book with several important additions by the author, has just been published by Mr. Griffin.

Wöhler is professor of chemistry in the Polytechnic School of Berlin ; he does not appear to have turned his attention to analytical chemistry, but rather towards extending our knowledge of the compounds which the different simple bodies are capable of forming with each other. His discovery of cyanic acid may be mentioned as a specimen. He is active and young ; much, therefore, may be expected from him.

Augustus Arfvedson has distinguished himself by the discovery of the new fixed alkali, lithia, in petalite and spodumene. It has been lately ascertained at Moscow, by M. R. Hermann, and the experiments have been repeated and confirmed by Berzelius, that lithia is a much lighter substance than it was found to be by Arfvedson, its atomic weight being only 1·75. We have from Arfvedson an important set of experiments on uranium and its oxides, and on the action of hydrogen on the metallic sulphurets. He has likewise analyzed a considerable number of minerals with great care ; but of late years he seems to have lost his activity. His analysis of chrysoberyl does not possess the accuracy of the rest : by some inadvertence, he has taken a compound of glucina and alumina for silica.

I ought to have included Walmstedt and Trollé-Wachmeister among the Swedish chemists who have contributed important papers towards the progress of analytical chemistry, the memoir of the former on chrysolite, and of the latter on the garnets, being peculiarly valuable. But it would extend this work to an almost interminable length, if I were to particularize every meritorious experimenter. This must plead my excuse for having omitted the names of Bucholz, Gehlen, Fuchs, Dumesnil, Dobereiner, Kupfer, and various other meritorious chemists who have contributed so much to the perfecting of the chemical analysis of the mineral kingdom. But it would be unpardonable to leave out the name of M. Mitcherlich, professor of chemistry in Berlin, and successor of Klaproth, who was also a pupil of Berzelius. He has opened a new branch of chemistry to our consideration. His papers on isomorphous bodies, on the crystalline forms of various sets of salts, on the artificial formation of various minerals, do him immortal honour, and will hand him down to posterity as a fit successor of his illustrious predecessors in the chemical chair of Berlin—a city in which an uninterrupted series of firstrate chemists have followed each other for more than a century; and where, thanks to the fostering care of the Prussian government, the number was never greater than at the present moment.

The most eminent analytical chemists at present in France are, Laugier, a nephew and successor of Fourcroy, as professor of chemistry in the Jardin du Roi, and Berthier, who has long had the superintendence of the laboratory of the School of Mines. Laugier has not published many analyses to the world, but those with which he has favoured us appear to have been made with great care, and are in general very accurate. Berthier is a much more active man; and has not merely given us many analyses, but has made various important improvements in the analytical processes. His mode of separating arsenic acid, and determining its weight, is now generally followed; and I can state from experience that his method of fusing minerals with oxide of lead, when the object is to detect an alkali, is both accurate and easy. Berthier is young, and active, and zealous; we may therefore expect a great deal from him hereafter.

The chemists in great Britain have never hitherto distinguished themselves much in analytical chemistry. This I conceive is owing to the mode of education which has been hitherto unhappily followed. Till within these very few years, practical chemistry has been nowhere taught. The consequence has been, that every chemist must discover processes for himself; and a long time elapses before he acquires the requisite dexterity and skill. About the beginning of the present century, Dr. Kennedy, of Edinburgh, was an enthusiastic and dexterous analyst; but unfortunately he was lost to the science by a premature death, after giving a very few, but these masterly, analyses to the public. About the same time, Charles Hatchett, Esq., was an active chemist, and published not a few very excellent analyses; but unfortunately this most amiable and accomplished man has been lost to science for more than a quarter of a century; the baneful effects of wealth, and the cares of a lucrative and extensive business, having completely weaned him from scientific pursuits. Mr. Gregor, of Cornwall, was an accurate man, and attended only to analytical chemistry : his analyses were not numerous, but they were in general excellent. Unfortunately the science was deprived of his services by a premature death. The same observation applies equally to Mr. Edward Howard, whose analyses of meteoric stones form an era in this branch of chemistry. He was not only a skilful chemist, but was possessed of a persevering industry which peculiarly fitted him for making a figure as a practical chemist. Of modern British analytical chemists, undoubtedly the first is Mr. Richard Philips; to whom we are indebted for not a few analyses, conducted with great chemical skill, and performed with great accuracy. Unfortunately, of late years he has done little, having been withdrawn from science by the necessity of providing for a large family, which can hardly be done, in this country,

except by turning one's attention to trade or manufactures.   The same remark applies to Dr. Henry,
who has contributed so much to our knowledge of
gaseous bodies, and whose analytical skill, had it
been wholly devoted to scientific investigations,
would have raised his reputation, as a discoverer,
much higher than it has attained; although the
celebrity of Dr. Henry, even under the disadvantages
of being a manufacturing chemist, is deservedly very
high.   Of the young chemists who have but recently
started in the path of analytical investigation, we
expect the most from Dr. Turner, of the London
University.   His analyses of the ores of manganese
are admirable specimens of skill and accuracy, and
have completely elucidated a branch of mineralogy
which, before his experiments, and the descriptions
of Haidinger appeared, was buried in impenetrable
darkness.

No man that Great Britain has produced was better fitted to have figured as an analytical chemist,
both by his uncommon chemical skill, and the
powers of his mind, which were of the highest order,
than Mr. Smithson Tennant, had he not been in
some measure prevented by a delicate frame of
body, which produced in him a state of indolence
somewhat similar to that of Dr. Black.   His discovery of osmium and iridium, and his analysis of
emery and magnesian limestone, may be mentioned
as proofs of what he could have accomplished had
his health allowed him a greater degree of exertion.
His experiments on the diamond first demonstrated
that it was composed of pure carbon ; while his discovery of phosphuret of lime has furnished lecturers
on chemistry with one of the most brilliant and
beautiful of those exhibitions which they are in the
habit of making to attract the attention of their
students.

Smithson Tennant was the only child of the Rev. Calvert Tennant, youngest son of a respectable family in Wensleydale, near Richmond, in Yorkshire, and vicar of Selby in that county. He was born on the 30th of November, 1761 : he had the misfortune to lose his father when he was only nine years of age ; and before he attained the age of manhood he was deprived likewise of his mother, by a very unfortunate accident : she was thrown from her horse while riding with her son, and killed on the spot. His education, after his father's death, was irregular, and apparently neglected ; he was sent successively to different schools in Yorkshire, at Scorton, Tadcaster, and Beverley. He gave many proofs while young of a particular turn for chemistry and natural philosophy, both by reading all books of that description which fell in his way, and by making various little experiments which the perusal of these books suggested. His first experiment was made at nine years of age, when he prepared a quantity of gunpowder for fireworks, according to directions contained in some scientific book to which he had access.

In the choice of a profession, his attention was naturally directed towards medicine, as being more nearly allied to his philosophical pursuits. He went accordingly to Edinburgh, about the year 1781, where he laid the foundation of his chemical knowledge under Dr. Black. In 1782 he was entered a member of Christ's College, Cambridge, where he began, from that time, to reside. He was first entered as a pensioner ; but disliking the ordinary discipline and routine of an academical life, he obtained an exemption from those restraints, by becoming a fellow commoner. During his residence at Cambridge his chief attention was bestowed on chemistry and botany ; though he made himself also acquainted with the elementary parts of mathematics, and had mastered the most important parts of Newton's Principia.

In 1784 he travelled into Denmark and Sweden, chiefly with the view of becoming personally acquainted with Scheele, for whom he had imbibed a high admiration. He was much gratified by what he saw of this extraordinary man, and was particularly struck with the simplicity of the apparatus with which his great experiments had been performed. On his return to England he took great pleasure in showing his friends at Cambridge various mineralogical specimens, which had been presented to him by Scheele, and in exhibiting several interesting experiments which he had learned from that great chemist. A year or two afterwards he went to France, to become personally acquainted with the most eminent of the French chemists. Thence he went to Holland and the Netherlands, at that time in a state of insurrection against Joseph II.

In 1786 he left Christ's College along with Professor Hermann, and removed with him to Emmanuel College. In 1788 he took his first degree as bachelor of physic, and soon after quitted Cambridge and came to reside in London. In 1791 he made his celebrated analysis of carbonic acid, which fully confirmed the opinions previously stated by Lavoisier respecting the constituents of this substance. His mode was to pass phosphorus through red-hot carbonate of lime. The phosphorus was acidified, and charcoal deposited. It was during these experiments that he discovered phosphuret of lime.

In 1792 he again visited Paris; but, from circumstances, being afraid of a convulsion, he was fortunate enough to leave that city the day before the memorable 10th of August. He travelled through Italy, and then passed through part of Germany.

On his return to Paris, in the beginning of 1793, he was deeply impressed with the gloom and desolation arising from the system of terror then beginning to prevail in that capital. On calling at the house of M. Delametherie, of whose simplicity and moderation he had a high opinion, he found the doors and windows closed, as if the owner were absent. Being at length admitted, he found his friend sitting in a back room, by candle-light, with the shutters closed in the middle of the day. On his departure, after a hurried and anxious conversation, his friend conjured him not to come again, as the knowledge of his being there might be attended with serious consequences to them both. To the honour of Delametherie, it deserves to be stated, that through all the inquisitions of the revolution, he preserved for his friend property of considerable value, which Mr. Tennant had intrusted to his care.

On his return from the continent, he took lodgings in the Temple, where he continued to reside during the rest of his life. He still continued the study of medicine, and attended the hospitals, but became more indifferent about entering into practice. He took, however, a doctor's degree at Cambridge in 1796; but resolved, as his fortune was independent, to relinquish all idea of practice, as not likely to contribute to his happiness. Exquisite sensibility was a striking feature in his character, and it would, as he very properly conceived, have made him peculiarly unfit for the exercise of the medical profession. It may be worth while to relate an example of his practical benevolence which happened about this time.

He had a steward in the country, in whom he had long placed implicit confidence, and who was considerably indebted to him. In consequence of this man's becoming embarrassed in his circumstances,

Mr. Tennant went into the country to examine his accounts.   A time and place were appointed for him produce his books, and show the extent of the deficiency; but the unfortunate steward felt himself unequal to the task of such an explanation, and in a fit of despair put an end to his existence.   Touched by this melancholy event, Mr. Tennant used his utmost exertions for the relief and protection of the family whom he had left, and not only forgave them the debt, but afforded them pecuniary assistance, and continued ever afterwards to be their friend and benefactor.

During the year 1796 he made his experiments to prove that the diamond is pure carbon.   His method was to heat it in a gold tube, with saltpetre.   The diamond was converted into carbonic acid gas, which combined with the potash from the saltpetre, and by the evolution of which the quantity of carbon, in a given weight of diamond, might be estimated.   A characteristic trait of Mr. Tennant occurred during the course of this experiment, which I relate on the authority of Dr. Wollaston, who was present as an assistant, and who related the fact to me.   Mr. Tennant was in the habit of taking a ride on horseback every day at a certain hour.   The tube containing the diamond and saltpetre were actually heating, and the experiment considerably advanced, when, suddenly recollecting that his hour for riding was come, he left the completion of the process to Dr. Wollaston, and went out as usual to take his ride.

In the year 1797, in consequence of a visit to a friend in Lincolnshire, where he witnessed the activity with which improvements in farming operations were at that time going on, he was induced to purchase some land in that country, in order to commence farming operations.   In 1799 he bought a considerable tract of waste land in Somersetshire, near the village of Cheddar, where he built a small house, in which, during the remainder of his life, he was in the habit of spending some months every summer, besides occasional visits at other times of the year. These farming speculations, as might have been anticipated from the indolent and careless habits of Mr. Tennant, were not very successful. Yet it appears from the papers which he left behind him, that he paid considerable attention to agriculture, that he had read the best books on the subject, and collected many facts on it during his different journeys through various parts of England. In the course of these inquiries he had discovered that there were two kinds of limestone known in the midland counties of England, one of which yielded a lime injurious to vegetation. He showed, in 1799, that the presence of carbonate of magnesia is the cause of the bad qualities of this latter kind of limestone. He found that the magnesian limestone forms an extensive stratum in the midland counties, and that it occurs also in primitive districts under the name of dolomite.

He infers from the slow solubility of this limestone in acids, that it is a double salt composed of carbonate of lime and carbonate of magnesia in chemical combination. He found that grain would scarcely germinate, and that it soon perished in moistened carbonate of magnesia: hence he concluded that magnesia is really injurious to vegetation. Upon this principle he accounted for the injurious effects of the magnesian limestone when employed as a manure.

In 1802 he showed that emery is merely a variety of corundum, or of the precious stone known by the name of sapphire.

During the same year, while endeavouring to make an alloy of lead with the powder which remains after treating crude platinum with aqua regia, he observed remarkable properties in this powder, and found that it contained a new metal. While he was engaged in the investigation, Descotils had turned his attention to the same powder, and had discovered that it contained a metal which gives a red colour to the ammoniacal precipitate of platinum. Soon after, Vauquelin, having treated the powder with alkali, obtained a volatile metallic oxide, which he considered as the same metal that had been observed by Descotils. In 1804 Mr. Tennant showed that this powder contains two new metals, to which he gave the name of *osmium* and *iridium*.

Mr. Tennant's health, by this time, had become delicate, and he seldom went to bed without a certain quantity of fever, which often obliged him to get up during the night and expose himself to the cold air. To keep himself in any degree in health, he found it necessary to take a great deal of exercise on horseback. He was always an awkward and a bad horseman, so that these rides were sometimes a little hazardous ; and I have more than once heard him say, that a fall from his horse would some day prove fatal to him. In 1809 he was thrown from his horse near Brighton, and had his collar-bone broken.

In the year 1812 he was prevailed upon to deliver a few lectures on the principles of mineralogy, to a number of his friends, among whom were many ladies, and a considerable number of men of science and information. These lectures were completely successful, and raised his reputation very much among his friends as a lecturer. He particularly excelled in the investigation of minerals by the blowpipe ; and I have heard him repeatedly say, that he was indebted for the first knowledge of the mode of using that valuable instrument to Assessor Gahn of Fahlun.

In 1813, a vacancy occurring in the chemical professorship at Cambridge, he was solicited to become a candidate. His friends exerted themselves in his favour with unexampled energy; and all opposition being withdrawn, he was elected professor in May, 1813.

After the general pacification in 1814 he went to France, and repaired to the southern provinces of that kingdom. He visited Lyons, Nismes, Avignon, Marseilles, and Montpellier. He returned to Paris in November, much gratified by his southern tour. He was to have returned to England about the latter end of the year; but he continued to linger on till the February following. On the 15th of that month he went to Calais; but the wind blew directly into Calais harbour, and continued unfavourable for several days. After waiting till the 20th he went to Boulogne, in order to take the chance of a better passage from that port. He embarked on board a vessel on the 22d, but the wind was still adverse, and blew so violently that the vessel was obliged to put back. When Mr. Tennant came ashore, he said that " it was in vain to struggle with the elements, and that he was not yet tired of life." It was determined, in case the wind should abate, to make another trial in the evening. During the interval Mr. Tennant proposed to his fellow-traveller, Baron Bulow, that they should hire horses and take a ride. They rode at first along the sea-side; but, on Mr. Tennant's suggestion, they went afterwards to Bonaparte's pillar, which stands on an eminence about a league from the sea-shore, and which, having been to see it the day before, he was desirous of showing to Baron Bulow. On their return from thence they deviated a little from the road, in order to look at a small fort near the pillar, the entrance to which was over a fosse twenty feet deep. On the side towards them, there was a standing bridge for some way, till it joined a drawbridge, which turned on a pivot. The end next the fort rested on the ground. On the side next to them it was usually fastened by a bolt; but the bolt had been stolen about a fortnight before, and was not replaced. As the bridge was too narrow for them to go abreast, the baron said he would go first, and attempted to ride over it; but perceiving that it was beginning to sink, he made an effort to pass the centre, and called out to warn his companion of his danger; but it was too late: they were both precipitated into the trench. The baron, though much stunned, fortunately escaped without any serious hurt; but on recovering his senses, and looking round for Mr. Tennant, he found him lying under his horse nearly lifeless. He was taken, however, to the Civil Hospital, as the nearest place ready to receive him. After a short interval, he seemed in some slight degree to recover his senses, and made an effort to speak, but without effect, and died within the hour. His remains were interred a few days after in the public cemetery at Boulogne, being attended to the grave by most of the English residents.

There is another branch of investigation intimately connected with analytical chemistry, the improvements in which have been attended with great advantage, both to mineralogists and chemists. I mean the use of the blowpipe, to make a kind of miniature analysis of minerals in the dry way; so far, at least, as to determine the nature of the constituents of the mineral under examination. This is attended with many advantages, as a preliminary to a rigid analysis by solution. By informing us of the nature of the constituents, it enables us to form a plan of the analysis beforehand, which, in many cases, saves the trouble and the tediousness of two separate analytical investigations; for when we set about analyzing a mineral, of the nature of which we are entirely ignorant, two separate sets of experiments are in most cases indispensable. We must examine the mineral, in the first place, to determine the nature of its constituents. These being known, we can form a plan of an analysis, by means of which we can separate and estimate in succession the amount of each constituent of the mineral. Now a judicious use of the blowpipe often enables us to determine the nature of the constituents in a few minutes, and thus saves the trouble of the preliminary analysis.

The blowpipe is a tube employed by goldsmiths in soldering. By means of it, they force the flame of a candle or lamp against any particular point which they wish to heat. This enables them to solder trinkets of various kinds, without affecting any other part except the portion which is required to be heated. Cronstedt and Engestroem first thought of applying this little instrument to the examination of minerals. A small fragment of the mineral to be examined, not nearly so large as the head of a small pin, was put upon a piece of charcoal, and the flame of a candle was made to play upon it by means of a blowpipe, so as to raise it to a white heat. They observed whether it decrepitated, or was dissipated, or melted; and whatever the effect produced was, they were enabled from it to draw consequences respecting the nature of the mineral under examination.

The importance of this instrument struck Bergman, and induced him to wish for a complete examination of the action of the heat of the blowpipe upon all different minerals, either tried *per se* upon charcoal, or mixed with various fluxes; for three different substances had been chosen as fluxes, namely, *carbonate of soda*, *borax*, and *biphosphate of soda;* or, at least, what was in fact an equivalent for this last substance, *ammonio-phosphate of soda,* or *microcosmic salt,* at that time extracted from urine. This salt is a compound of one integrant particle of phosphate of soda, and one integrant particle of phosphate of ammonia. When heated before the blowpipe it fuses, and the water of crystallization, together with the ammonia, are gradually dissipated, so that at last nothing remains but biphosphate of soda. These fluxes have been found to act with considerable energy on most minerals. The carbonate of soda readily fuses with those that contain much silica, while the borax and biphosphate of soda act most powerfully on the bases, not sensibly affecting the silica, which remains unaltered in the fused bead. A mixture of borax and carbonate of soda upon charcoal in general enables us to reduce the metallic oxides to the state of metals, provided we understand the way of applying the flame properly. Bergman employed Gahn, who was at that time his pupil, and whose skill he was well acquainted with, to make the requisite experiments. The result of these experiments was drawn up into a paper, which Bergman sent to Baron Born in 1777, and they were published by him at Vienna in 1779. This valuable publication threw a new light upon the application of the blowpipe to the assaying of minerals ; and for every thing new which it contained Bergman was indebted to Gahn, who had made the experiments.

John Gottlieb Gahn, the intimate friend of Bergman and of Scheele, was one of the best-informed men, and one whose manners were the most simple, unaffected, and pleasing, of all the men of science with whom I ever came in contact. I spent a few days with him at Fahlun, in 1812, and they were some of the most delightful days that I ever passed in my life. His fund of information was inexhaustible, and was only excelled by the charming simplicity of his manners, and by the benevolence and goodness of heart which beamed in his countenance. He was born on the 17th of August, 1745, at the Woxna iron-works, in South Helsingland, where his father, Hans Jacob Gahn, was treasurer to the government of Stora Kopperberg. His grandfather, or great-grandfather, he told me, had emigrated from Scotland; and he mentioned several families in Scotland to which he was related. After completing his school education at Westerås, he went, in the year 1760, to the University of Upsala. He had already shown a decided bias towards the study of chemistry, mineralogy, and natural philosophy; and, like most men of science in Sweden, where philosophical instrument-makers are scarcely to be found, he had accustomed himself to handle the different tools, and to supply himself in that manner with all the different pieces of apparatus which he required for his investigations. He seems to have spent nearly ten years at Upsala, during which time he acquired a very profound knowledge in chemistry, and made various important discoveries, which his modesty or his indifference to fame made him allow others to pass as their own. The discovery of the rhomboidal nucleus of carbonate of lime in a six-sided prism of that mineral, which he let fall, and which was accidentally broken, constitutes the foundation of Hauy's system of crystallization. He communicated the fact to Bergman, who published it as his own in the second volume of his Opuscula, without any mention of Gahn's name.

The earth of bones had been considered as a peculiar simple earth; but Gahn ascertained, by analysis, that it was a compound of phosphoric acid and lime; and this discovery he communicated to Scheele, who, in his paper on fluor spar, published in 1771,

R 2

observed, in the seventeenth section, in which he is describing the effect of phosphoric acid on fluor spar, " It has lately been discovered that the earth of bones, or of horns, is calcareous earth combined with phosphoric acid." In consequence of this remark, in which the name of Gahn does not appear, it was long supposed that Scheele, and not Gahn, was the author of this important discovery.

It was during this period that he demonstrated the metallic nature of manganese, and examined the properties of the metal. This discovery was announced as his, at the time, by Bergman, and was almost the only one of the immense number of new facts which he had ascertained that was publicly known to be his.

On the death of his father he was left in rather narrow circumstances, which obliged him to turn his immediate attention to mining and metallurgy. To acquire a practical knowledge of mining he associated with the common miners, and continued to work like them till he had acquired all the practical dexterity and knowledge which actual labour could give. In 1770 he was commissioned by the College of Mines to institute a course of experiments, with a view to improve the method of smelting copper, at Fahlun. The consequence of this investigation was a complete regeneration of the whole system, so as to save a great deal both of time and fuel.

Sometime after, he became a partner in some extensive works at Stora Kopperberg, where he settled as a superintendent. From 1770, when he first settled at Fahlun, down to 1785, he took a deep interest in the improvement of the chemical works in that place and neighbourhood. He established manufactories of sulphur, sulphuric acid, and red ochre.

In 1780 the Royal College of Mines, as a testimony of their sense of the value of Gahn's improvements, presented him with a gold medal of merit. In 1782 he received a royal patent as mining master. In 1784 he was appointed assessor in the Royal College of Mines, in which capacity he officiated as often as his other vocations permitted him to reside in Stockholm. The same year he married Anna Maria Bergstrom, with whom he enjoyed for thirty-one years a life of uninterrupted happiness. By his wife he had a son and two daughters.

In the year 1773 he had been elected chemical stipendiary to the Royal College of Mines, and he continued to hold this appointment till the year 1814. During the whole of this period the solution of almost every difficult problem remitted to the college devolved upon him. In 1795 he was chosen a member of the committee for directing the general affairs of the kingdom. In 1810 he was made one of the committee for the general maintenance of the poor. In 1812 he was elected an active associate of the Royal Academy for Agriculture; and in 1816 he became a member of the committee for organizing the plan of a Mining Institute. In 1818 he was chosen a member of the committee of the Mint; but from this situation he was shortly after, at his own request, permitted to withdraw.

His wife died in 1815, and from that period his health, which had never been robust, visibly declined. Nature occasionally made an effort to shake off the disease; but it constantly returned with increasing strength, until, in the autumn of 1818, the decay became more rapid in its progress, and more decided in its character. He became gradually weaker, and on the 8th of December, 1818, died without a struggle, and seemingly without pain.

Ever after the experiments on the blowpipe which Gahn performed at the request of Bergman, his attention had been turned to that piece of apparatus;

and during the course of a long life he had introduced so many improvements, that he was enabled,
by means of the blowpipe, to determine in a few
minutes the constituents of almost any mineral. He
had gone over almost all the mineral kingdom, and
determined the behaviour of almost every mineral
before the blowpipe, both by itself and when mixed
with the different fluxes and reagents which he had
invented for the purpose of detecting the different
constituents; but, from his characteristic unwilling-
ness to commit his observations and experiments to
writing, or to draw them up into a regular memoir,
had not Berzelius offered himself as an assistant,
they would probably have been lost. By his means
a short treatise on the blowpipe, with minute directions how to use the different contrivances which
he had invented, was drawn up and inserted in the
second volume of Berzelius's Chemistry. Berzelius
and he afterwards examined all the minerals known,
or at least which they could procure, before the blowpipe; and the result of the whole constituted the
materials of Berzelius's treatise on the blowpipe,
which has been translated into German, French, and
English. It may be considered as containing the
sum of all the improvements which Gahn had made
on the use of the blowpipe, together with all the
facts that he had collected respecting the phenomena exhibited by minerals before the blowpipe. It
constitutes an exceedingly useful and valuable book,
and ought to make a part of the library of every
analytical chemist.

Dr. Wollaston had paid as much attention to the
blowpipe as Gahn, and had introduced so many improvements into its use, that he was able, by means
of it, to determine the nature of the constituents of
any mineral in the course of a few minutes. He
was fond of such analytical experiments, and was

generally applied to by every person who thought himself possessed of a new mineral, in order to be enabled to state what its constituents were. The London mineralogists if the race be not extinct, must sorely feel the want of the man to whom they were in the habit of applying on all occasions, and to whom they never applied in vain.

Dr. William Hyde Wollaston, was the son of the Reverend Dr. Wollaston, a clergyman of some rank in the church of England, and possessed of a competent fortune. He was a man of abilities, and rather eminent as an astronomer. His grandfather was the celebrated author of the Religion of Nature delineated. Dr. William Hyde Wollaston was born about the year 1767, and was one of fifteen children, who all reached the age of manhood. His constitution was naturally feeble ; but by leading a life of the strictest sobriety and abstemiousness he kept himself in a state fit for mental exertion. He was educated at Cambridge, where he was at one time a fellow. After studying medicine by attending the hospitals and lectures in London, and taking his degree of doctor at Cambridge, he settled at Bury St. Edmund's, where he practised as a physician for some years. He then went to London, became a fellow of the Royal College of Physicians, and commenced practitioner in the metropolis. A vacancy occurring in St. George's Hospital, he offered himself for the place of physician to that institution; but another individual, whom he considered his inferior in knowledge and science, having been preferred before him, he threw up the profession of medicine altogether, and devoted the rest of his life to scientific pursuits. His income, in consequence of the large family of his father, was of necessity small. In order to improve it he turned his thoughts to the manufacture of platinum, in which he succeeded so well, that he must have, by means of it, realized considerable sums. It was he who first succeeded in reducing it into ingots in a state of purity and fit for every kind of use : it was employed, in consequence, for making vessels for chemical purposes ; and it is to its introduction that we are to ascribe the present accuracy of chemical investigations. It has been gradually introduced into the sulphuric acid manufactories, as a substitute for glass retorts.

Dr. Wollaston had a particular turn for contriving pieces of apparatus for scientific purposes. His reflecting goniometer was a most valuable present to mineralogists, and it is by its means that crystallography has acquired the great degree of perfection which it has recently exhibited. He contrived a very simple apparatus for ascertaining the power of various bodies to refract light. His camera lucida furnished those who were ignorant of drawing with a convenient method of delineating natural objects. His periscopic glasses must have been found useful, for they sold rather extensively : and his sliding rule for chemical equivalents furnished a ready method for calculating the proportions of one substance necessary to decompose a given weight of another.

Dr. Wollaston's knowledge was more varied, and his taste less exclusive than any other philosopher of his time, except Mr. Cavendish : but optics and chemistry are the two sciences which lie under the greatest obligations to him. His first chemical paper on urinary calculi at once added a vast deal to what had been previously known. He first pointed out the constituents of the mulberry calculi, showing them to be composed of oxalate of lime and animal matter. He first distinguished the nature of the triple phosphates. It was he who first ascertained the nature of the cystic oxides, and of the chalk-stones, which appear occasionally in the joints of gouty patients. To him we owe the first demonstration of the identity of galvanism and common electricity; and the first explanation of the cause of the different phenomena exhibited by galvanic and common electricity. To him we are indebted for the discovery of palladium and rhodium, and the first account of the properties and characters of these two metals. He first showed that oxalic acid and potash unite in three different proportions, constituting oxalate, binoxalate, and quadroxalate of potash. Many other chemical facts, first ascertained by him, are to be found in the numerous papers of his scattered over the last forty volumes of the Philosophical Transactions: and perhaps not the least valuable of them is his description of the mode of reducing platinum from the raw state, and bringing it into the state of an ingot.

Dr. Wollaston died in the month of January, 1829, in consequence of a tumour formed in the brain, near, if I remember right, the thalami nervorum opticorum. There is reason to suspect that this tumour had been some time in forming. He had, without exception, the sharpest eye that I have ever seen: he could write with a diamond upon glass in a character so small, that nothing could be distinguished by the naked eye but a ragged line; yet when the letters were viewed through a microscope, they were beautifully regular and quite legible. He retained his senses to almost the last moment of his life: when he lay apparently senseless, and his friends were anxiously solicitous whether he still retained his understanding, he informed them, by writing, that his senses were still perfectly entire. Few individuals ever enjoyed a greater share of general respect and confidence, or had fewer enemies, than Dr. Wollaston. He was at first shy and distant, and remarkably circumspect, but he grew insensibly more and more agreeable as you got better acquainted with him, till at last you formed for him the most sincere friendship, and your acquaintance ended in the warmest and closest attachment.

CHAPTER V.

OF ELECTRO-CHEMISTRY.

ELECTRICITY, like chemistry, is a modern science; for it can scarcely claim an older origin than the termination of the first quarter of the preceding century; and during the last half of that century, and a small portion of the present, it participated with chemistry in the zeal and activity with which it was cultivated by the philosophers of Europe and America. For many years it was not suspected that any connexion existed between chemistry and electricity; though some of the meteorological phenomena, especially the production of clouds and the formation of rain, which are obviously connected with chemistry, seem likewise to claim some connexion with the agency of electricity.

The discovery of the intimate relation between chemistry and electricity was one of the consequences of a controversy carried on about the year 1790 between Galvani and Volta, two Italian philosophers, whose discoveries will render their names immortal. Galvani, who was a professor of anatomy, was engaged in speculations respecting muscular motion. He was of opinion that a peculiar fluid was secreted in the brain, which was sent along the nerves to all the different parts of the body. This nervous fluid possessed many characters analogous to those of electricity : the muscles were capable of being charged with it somewhat like a Leyden phial ; and it was by the discharge of this accumulation, by the voluntary power of the nerves, that muscular motion was produced. He accidently discovered, that if the crural nerve going into the muscles of a frog, and the crural muscles, be laid bare immediately after death, and a piece of zinc be placed in contact with the nerve, and a piece of silver or copper with the muscle ; when these two pieces of metal are made to touch each other, violent convulsions are produced in the muscle, which cause the limb to move. He conceived that these convulsions were produced by the discharge of the nervous energy from the muscles, in consequence of the conducting power of the metals.

Volta, who repeated these experiments, explained them in a different manner. According to him, the convulsions were produced by the passage of a current of common electricity through the limb of the frog, which was thrown into a state of convulsion merely in consequence of its irritability. This irritability vanishes after the death of the muscle ; accordingly it is only while the principle of life remains that the convulsions can be produced. Every metallic conductor, according to him, possesses a certain electricity which is peculiar to it, either positive or negative, though the quantity is so small, as to be imperceptible, in the common state of the metal. But if a metal, naturally positive, be placed in contact, while insulated, with a metal naturally negative, the charge of electricity in both is increased by induction, and becomes perceptible when the two metals are separated and presented to a sufficiently delicate electrometer. Thus zinc is naturally positive, and copper and silver naturally negative. If we take two discs of copper and zinc, to the centre of each of which a varnished glass handle is cemented, and after keeping them for a short time in contact, separate them by the handles, and apply each to a sufficiently delicate electrometer, we shall find that the zinc is positive, and the silver or copper disc negative. When the silver and copper are placed in contact while lying on the nerve and muscles of the leg of a frog, the zinc becomes positive, and the silver negative, by induction; but, as the animal substance is a conductor, this state cannot continue: the two electricities pass through the conducting muscles and nerve, and neutralize one another. And it is this current which occasions the convulsions.

Such was Volta's simple explanation of the convulsions produced in galvanic experiments in the limb of a frog. Galvani was far from allowing the accuracy of it; and, in order to obviate the objection to his reasoning advanced by Volta from the necessity of employing two metals, he showed that the convulsions might, in certain cases, be produced by one metal. Volta showed that a very small quantity of one metal, either alloyed with, or merely in contact with another, were capable of inducing the two electricities. But in order to prove in the most unanswerable manner that the two electricities were induced when two different metals were placed in contact, he contrived the following piece of apparatus:

He procured a number (say 50) of pieces of zinc, about the size of a crown-piece, and as many pieces of copper, and thirdly, the same number of pieces of card of the same size. The cards were steeped in a solution of salt, so as to be moist. He lays upon the table a piece of zinc, places over it a piece of copper, and then a piece of moist card. Over the card is placed a second piece of zinc, then a piece of copper, then a piece of wet card. In this way

all the pieces are piled upon each other in exactly
the same order, namely, zinc, copper, card ; zinc,
copper, card ; zinc, copper, card. So that the lowest
plate is zinc and the uppermost is copper (for the
last wet card may be omitted). In this way there
are fifty pairs of zinc and copper plates in contact,
each separated by a piece of wet card, which is a
conductor of electricity. If you now moisten a
finger of each hand with water, and apply one wet
finger to the lowest zinc plate, and the other to the
highest copper plate, the moment the fingers come
in contact with the plates an electric shock is felt,
the intensity of which increases with the number of
pairs of plates in the pile. This is what is called
the Galvanic, or rather the Voltaic pile. It was
made known to the public in a paper by Volta, inserted in the Philosophical Transactions for 1800.
This pile was gradually improved, by substituting
troughs, first of baked wood, and afterwards of
porcelain, divided into as many cells as there were
pairs of plates. The size of the plates was increased;
they were made square, and instead of all being in
contact, it was found sufficient if they were soldered
together by means of metallic slips rising from one
side of each square. The two plates thus soldered
were slipped over the diaphragm separating the
contiguous cells, so that the zinc plate was in one
cell and the copper in the other. Care was taken
that the pairs were introduced all looking one way,
so that a copper plate had always a zinc plate immediately opposite to it. The cells were filled with
conducting liquid : brine, or a solution of salt in
vinegar, or dilute muriatic, sulphuric, or nitric acid,
might be employed ; but dilute nitric acid was found
to answer best, and the energy of the battery is
directly proportional to the strength of the nitric
acid employed.

Messrs. Nicholson and Carlisle were the first persons who repeated Volta's experiments with this apparatus, which speedily drew the attention of all Europe. They ascertained that the zinc end of the pile was positive and the copper end negative. Happening to put a drop of water on the uppermost plate, and to put into it the extremity of a gold wire connected with the undermost plate, they observed an extrication of air-bubbles from the wire. This led them to suspect that the water was decomposed. To determine the point, they collected a little of the gas extricated and found it hydrogen. They then attached a gold wire to the zinc end of the pile, and another gold wire to the copper end, and plunged the two wires into a glass of water, taking care not to allow them to touch each other. Gas was extricated from both wires. On collecting that from the wire attached to the zinc end, it was found to be *oxygen gas*, while that from the copper end was hydrogen gas. The volume of hydrogen gas extricated was just double that of the oxygen gas; and the two gases being mixed, and an electric spark passed through them, they burnt with an explosion, and were completely converted into water. Thus it was demonstrated that water was decomposed by the action of the pile, and that the oxygen was extricated from the positive pile and the hydrogen from the negative. This held when the communicating wires were gold or platinum; but if they were of copper, silver, iron, lead, tin, or zinc, then only hydrogen gas was extricated from the negative wire. The positive wire extricated little or no gas; but it was rapidly oxidized. Thus the connexion between chemical decompositions and electrical currents was first established.

It was soon after observed by Henry, Haldane, Davy, and other experimenters, that other chemical compounds were decomposed by the electrical currents as well as water. Ammonia, for example, nitric acid, and various salts, were decomposed by it. In the year 1803 an important set of experiments was published by Berzelius and Hisinger. They decomposed eleven different salts, by exposing them to the action of a current of electricity. The salts were dissolved in water, and iron or silver wires from the two poles of the pile were plunged into the solution. In every one of these decompositions, the acid was deposited round the positive wire, and the base of the salt round the negative wire. When ammonia was decomposed by the action of galvanic electricity, the azotic gas separated from the positive wire, and the hydrogen gas from the negative.

But it was Davy that first completely elucidated the chemical decompositions produced by galvanic electricity, who first explained the laws by which these decompositions were regulated, and who employed galvanism as an instrument for decomposing various compounds, which had hitherto resisted all the efforts of chemists to reduce them to their elements. These discoveries threw a blaze of light upon the obscurest parts of chemistry, and secured for the author of them an immortal reputation.

Humphry Davy, to whom these splendid discoveries were owing, was born at Penzance, in Cornwall, in the year 1778. He displayed from his very infancy a spirit of research, and a brilliancy of fancy, which augured, even at that early period, what he was one day to be. When very young, he was bound apprentice to an apothecary in his native town. Even at that time, his scientific acquirements were so great, that they drew the attention of Mr. Davis Gilbert, the late distinguished president of the Royal Society. It was by his advice that he resolved to devote himself to chemistry, as the pursuit best calculated to procure him celebrity. About this time Mr. Gregory Watt, youngest son of the celebrated improver of the steam-engine, happening to be at Penzance, met with young Davy, and was delighted with the uncommon knowledge which he displayed, at the brilliancy of his fancy, and the great dexterity and ardour with which, under circumstances the most unfavourable, he was prosecuting his scientific investigations. These circumstances made an indelible impression on his mind, and led him to recommend Davy as the best person to superintend the Bristol Institution for trying the medicinal effects of the gases.

After the discovery of the different gases, and the investigation of their properties by Dr. Priestley, it occurred to various individuals, nearly about the same time, that the employment of certain gases, or at least of mixtures of certain gases, with common air in respiration, instead of common air, might be powerful means of curing diseases. Dr. Beddoes, at that time professor of chemistry at Oxford, was one of the keenest supporters of these opinions. Mr. Watt, of Birmingham, and Mr. Wedgewood, entertained similar sentiments. About the beginning of the present century, a sum of money was raised by subscription, to put these opinions to the test of experiment; and, as Dr. Beddoes had settled as a physician in Bristol, it was agreed upon that the experimental investigation should take place at Bristol. But Dr. Beddoes was not qualified to superintend an institution of the kind : it was necessary to procure a young man of zeal and genius, who would take such an interest in the investigation as would compensate for the badness of the apparatus and the defects of the arrangements. The greatest part of the money had been subscribed by Mr. Wedgewood and Mr. Watt : their influence of course would

be greatest in recommending a proper superintendent. Gregory Watt thought of Mr. Davy, whom
he had lately been so highly pleased with, and recommended him with much zeal to superintend the
undertaking. This recommendation being seconded
by that of Mr. Davis Gilbert, who was so well acquainted with the scientific acquirements and genius
of Davy, proved successful, and Davy accordingly
got the appointment. At Bristol he was employed
about a year in investigating the effects of the gases
when employed in respiration. But he did not by
any means confine himself to this, which was the
primary object of the institution; but investigated
the properties and determined the composition of
nitric acid, ammonia, protoxide of azote and deutoxide of azote. The fruit of his investigations was
published in 1800, in a volume entitled, " Researches, Chemical and Philosophical; chiefly concerning Nitrous Oxide, or Dephlogisticated Nitrous
Air, and its Respiration." This work gave him at
once a high reputation as a chemist, and was really
a wonderful performance, when the circumstances
under which it was produced are taken into consideration. He had discovered the intoxicating effects
which protoxide of azote (nitrous oxide) produces
when breathed, and had tried their effects upon a
great number of individuals. This fortunate discovery perhaps contributed more to his celebrity, and
to his subsequent success, than all the sterling merit
of the rest of his researches—so great is the effect of
display upon the greater part of mankind.

A few years before, a philosophical institution had
been established in London, under the auspices of
Count Rumford, which had received the name of
the Royal Institution. Lectures on chemistry and
natural philosophy were delivered in this institution;
a laboratory was provided, and a library established.

The first professor appointed to this institution, Dr. Garnet, had been induced, in consequence of some disagreement between him and Count Rumford, to throw up his situation. Many candidates started for it; but Davy, in consequence of the celebrity which he had acquired by his researches, or perhaps by the intoxicating effects of protoxide of azote, which he had discovered, was, fortunately for the institution and for the reputation of England, preferred to them all. He was appointed professor of chemistry, and Dr. Thomas Young professor of natural philosophy, in the year 1801. Davy, either from the more popular nature of his subject, or from his greater oratorical powers, became at once a popular lecturer, and always lectured to a crowded room; while Dr. Young, though both a profound and clear lecturer, could scarcely command an audience of a dozen. It was here that Davy laboured with unwearied industry during eleven years, and acquired by his discoveries the highest reputation of any chemist in Europe.

In 1811 he was knighted, and soon after married Mrs. Apreece, a widow lady, daughter of Mr. Ker, who had been secretary to Lord Rodney, and had made a fortune in the West Indies. He was soon after created a baronet. About this time he resigned his situation as professor of chemistry in the Royal Institution, and went to the continent. He remained for some years in France and Italy. In the year 1821, when Sir Joseph Banks died, a very considerable number of the fellows offered their votes to Dr. Wollaston; but he declined standing as a candidate for the president's chair. Sir Humphry Davy, on the other hand, was anxious to obtain that honourable situation, and was accordingly elected president by a very great majority of votes on the 30th of November, 1821. This honourable situa-

tion he filled about seven years; but his health declining, he was induced to resign in 1828, and to
go to Italy. Here he continued till 1829, when
feeling himself getting worse, and wishing to draw
his last breath in his own country, he began to turn
his way homewards; but at Geneva he felt himself
so ill, that he was unable to proceed further: here
he took to his bed, and here he died on the 29th of
May, 1829.

It was his celebrated paper " On some chemical
Agencies of Electricity," inserted in the Philosophical Transactions for 1807, that laid the foundation of the high reputation which he so deservedly
acquired. I consider this paper not merely as the
best of all his own productions, but as the finest and
completest specimen of inductive reasoning which
appeared during the age in which he lived. It had
been already observed, that when two platinum wires
from the two poles of a galvanic pile are plunged
each into a vessel of water, and the two vessels
united by means of wet asbestos, or any other conducting substance, an *acid* appeared round the positive wire and an *alkali* round the negative wire.
The alkali was said by some to be *soda*, by others to
be *ammonia*. The acid was variously stated to be
nitric acid, muriatic acid, or even *chlorine*. Davy
demonstrated, by decisive experiments, that in all
cases the acid and alkali are derived from the decomposition of some salt contained either in the water
or in the vessel containing the water. Most commonly the salt decomposed is common salt, because
it exists in water and in agate, basalt, and various
other stony bodies, which he employed as vessels.
When the same agate cup was used in successive
experiments, the quantity of acid and alkali evolved
diminished each time, and at last no appreciable
quantity could be perceived. When glass vessels

were used, soda was disengaged at the expense of the glass, which was sensibly corroded. When the water into which the wires were dipped was perfectly pure, and when the vessel containing it was free from every trace of saline matter, no acid or alkali made its appearance, and nothing was evolved except the constituents of water, namely, oxygen and hydrogen; the oxygen appearing round the positive wire, and the hydrogen round the negative wire.

When a salt was put into the vessel in which the positive wire dipped, the vessel into which the negative wire dipped being filled with pure water, and the two vessels being united by means of a slip of moistened asbestos, the acid of the salt made its appearance round the positive wire, and the alkali round the negative wire, before it could be detected in the intermediate space; but if an intermediate vessel, containing a substance for which the alkali has a strong affinity, be placed between these two vessels, the whole being united by means of slips of asbestos, then great part, or even the whole of the alkali, was stopped in this intermediate vessel. Thus, if the salt was nitrate of barytes, and sulphuric acid was placed in the intermediate vessel, much sulphate of barytes was deposited in the intermediate vessel, and very little or even no barytes made its appearance round the negative wire. Upon this subject a most minute, extensive, and satisfactory series of experiments was made by Davy, leaving no doubt whatever of the accuracy of the fact.

The conclusions which he drew from these experiments are, that all substances which have a chemical affinity for each other, are in different states of electricity, and that the degree of affinity is proportional to the intensity of these opposite states.

When such a compound body is placed in contact
with the poles of a galvanic battery, the positive
pole attracts the constituent, which is negative, and
repels the positive.   The negative acts in the opposite way, attracting the positive constituent and repelling the negative.   The more powerful the battery, the greater is the force of these attractions and
repulsions.   We may, therefore, by increasing the
energy of a battery sufficiently, enable it to decompose any compound whatever, the negative constituent being attracted by the positive pole, and the
positive constituent by the negative pole.   Oxygen,
chlorine, bromine, iodine, cyanogen, and acids, are
negative bodies; for they always appear round the
positive pole of the battery, and are therefore attracted to it: while hydrogen, azote, carbon, selenium, metals, alkalies, earths, and oxide bases, are
deposited round the negative pole, and consequently
are *positive*.

According to this view of the subject, chemical
affinity is merely a case of the attractions exerted
by bodies in different states of electricity.   Volta
first broached the idea, that every body possesses
naturally a certain state of electricity.   Davy went a
step further, and concluded, that the attractions
which exist between the atoms of different bodies are
merely the consequence of these different states of
electricity.   The proof of this opinion is founded on
the fact, that if we present to a compound, sufficiently strong electrical poles, it will be separated into
its constituents, and one of these constituents will
invariably make its way to the positive and the other
to the negative pole.   Now bodies in a state of electrical excitement always attract those that are in the
opposite state.

If electricity be considered as consisting of two
distinct fluids, which attract each other with a force

inversely, as the square of the distance, while the particles of each fluid repel each other with a force varying according to the same law, then we must conclude that the atoms of each body are covered externally with a coating of some one electric fluid to a greater or smaller extent. Oxygen and the other supporters of combustion are covered with a coating of negative electricity; while hydrogen, carbon, and the metals, are covered with a coating of positive electricity. What is the cause of the adherence of the electricity to these atoms we cannot explain. It is not owing to an attraction similar to gravitation; for electricity never penetrates into the interior of bodies, but spreads itself only on the surface, and the quantity of it which can accumulate is not proportional to the quantity of matter but to the extent of surface. But whatever be the cause, the adhesion is strong, and seemingly cannot be overcome. If we were to suppose an atom of any body, of oxygen for example, to remain uncombined with any other body, but surrounded by electricity, it is obvious that the coating of negative electricity on its surface would be gradually neutralized by its attracting and combining with positive electricity. But let us suppose an atom of oxygen and an atom of hydrogen to be united together, it is obvious that the positive electricity of the one atom would powerfully attract the negative electricity of the other, and vice versâ. And if these respective electricities cannot leave the atoms, the two atoms will remain firmly united, and the opposite electrical intensities will in some measure neutralize each other, and thus prevent them from being neutralized by electricity from any other quarter. But a current of the opposite electricities passing through such a compound, might neutralize the electricity in each, and thus putting an end to their attractions, occasion decomposition.

Such is a very imperfect outline of the electrical theory of affinity first proposed by Davy to account for the decompositions produced by electricity. It has been universally adopted by chemists; and some progress has been made in explaining and accounting for the different phenomena. It would be improper, in a work of this kind, to enter further into the subject. Those who are interested in such discussions will find a good deal of information in the first volume of Berzelius's Treatise on Chemistry, in the introduction to the Traité de Chimie appliqué aux Arts, by Dumas, or in the introduction to my System of Chemistry, at present in the press.

Davy having thus got possession of an engine, by means of which the compounds, whose constituents adhered to each other might be separated, immediately applied it to the decomposition of potash and soda; bodies which were admitted to be compounds, though all attempts to analyze them had hitherto failed. His attempt was successful. When a platinum wire from the negative pole of a strong battery in full action was applied to a lump of potash, slightly moistened, and lying on a platinum tray attached to the positive pole of the battery, small globules of a white metal soon appeared at its extremity. This white metal he speedily proved to be the basis of potash. He gave it the name of *potassium*, and very soon proved, that potash is a compound of five parts by weight of this metal and one part of oxygen. Potash, then, is a metallic oxide. He proved soon after that soda is a compound of oxygen and another white metal, to which he gave the name of *sodium*. Lime is a compound of *calcium* and oxygen, magnesia of *magnesium* and oxygen, barytes of *barium* and oxygen, and strontian of *strontium* and oxygen. In short, the fixed alkalies and alkaline earths, are metallic oxides. When *lithia* was afterwards discovered by Arfvedson, Davy succeeded in decomposing it also by the galvanic battery, and resolving it into oxygen and a white metal, to which the name of *lithium* was given.

Davy did not succeed so well in decomposing alumina, glucina, yttria, and zirconia, by the galvanic battery: they were not sufficiently good conductors of electricity; but nobody entertained any doubt that they also were metallic oxides. They have been all at length decomposed, and their bases obtained by the joint action of chlorine and potassium, and it has been demonstrated, that they also are metallic oxides. Thus it has been ascertained, in consequence of Davy's original discovery of the powers of the galvanic battery, that all the bases formerly distinguished into the four classes of alkalies, alkaline earths, earths proper, and metallic oxides, belong in fact only to one class, and are all metallic oxides.

Important as these discoveries are, and sufficient as they would have been to immortalize the author of them, they are not the only ones for which we are indebted to Sir Humphry Davy. His experiments on *chlorine* are not less interesting or less important in their consequences. I have already mentioned in a former chapter, that Berthollet made a set of experiments on chlorine, from which he had drawn as a conclusion, that it is a compound of oxygen and muriatic acid, in consequence of which it got the name of *oxymuriatic acid*. This opinion of Berthollet had been universally adopted by chemists, and admitted by them as a fundamental principle, till Gay-Lussac and Thenard endeavoured, unsuccessfully, to decompose this gas, or to resolve it into muriatic acid and chlorine. They showed, in the clearest manner, that such a resolution was impossible, and that no direct evidence could be ad-

duced to prove that oxygen was one of its constituents.   The conclusion to which they came was,
that muriatic acid gas contained water as an essential
constituent; and they succeeded by this hypothesis
in accounting for all the different phenomena which
they had observed.   They even made an experiment
to determine the quantity of water thus combined.
They passed muriatic acid through hot litharge
(protoxide of lead); muriate of lead was formed,
and abundance of water made its appearance and
was collected.   They did not attempt to determine
the proportions; but we can now easily calculate the
quantity of water which would be deposited when
a given weight of muriatic acid gas is absorbed by a
given weight of litharge.   Suppose we have fourteen
parts of oxide of lead : to convert it into muriate of
lead, 4·625 parts (by weight) of muriatic acid would
be necessary, and during the formation of the muriate of lead there would be deposited 1 125 parts
of water.   So that from this experiment it might be
concluded, that about one-fourth of the weight of
muriatic acid gas is water.

The very curious and important facts respecting
chlorine and muriatic acid gas which they had ascertained, were made known by Gay-Lussac and Thenard to the Institute, on the 27th of February, 1809,
and an abstract of them was published in the second
volume of the Mémoires d'Arcueil.   There can
be little doubt that it was in consequence of these
curious and important experiments of the French
chemists that Davy's attention was again turned to
muriatic acid gas.   He had already, in 1808, shown
that when potassium is heated in muriatic acid gas,
muriate of potash is formed, and a quantity of hydrogen gas evolved, amounting to more than one-
third of the muriatic acid gas employed, and he had
shown, that in no case can muriatic acid be obtained

from chlorine, unless water or its elements be present. This last conclusion had been amply confirmed by the new investigations of Gay-Lussac and Thenard. In 1810 Davy again resumed the examination of the subject, and in July of that year read a paper to the Royal Society, to prove that *chlorine* is a simple substance, and that muriatic acid is a compound of *chlorine* and *hydrogen*.

This was introducing an alteration in chemical theory of the same kind, and nearly as important, as was introduced by Lavoisier, with respect to the action of oxygen in the processes of combustion and calcination. It had been previously supposed that sulphur, phosphorus, charcoal, and metals, were compounds; one of the constituents of which was phlogiston, and the other the acids or oxides which remained after the combustion or calcination had taken place. Lavoisier showed that the sulphur, phosphorus, charcoal, and metals, were simple substances; and that the acids or calces formed were compounds of these simple bodies and oxygen. In like manner, Davy showed that chlorine, instead of being a compound of muriatic acid and oxygen, was, in fact, a simple substance, and muriatic acid a compound of chlorine and hydrogen. This new doctrine immediately overturned the Lavoisierian hypothesis respecting oxygen as the acidifying principle, and altered all the previously received notions respecting the muriates. What had been called *muriates* were, in fact, combinations of chlorine with the combustible or metal, and were analagous to oxides. Thus, when muriatic acid gas was made to act upon hot litharge, a double decomposition took place, the chlorine united to the lead, while the hydrogen of the muriatic acid united with the oxygen of the litharge, and formed water. Hence the reason of the appearance of water in this case; and hence it was obvious that what had been called muriate of lead, was, in reality, a compound of chlorine and metallic lead. It ought, therefore, to be called, not muriate of lead, but chloride of lead.

It was not likely that this new opinion of Davy should be adopted by chemists in general, without a struggle to support the old opinions. But the feebleness of the controversy which ensued, affords a striking proof how much chemistry had advanced since the days of Lavoisier, and how free from prejudices chemists had become. One would have expected that the French chemists would have made the greatest resistance to the admission of these new opinions; because they had a direct tendency to diminish the reputation of two of their most eminent chemists, Lavoisier and Berthollet. But the fact was not so: the French chemists showed a degree of candour and liberality which redounds highly to their credit. Berthollet did not enter at all into the controversy. Gay-Lussac and Thenard, in their Recherches Physico-chimiques, published in 1811, state their reasons for preferring the old hypothesis to the new, but with great modesty and fairness; and, within less than a year after, they both adopted the opinion of Davy, that chlorine is a simple substance, and muriatic acid a compound of hydrogen and chlorine.

The only opponents to the new doctrine who appeared against it, were Dr. John Murray, of Edinburgh, and Professor Berzelius, of Stockholm. Dr. Murray was a man of excellent abilities, and a very zealous cultivator of chemistry; but his health had been always very delicate, which had prevented him from dedicating so much of his time to experimenting as he otherwise would have been inclined to do. The only experimental investigations into which he entered was the analysis of some mineral waters.

His powers of elocution were great. He was, in consequence, a popular and very useful lecturer. He published animadversions upon the new doctrine respecting *chlorine*, in Nicholson's Journal ; and his observations were answered by Dr. John Davy.

Dr. John Davy was the brother of Sir Humphry, and had shown, by his paper on fluoric acid and on the chlorides, that he possessed the same dexterity and the same powers of inductive reasoning, which had given so much celebrity to his brother. The controversy between him and Dr. Murray was carried on for some time with much spirit and ingenuity on both sides, and was productive of some advantage to the science of chemistry, by the discovery of phosgene gas or chlorocarbonic acid, which was made by Dr. Davy. It is needless to say to what side the victory fell. The whole chemical world has for several years unanimously adopted the theory of Davy ; showing sufficiently the opinion entertained respecting the arguments advanced by either party. Berzelius supported the old opinion respecting the compound nature of chlorine, in a paper which he published in the Annals of Philosophy. No person thought it worth while to answer his arguments, though Sir Humphry Davy made a few animadversions upon one or two of his experiments. The discovery of iodine, which took place almost immediately after, afforded so close an analogy with chlorine, and the nature of the compounds which it forms was so obvious and so well made out, that chemists were immediately satisfied ; and they furnished so satisfactory an answer to all the objections of Berzelius, that I am not aware of any person, either in Great Britain or in France, who adopted his opinions. I have not the same means of knowing the impression which his paper made upon the chemists of Germany and Sweden. Berzelius con-

tinued for several years a very zealous opponent to
the new doctrine, that chlorine is a simple substance.
But he became at last satisfied of the futility of his
own objections, aud the inaccuracy of his reasoning.
About the year 1820 he embraced the opinion of
Davy, and is now one of its most zealous defenders.
Dr. Murray has been dead for many years, and Berzelius has renounced his notion, that muriatic acid is
a compound of oxygen and an unknown combustible basis.   We may say then, I believe with justice, that at present all the chemical world adopts
the notion that chlorine is a simple substance, and
muriatic acid a compound of chlorine and hydrogen.

The recent discovery of bromine, by Balard, has
added another strong analogy in favour of Davy's
theory; as has likewise the discovery by Gay-
Lussac respecting prussic acid.   At present, then,
(not reckoning sulphuretted and telluretted hydrogen
gas), we are acquainted with four acids which contain no oxygen, but are compounds of hydrogen
with another negative body.   These are

  Muriatic acid, composed of chlorine and hydrogen
  Hydriodic acid . . . iodine and hydrogen
  Hydrobromic acid . . bromine and hydrogen
  Prussic acid . . . . cyanogen and hydrogen.

So that even if we were to leave out of view the
chlorine acids, the sulphur acids, &c., no doubt
can be entertained that many acids exist which contain no oxygen.   Acids are compounds of electro-
negative bodies and a base, and in them all the
electro-negative electricity continues to predomi-
nate.

Next to Sir Humphry Davy, the two chemists
who have most advanced electro-chemistry are Gay-
Lussac and Thenard.   About the year 1808, when
the attention of men of science was particularly
drawn towards the galvanic battery, in consequence

of the splendid discoveries of Sir Humphry Davy, Bonaparte, who was at that time Emperor of France, consigned a sufficient sum of money to Count Cessac, governor of the Polytechnic School, to construct a powerful galvanic battery; and Gay-Lussac and Thenard were appointed to make the requisite experiments with this battery. It was impossible that a better choice could have been made. These gentlemen undertook a most elaborate and extensive set of experiments, the result of which was published in 1811, in two octavo volumes, under the title of " Recherches Physico-chimiques, faites sur la Pile; sur la Preparation chimique et les Propriétés du Potassium et du Sodium; sur la Décomposition de l'Acide boracique; sur les Acides fluorique, muriatique, et muriatique oxygené; sur l'Action chimique de la Lumière; sur l'Analyse vegetale et animale, &c." It would be difficult to name any chemical book that contains a greater number of new facts, or which contains so great a collection of important information, or which has contributed more to the advancement of chemical science.

The first part contains a very minute and interesting examination of the galvanic battery, and upon what circumstances its energy depends. They tried the effect of various liquid conductors, varied the strength of the acids and of the saline solutions. This division of their labours contains much valuable information for the practical electro-chemist, though it would be inconsistent with the plan of this work to enter into details.

The next division of the work relates to potassium. Davy had hitherto produced that metal only in minute quantities by the action of the galvanic battery upon potash. But Gay-Lussac and Thenard contrived a process by which it can be prepared on a large scale by chemical decomposition. Their

method was, to put into a bent gun-barrel, well
coated externally with clay, and passed through a
furnace, a quantity of clean iron-filings. To one
extremity of this barrel was fitted a tube containing
a quantity of caustic potash. This tube was either
shut at one end by a stopper, or by a glass tube
luted to it, and plunged under the surface of mercury. To the other extremity of the gun-barrel
was also luted a tube, which plunged into a vessel
containing mercury. Heat was applied to the gun-
barrel till it was heated to whiteness ; then, by means
of a choffer, the caustic potash was melted and made
to trickle slowly into the white-hot iron-filings.
At this temperature the potash undergoes decomposition, the iron uniting with its oxygen. The
potassium is disengaged, and being volatile is deposited at a distance from the hot part of the tube,
where it is collected after the process is finished.

Being thus in possession, both of potassium and
sodium in considerable quantities, they were enabled to examine its properties more in detail than
Davy had done : but such was the care and industry with which Davy's experiments had been
made that very little remained to be done. The
specific gravity of the two metals was determined
with more precision than it was possible for Davy to
do. They determined the action of these metals on
water, and measured the quantity of hydrogen gas
given out with more precision than Davy could.
They discovered also, by heating these metals in
oxygen gas, that they were capable of uniting with
an additional dose of oxygen, and of forming per-
oxides of potassium and sodium. These oxides
have a yellow colour, and give out the surplus
oxygen, and are brought back to the state of potash
and soda when they are plunged into water. They
exposed a great variety of substances to the action

of potassium, and brought to light a vast number of curious and important facts, tending to throw new light on the properties and characters of that curious metallic substance.

By heating together anhydrous boracic acid and potassium in a copper tube, they succeeded in decomposing the acid, and in showing it to be a compound of oxygen, and a black matter like charcoal, to which the name of *boron* has been given. They examined the properties of boron in detail, but did not succeed in determining with exactness the proportions of the constituents of boracic acid. The subsequent experiments of Davy, though not exact, come a good deal nearer the truth.

Their experiments on fluoric acid are exceedingly valuable. They first obtained that acid in a state of purity, and ascertained its properties. Their attempts to decompose it as well as those of Davy, ended in disappointment. But Ampere conceived the idea that this acid, like muriatic acid, is a compound of hydrogen with an unknown supporter of combustion, to which the name *fluorine* was given. This opinion was adopted by Davy, and his experiments, though they do not absolutely prove the truth of the opinion, give it at least considerable probability, and have disposed chemists in general to adopt it. The subsequent researches of Berzelius, while they have added a great deal to our former knowledge respecting fluoric acid and its compounds, have all tended to confirm and establish the doctrine that it is a hydracid, and similar in its nature to the other hydracids. But such is the tendency of fluorine to combine with every substance, that hitherto it has been impossible to obtain it in an insulated state. We want therefore, still, a decisive proof of the accuracy of the opinion.

To the experiments of Gay-Lussac and Thenard

on chlorine and muriatic acid, I have already alluded in a former part of this chapter. It was
during their investigations connected with this subject, that they discovered *fluoboric* acid gas, which
certainly adds considerably to the probability of the
theory of Ampere respecting the nature of fluoric
acid.

I pass over a vast number of other new and important facts and observations contained in this admirable work, which ought to be studied with minute attention by every person who aspires at becoming a chemist.

Besides the numerous discoveries contained in the
Recherches Physico-chimique, Gay-Lussac is the
author of two of so much importance that it would
be wrong to omit them. He showed that cyanogen
is one of the constituents of prussic acid; succeeded
in determining the composition of cyanogen, and
showing it to be a compound of two atoms of carbon
and one atom of azote. Prussic acid is a compound of one atom of hydrogen and one atom of
cyanogen. Sulpho-cyanic acid, discovered by Mr.
Porrett, is a compound of one atom sulphuric, and
one atom cyanogen; chloro-cyanic acid, discovered
by Berthollet, is a compound of one atom chlorine
and one atom cyanogen; while cyanic acid, discovered by Wöhler, is a compound of one atom
oxygen and one atom cyanogen. I take no notice
of the fulminic acid; because, although Gay-Lussac's experiments are exceedingly ingenious, and
his reasoning very plausible, it is not quite convincing; especially as the results obtained by Mr.
Edmund Davy, and detailed by him in his late interesting memoir on this subject, are somewhat different.

The other discovery of Gay-Lussac is his demonstration of the peculiar nature of iodine, his account of iodic and hydriodic acids, and of many

other compounds into which that curious substance enters as a constituent. Sir H. Davy was occupied with iodine at the same time with Gay-Lussac; and his sagacity and inventive powers were too great to allow him to work upon such a substance without discovering many new and interesting facts.

To M. Thenard we are indebted for the discovery of the important fact, that hydrogen is capable of combining with twice as much oxygen as exists in water, and determining the properties of this curious liquid which has been called deutoxide of hydrogen. It possesses bleaching properties in perfection, and I think it likely that chlorine owes its bleaching powers to the formation of a little deutoxide of hydrogen in consequence of its action on water.

The mantle of Davy seems in some measure to have descended on Mr. Faraday, who occupies his old place at the Royal Institution. He has shown equal industry, much sagacity, and great powers of invention. The most important discovery connected with electro-magnetism, next to the great fact, for which we are indebted to Professor Œrstedt of Copenhagen, is due to Mr. Faraday; I mean the rotation of the electric wires round the magnet. To him we owe the knowledge of the fact, that several of the gases can be condensed into liquids by the united action of pressure and cold, which has removed the barrier that separated gaseous bodies from vapours, and shown us that all owe their elasticity to the same cause. To him also we owe the knowledge of the important fact, that chlorine is capable of combining with carbon. This has considerably improved the history of chlorine and served still further to throw new light on the analogy which exists between all the supporters of combustion. They are doubtless all of them capable of combining with every one of the other simple bodies, and of forming compounds with them. For they are all negative bodies; while the other simple substances without exception, when compared to them, possess positive properties. We must therefore view the history of chemistry as incomplete, till we have become acquainted with the compounds of every supporter with every simple base.

CHAPTER VI.

OF THE ATOMIC THEORY.

I COME now to the last improvement which chemistry has received—an improvement which has given a degree of accuracy to chemical experimenting almost approaching to mathematical precision, which has simplified prodigiously our views respecting chemical combinations; which has enabled manufacturers to introduce theoretical improvements into their processes, and to regulate with almost perfect precision the relative quantities of the various constituents necessary to produce the intended effects. The consequence of this is, that nothing is wasted, nothing is thrown away. Chemical products have become not only better in quality, but more abundant and much cheaper. I allude to the atomic theory still only in its infancy, but already productive of the most important benefits. It is destined one day to produce still more wonderful effects, and to render chemistry not only the most delightful, but the most useful and indispensable, of all the sciences.

Like all other great improvements in science, the atomic theory developed itself by degrees, and several of the older chemists ascertained facts which might, had they been aware of their importance, have led them to conclusions similar to those of the moderns. The very attempt to analyze the salts was an acknowledgment that bodies united with each other in definite proportions : and these definite proportions, had they been followed out, would have led ultimately to the doctrine of atoms. For how could it be, that six parts of potash were always saturated by five parts of sulphuric acid and 6·75 parts of nitric acid? How came it that five of sulphuric acid always went as far in saturating potash as 6·75 of nitric acid? It was known, that in chemical combinations it was the ultimate particles of matter that combined. The simple explanation, therefore, would have been—that the weight of an ultimate particle of sulphuric acid was only five, while that of an ultimate particle of nitric acid was 6·75. Had such an inference been drawn, it would have led directly to the atomic theory.

The atomic theory in chemistry has many points of resemblance to the fluxionary calculus in mathematics. Both give us the ratios of quantities; both reduce investigations that would be otherwise extremely difficult, or almost impossible, to the utmost simplicity; and what is still more curious, both have been subjected to the same kind of ridicule by those who have not put themselves to the trouble of studying them with such attention as to understand them completely. The minute philosopher of Berkeley, *mutatis mutandis*, might be applied to the atomic theory with as much justice as to the fluxionary calculus; and I have heard more than one individual attempt to throw ridicule upon the atomic theory by nearly the same kind of arguments.

The first chemists, then, who attempted to analyze the salts may be considered as contributing towards laying the foundation of the atomic theory, though they were not themselves aware of the importance of the structure which might have been raised upon their experiments, had they been made with the requisite precision.

Bergman was the first chemist who attempted regular analyses of salts. It was he that first tried to establish regular formulas for the analyses of mineral waters, stones, and ores, by the means of solution and precipitation. Hence a knowledge of the constituents of the salts was necessary, before his formulas could be applied to practice. It was to supply this requisite information that he set about analyzing the salts, and his results were long considered by chemists as exact, and employed by them to determine the results of their analyses. We now know that these analytical results of Bergman are far from accurate; they have accordingly been laid aside as useless : but this knowledge has been derived from the progress of the atomic theory.

The first accurate set of experiments to analyze the salts was made by Wenzel, and published by him in 1777, in a small volume entitled " Lehre von der Verwandschaft der Körper," or, " Theory of the Affinities of Bodies." These analyses of Wenzel are infinitely more accurate than those of Bergman, and indeed in many cases are equally precise with the best which we have even at the present day. Yet the book fell almost dead-born from the press ; Wenzel's results never obtained the confidence of chemists, nor is his name ever quoted as an authority. Wenzel was struck with a phenomenon, which had indeed been noticed by preceding chemists; but they had not drawn the advantages from it which it was capable of affording. There are several saline solutions which, when mixed with each other, completely decompose each other, so that two new salts are produced. Thus, if we mix together solutions of nitrate of lead and sulphate of soda in the requisite proportions, the sulphuric acid of the latter salt will combine with the oxide of lead of the former, and will form with it sulphate of lead, which will precipitate to the bottom in the state of an insoluble powder, while the nitric acid formerly united to the oxide of lead, will combine with the soda formerly in union with the sulphuric acid, and form nitrate of soda, which being soluble, will remain in solution in the liquid. Thus, instead of the two old salts,

> Sulphate of soda
> Nitrate of lead,

we obtain the two new salts,

> Sulphate of lead
> Nitrate of soda.

If we mix the two salts in the requisite proportions, the decomposition will be complete ; but if there be an excess of one of the salts, that excess will still remain in solution without affecting the result. If we suppose the two salts anhydrous, then the proportions necessary for complete decomposition are,

> Sulphate of soda   9
> Nitrate of lead   20·75
> ————
> 29·75

and the quantities of the new salts formed will be

> Sulphate of lead 19
> Nitrate of soda   10·75
> ————
> 29·75

We see that the absolute weights of the two sets of salts are the same : all that has happened is, that both the acids and both the bases have exchanged situations. Now if, instead of mixing these two salts together in the preceding proportions, we employ

> Sulphate of soda   9
> Nitrate of lead   25·75

That is to say, if we employ 5 parts of nitrate of

lead more than is sufficient for the purpose; we shall have exactly the same decompositions as before; but the 5 of excess of nitrate of lead will remain in solution, mixed with the nitrate of soda. There will be precipitated as before,

Sulphate of lead 19

and there will remain in solution a mixture of

Nitrate of soda 10·75

Nitrate of lead 5

The phenomena are precisely the same as before; the additional 5 of nitrate of lead have occasioned no alteration; the decomposition has gone on just as if they had not been present.

Now the phenomena which drew the particular attention of Wenzel is, that if the salts were neutral before being mixed, the neutrality was not affected by the decomposition which took place on their mixture.* A salt is said to be neutral when it neither possesses the characters of an acid or an alkali. Acids *redden* vegetable *blues*, while alkalies render them *green*. A neutral salt produces no effect whatever upon vegetable blues. This observation of Wenzel is very important : it is obvious that the salts, after their decomposition, could not have remained neutral unless the elements of the two salts had been such that the bases in each just saturated the acids in either of the salts.

The constituents of the two salts are as follows :

9 sulphate of soda $\begin{cases} 5 & \text{sulphuric acid} \\ 4 & \text{soda,} \end{cases}$

20·75 nitrate of lead $\begin{cases} 6\cdot75 & \text{nitric acid} \\ 14 & \text{oxide of lead.} \end{cases}$

---

* This observation is not without exception. It does not hold when one of the salts is a phosphate or an arseniate, and this is the cause of the difficulty attending the analysis of these genera of salts.

Now it is clear, that unless 5 sulphuric acid were just saturated by 4 soda and by 14 oxide of lead; and 6·75 of nitric acid by the same 4 soda and 14 oxide of lead, the salts, after their decomposition, could not have preserved their neutrality. Had 4 soda required only 5·75 of nitric acid, or had 14 oxide of lead required only 4 sulphuric acid, to saturate them, the liquid, after decomposition, would have contained an excess of acid. As no such excess exists, it is clear that in saturating an acid, 4 soda goes exactly as far as 14 oxide of lead; and that, in saturating a base, 5 sulphuric acid goes just as far as 6·75 nitric acid.

Nothing can exhibit in a more striking point of view, the almost despotic power of fashion and authority over the minds even of men of science, and the small number of them that venture to think for themselves, than the fact, that this most important and luminous explanation of Wenzel, confirmed by much more accurate experiments than any which chemistry had yet seen, is scarcely noticed by any of his contemporaries, and seems not to have attracted the smallest attention. In science, it is as unfortunate for a man to get before the age in which he lives, as to continue behind it. The admirable explanation of combustion by Hooke, and the important experiments on combustion and respiration by Mayow, were lost upon their contemporaries, and procured them little or no reputation whatever; while the same theory, and the same experiments, advanced by Lavoisier and Priestley, a century later, when the minds of men of science were prepared to receive them, raised them to the very first rank among philosophers, and produced a revolution in chemistry. So much concern has fortune, not merely in the success of kings and conquerors, but in the reputation acquired by men of science.

In the year 1792 another labourer, in the same department of chemistry, appeared : this was Jeremiah Benjamin Richter, a Prussian chemist, of whose history I know nothing more than that his publications were printed and published in Breslau, from which I infer that he was a native of, or at least resided in, Silesia. He calls himself Assessor of the Royal Prussian Mines and Smeltinghouses, and Arcanist of the Commission of Berlin Porcelain Manufacture. He died in the prime of life, on the 4th of May, 1807. In the year 1792 he published a work entitled " Anfansgründe der Stochyometrie ; oder, Messkunst Chymischer Elemente " (Elements of Stochiometry; or, the Mathematics of the Chemical Elements). A second and third volume of this work appeared in 1793, and a fourth volume in 1794. The object of this book was a rigid analysis of the different salts, founded on the fact just mentioned, that when two salts decompose each other, the salts newly formed are neutral as well as those which have been decomposed. He took up the subject nearly in the same way as Wenzel had done, but carried the subject much further ; and endeavoured to determine the capacity of saturation of each acid and base, and to attach numbers to each, indicating the weights which mutually saturate each other. He gave the whole subject a mathematical dress, and endeavoured to show that the same relation existed, between the numbers representing the capacity of saturation of these bodies, as does between certain classes of figurate numbers. When we strip the subject of the mystical form under which he presented it, the labours of Richter may be exhibited under the two following tables, which represent the capacity of saturation of the acids and bases, according to his experiments.

1. ACIDS.		2. BASES.	
Fluoric acid . .	427	Alumina . . .	525
Carbonic . . .	577	Magnesia . .	615
Sebacic . . .	706	Ammonia . .	672
Muriatic . . .	712	Lime	793
Oxalic . . .	755	Soda	859
Phosphoric . .	979	Strontian . . .	1329
Formic . . .	988	Potash . . .	1605
Sulphuric . .	1000	Barytes . . .	2222
Succinic . . .	1209		
Nitric	1405		
Acetic	1480		
Citric	1683		
Tartaric . . .	1694		

To understand this table, it is only necessary to observe, that if we take the quantity of any of the acids placed after it in the table, that quantity will be exactly saturated by the weight of each base put after it in the second column: thus, 1000 of sulphuric acid will be just saturated by 525 alumina, 615 magnesia, 672 ammonia, 793 lime, and so on. On the other hand, the quantity of any base placed after its name in the second column, will be just saturated by the weight of each acid placed after its name in the first column : thus, 793 parts of lime will be just saturated by 427 of fluoric acid, 577 of carbonic acid, 706 of sebacic acid, and so on.

This work of Richter was followed by a periodical work entitled " Ueber die neuern Gegenstande der Chymie " (On the New Objects of Chemistry). This work was begun in the year 1792, and continued in twelve different numbers, or volumes, to the time of his death in 1807.*

---

* I have only seen eleven parts of this work, the last of which appeared in 1802 ; but I believe that a twelfth part was published afterwards.

Richter's labours in this important field produced as little attention as those of Wenzel. Gehlen wrote a short panegyric upon him at his death, praising his views and pointing out their importance; but I am not aware of any individual, either in Germany or elsewhere, who adopted Richter's opinions during his lifetime, or even seemed aware of their importance, unless we are to except Berthollet, who mentions them with approbation in his Chemical Statics. This inattention was partly owing to the great want of accuracy which it is impossible not be sensible of in Richter's experiments. He operated upon too large quantities of matter, which indeed was the common defect of the times, and was first checked by Dr. Wollaston. The dispute between the phlogistians and the antiphlogistians, which was not fully settled in Richter's time, drew the attention of chemists to another branch of the subject. Richter in some measure went before the age in which he lived, and had his labours not been recalled to our recollection by the introduction of atomic theory, he would probably have been forgotten, like Hooke and Mayow, and only brought again under review after the new discoveries in the science had put it in the power of chemists in general to appreciate the value of his labours.

It is to Mr. Dalton that we are indebted for the happy and simple idea from which the atomic theory originated.

John Dalton, to whose lot it has fallen to produce such an alteration and improvement in chemistry, was born in Westmorland, and belongs to that small and virtuous sect known in this country by the name of Quakers. When very young he lived with Mr. Gough of Kendal, a blind philosopher, to whom he read, and whom he assisted in his philosophical investigations   It was here, probably, that he

acquired a considerable part of his education, particularly his taste for mathematics. For Mr. Gough
was remarkably fond of mathematical investigations,
and has published several mathematical papers that
do him credit. From Kendal Mr. Dalton went to
Manchester, about the beginning of the present
century, and commenced teaching elementary mathematics to such young men as felt inclined to
acquire some knowledge of that important subject.
In this way, together with a few courses of lectures
on chemistry, which he has occasionally given at
the Royal Institution in London, at the Institution
in Birmingham, in Manchester, and once in Edinburgh and in Glasgow, he has contrived to support
himself for more than thirty years, if not in affluence,
at least in perfect independence. And as his de-
sires have always been of the most moderate kind,
his income has always been equal to his wants. In
a country like this, where so much wealth abounds,
and where so handsome a yearly income was subscribed to enable Dr. Priestley to prosecute his
investigations undisturbed and undistracted by the
necessity of providing for the daily wants of his
family, there is little doubt that Mr. Dalton, had
he so chosen it, might, in point of pecuniary circumstances, have exhibited a much more brilliant
figure. But he has displayed a much nobler mind
by the career which he has chosen—equally regardless of riches as the most celebrated sages of antiquity, and as much respected and beloved by his
friends, even in the rich commercial town of Manchester, as if he were one of the greatest and most
influential men in the country. Towards the end
of the last century, a literary and scientific society
had been established in Manchester, of which Mr.
Thomas Henry, the translator of Lavoisier's Essays,
and who distinguished himself so much in promoting

the introduction of the new mode of bleaching into Lancashire, was long president. Mr. Dalton, who had already distinguished himself by his meteorological observations, and particularly by his account of the Aurora Borealis, soon became a member of the society; and in the fifth volume of their Memoirs, part II., published in 1802, six papers of his were inserted, which laid the foundation of his future celebrity. These papers were chiefly connected with meteorological subjects; but by far the most important of them all was the one entitled " Experimental Essays on the Constitution of mixed Gases; on the Force of Steam or Vapour from water and other liquids in different temperatures, both in a torricellian vacuum and in air; on Evaporation; and on the Expansion of Gases by Heat."

From a careful examination of all the circumstances, he considered himself as entitled to infer, that when two elastic fluids or gases, A and B, are mixed together, there is no mutual repulsion among their particles; that is, the particles of A do not repel those of B, as they do one another. Consequently, the pressure or whole weight upon any one particle arises solely from those of its own kind. This doctrine is of so startling a nature and so contrary to the opinions previously received, that chemists have not been much disposed to admit it. But at the same time it must be confessed, that no one has hitherto been able completely to refute it. The consequences of admitting it are obvious: we should be able to account for a fact which has been long known, though no very satisfactory reason for it had been assigned; namely, that if two gases be placed in two separate vessels, communicating by a narrow orifice, and left at perfect rest in a place where the temperature never varies,

if we examine them after a certain interval of time
we shall find both equally diffused through both
vessels. If we fill a glass phial with hydrogen
gas and another phial with common air or carbonic
acid gas and unite the two phials by a narrow glass
tube two feet long, filled with common air, and
place the phial containing the hydrogen gas uppermost, and the other perpendicularly below it, the
hydrogen, though lightest, will not remain in the
upper phial, nor the carbonic acid, though heaviest,
in the undermost phial; but we shall find both
gases equally diffused through both phials.

But the second of these essays is by far the most
important. In it he establishes, by the most unexceptionable evidence, that water, when it evaporates, is always converted into an elastic fluid,
similar in its properties to air. But that the distance between the particles is greater the lower the
temperature is at which the water evaporates. The
elasticity of this vapour increases as the temperature
increases. At 32° it is capable of balancing a column of mercury about half an inch in height, and
at 212° it balances a column thirty inches high, or
it is then equal to the pressure of the atmosphere.
He determined the elasticity of vapour at all temperatures from 32° to 212°, pointed out the method
of determining the quantity of vapour that at any
time exists in the atmosphere, the effect which it
has upon the volume of air, and the mode of determining its quantity.   Finally, he determined, experimentally, the rate of evaporation from the surface of water at all temperatures from 32° to 212°.
These investigations have been of infinite use to chemists in all their investigations respecting the specific
gravity of gases, and have enabled them to resolve
various interesting problems, both respecting specific
gravity, evaporation, rain and respiration, which,

had it not been for the principles laid down in this essay, would have eluded their grasp.

In the last essay contained in this paper he has shown that all elastic fluids expand the same quantity by the same addition of heat, and this expansion is very nearly 1-480th part for every degree of Fahrenheit's thermometer. In this last branch of the subject Mr. Dalton was followed by Gay-Lussac, who, about half a year after the appearance of his Essays, published a paper in the Annales de Chimie, showing that the expansion of all elastic fluids, when equally heated, is the same. Mr. Dalton concluded that the expansion of all elastic fluids by heat is equable. And this opinion has been since confirmed by the important experiments of Dulong and Petit, which have thrown much additional light on the subject.

In the year 1804, on the 26th of August, I spent a day or two at Manchester, and was much with Mr. Dalton. At that time he explained to me his notions respecting the composition of bodies. I wrote down at the time the opinions which he offered, and the following account is taken literally from my journal of that date:

The ultimate particles of all simple bodies are *atoms* incapable of further division. These atoms (at least viewed along with their atmospheres of heat) are all spheres, and are each of them possessed of particular weights, which may be denoted by numbers. For the greater clearness he represented the atoms of the simple bodies by symbols. The following are his symbols for four simple bodies, together with the numbers attached to them by him in 1804:

Relative weights.

O Oxygen . . . . . . 6·5

⊙ Hydrogen 1

Relative weights.

● Carbon 5
◑ Azote 5

The following symbols represent the way in which he thought these atoms were combined to form certain binary compounds, with the weight of an integrant particle of each compound:

Weights.

○⊙ Water . . . . . . 7·5
○◑ Nitrous gas . . . . 11·5
●⊙ Olefiant gas 6
◑⊙ Ammonia 6
○● Carbonic oxide . . . 11·5

The following were the symbols by which he represented the composition of certain tertiary compounds:

Weights.

○●○ Carbonic acid . . . 18
○◑○ Nitrous oxide . . . 16·5
●⊙● Ether 11
⊙●⊙ Carburetted hydrogen 7
○◑○ Nitric acid 18

A quaternary compound:

○◑○
○ Oxynitric acid . . 24·5

A quinquenary compound:

○
◑○◑○ Nitrous acid . . 29·5
○

A sextenary compound:

●○●
⊙●⊙ Alcohol . . . . . 23·5

These symbols are sufficient to give the reader an idea of the notions entertained by Dalton respecting the nature of compounds. Water is a compound of one atom oxygen and one atom hydrogen as is obvious from the symbol ○⊙. Its weight 7·5 is that of an atom of oxygen and an atom of hydrogen united together. In the same way carbonic oxide is a compound of one atom oxygen and one atom carbon. Its symbol is ○●, and its weight 11·5 is equal to an atom of oxygen and an atom of carbon added together. Carbonic acid is a tertiary compound, or it consists of three atoms united together; namely, two atoms of oxygen and one atom of carbon. Its symbol is ○●○, and its weight 18. A bare inspection of the symbols and weights will make Mr. Dalton's notions respecting the constitution of every body in the table evident to every reader.

It was this happy idea of representing the atoms and constitution of bodies by symbols that gave Mr. Dalton's opinions so much clearness. I was delighted with the new light which immediately struck my mind, and saw at a glance the immense importance of such a theory, when fully developed. Mr. Dalton informed me that the atomic theory first occurred to him during his investigations of olefiant gas and carburetted hydrogen gases, at that time imperfectly understood, and the constitution of which was first fully developed by Mr. Dalton himself. It was obvious from the experiments which he made upon them, that the constituents of both were carbon and hydrogen, and nothing else. He found further, that if we reckon the carbon in each the same, then carburetted hydrogen gas contains exactly twice as much hydrogen as olefiant gas does. This determined him to state the ratios of these constituents in numbers, and to consider the olefiant gas as a compound of one atom of carbon and one atom of hydrogen; and carburetted hydrogen of one atom of carbon and two atoms of hydrogen. The idea thus conceived was applied to carbonic oxide, water ammonia, &c. ; and numbers representing the atomic weights of oxygen, azote, &c., deduced from the best analytical experiments which chemistry then possessed.

Let not the reader suppose that this was an easy task. Chemistry at that time did not possess a single analysis which could be considered as even approaching to accuracy. A vast number of facts had been ascertained, and a fine foundation laid for future investigation ; but nothing, as far as weight and measure were concerned, deserving the least confidence, existed. We need not be surprised, then, that Mr. Dalton's first numbers were not exact. It required infinite sagacity, and not a little labour, to come so near the truth as he did. How could accurate analyses of gases be made when there was not a single gas whose specific gravity was known, with even an approach to accuracy ; the preceding investigations of Dalton himself paved the way for accuracy in this indispensable department ; but still accurate results had not yet been obtained.

In the third edition of my System of Chemistry, published in 1807, I introduced a short sketch of Mr. Dalton's theory, and thus made it known to the chemical world. The same year a paper of mine on *oxalic acid* was published in the Philosophical Transactions, in which I showed that oxalic acid unites in two proportions with strontian, forming an *oxalate* and *binoxalate;* and that, supposing the strontian in both salts to be the same, the oxalic acid in the latter is exactly twice as much as in the former. About the same time, Dr. Wollaston showed that bicarbonate of potash contains just twice the quantity of carbonic acid that exists in carbonate of potash; and that there are three oxalates of potash; viz., *oxalate, binoxalate,* and *quadroxalate;* the weight of acids in each of which are as the numbers 1, 2, 4.

These facts gradually drew the attention of chemists to Mr. Dalton's views. There were, however, some of our most eminent chemists who were very hostile to the atomic theory. The most conspicuous of these was Sir Humphry Davy. In the autumn of 1807 I had a long conversation with him at the Royal Institution, but could not convince him that there was any truth in the hypothesis. A few days after I dined with him at the Royal Society Club, at the Crown and Anchor, in the Strand. Dr. Wollaston was present at the dinner. After dinner every member of the club left the tavern, except Dr. Wollaston, Mr. Davy, and myself, who staid behind and had tea. We sat about an hour and a half together, and our whole conversation was about the atomic theory. Dr. Wollaston was a convert as well as myself; and we tried to convince Davy of the inaccuracy of his opinions; but, so far from being convinced, he went away, if possible, more prejudiced against it than ever. Soon after, Davy met Mr. Davis Gilbert, the late distinguished president of the Royal Society; and he amused him with a caricature description of the atomic theory, which he exhibited in so ridiculous a light, that Mr. Gilbert was astonished how any man of sense or science could be taken in with such a tissue of absurdities. Mr. Gilbert called on Dr. Wollaston (probably to discover what could have induced a man of Dr. Wollaston's sagacity and caution to adopt such opinions), and was not sparing in laying the absurdities of the theory, such as they had been represented to him by Davy, in the broadest point of view. Dr. Wollaston begged Mr. Gilbert to sit down, and listen to a few facts which he would state to him. He then went over all the principal facts at that time known respecting the salts; mentioned the alkaline carbonates and bicarbonates, the oxalate, binoxalate, and quadroxalate of potash, carbonic oxide and carbonic acid, olefiant gas, and carburetted hydrogen; and doubtless many other similar compounds, in which the proportion of one of the constituents increases in a regular ratio. Mr. Gilbert went away a convert to the truth of the atomic theory; and he had the merit of convincing Davy that his former opinions on the subject were wrong. What arguments he employed I do not know; but they must have been convincing ones, for Davy ever after became a strenuous supporter of the atomic theory. The only alteration which he made was to substitute *proportion* for Dalton's word, *atom*. Dr. Wollaston substituted for it the term *equivalent*. The object of these substitutions was to avoid all theoretical annunciations. But, in fact, these terms, *proportion, equivalent*, are neither of them so convenient as the term *atom*: and, unless we adopt the hypothesis with which Dalton set out, namely, that the ultimate particles of bodies are *atoms* incapable of further division, and that chemical combination consists in the union of these atoms with each other, we lose all the new light which the atomic theory throws upon chemistry, and bring our notions back to the obscurity of the days of Bergman and of Berthollet.

In the year 1808 Mr. Dalton published the first volume of his New System of Chemical Philosophy. This volume consists chiefly of two chapters: the first, on *heat*, occupies 140 pages. In it he treats of all the effects of heat, and shows the same sagacity and originality which characterize all his writings. Even when his opinions on a subject are not correct, his reasoning is so ingenious and original, and the new facts which he contrives to bring forward so important, that we are always pleased and always instructed. The second chapter, on the *constitution of bodies*, occupies 70 pages. The chief object of it is to combat the peculiar notions respecting elastic fluids, which had been advanced by Berthollet, and supported by Dr. Murray, of Edinburgh. In the third chapter, on *chemical synthesis*, which occupies only a few pages, he gives us the outlines of the atomic theory, such as he had conceived it. In a plate at the end of the volume he exhibits the symbols and atomic weights of thirty-seven bodies, twenty of which were then considered as simple, and the other seventeen as compound. The following table shows the atomic weight of the simple bodies, as he at that time had determined them from the best analytical experiments that had been made:

	Weight of atom.		Weight of atom.
Hydrogen . .	1	Strontian . .	46
Azote . . .	5	Barytes . .	68
Carbon . .	5	Iron . . .	38
Oxygen . .	7	Zinc . . .	56
Phosphorus .	9	Copper . .	56
Sulphur . .	13	Lead . . .	95
Magnesia . .	20	Silver . . .	100
Lime . . .	23	Platinum . .	100
Soda . . .	28	Gold . . .	140
Potash . . .	42	Mercury . .	167

He had made choice of hydrogen for unity, because it is the lightest of all bodies. He was of opinion that the atomic weights of all other bodies are multiples of hydrogen; and, accordingly, they are all expressed in whole numbers. He had raised the atomic weight of oxygen from 6·5 to 7, from a more careful examination of the experiments on the component parts of water. Davy, from a more accurate set of experiments, soon after raised the number for oxygen to 7·5: and Dr. Prout, from a still more careful investigation of the relative specific gravities of oxygen and hydrogen, showed that if the atom of hydrogen be 1, that of oxygen must be 8. Every thing conspires to prove that this is the true ratio between the atomic weights of oxygen and hydrogen.

In 1810 appeared the second volume of Mr. Dalton's New System of Chemical Philosophy. In it he examines the elementary principles, or simple bodies, namely, oxygen, hydrogen, azote, carbon, sulphur, phosphorus, and the metals; and the compounds consisting of two elements, namely, the compounds of oxygen with hydrogen, azote, carbon, sulphur, phosphorus; of hydrogen with azote, carbon, sulphur, phosphorus. Finally he treats of the fixed alkalies and earths. All these combinations are treated of with infinite sagacity; and he endeavours to determine the atomic weights of the different elementary substances. Nothing can exceed the ingenuity of his reasoning. But unfortunately at that time very few accurate chemical analyses existed; and in chemistry no reasoning, however ingenious, can compensate for this indispensable datum. Accordingly his table of atomic weights at the end this second volume, though much more complete than that at the end of the first volume, is still exceedingly defective; indeed no one number can be considered as perfectly correct.

The third volume of the New System of Chemical Philosophy was only published in 1827; but the greatest part of it had been printed nearly ten years before. It treats of the metallic oxides, the sulphurets, phosphurets, carburets, and alloys. Doubtless many of the facts contained in it were new when the sheets were put to the press; but during the interval between the printing and publication, almost the whole of them had not merely been anticipated, but the subject carried much further. By far the most important part of the volume is the Appendix, consisting of about ninety pages, in which he discusses, with his usual sagacity, various important points connected with heat and vapour. In page 352 he gives a new table of the atomic weights of bodies, much more copious than those contained in the two preceding volumes ; and into which he has introduced the corrections necessary from the numerous correct analyses which had been made in the interval.   He still adheres to the ratio 1 : 7 as the correct difference between the weights of the atoms of hydrogen and oxygen.   This shows very clearly that he has not attended to the new facts which have been brought forward on the subject.   No person who has attended to the experiments made on the specific gravity of these two gases during the last twelve years, could admit that these specific gravities are to each other as 1 to 14.   If 1 to 16 be not the exact ratio, it will surely be admitted on all hands that it is infinitely near it.

Mr. Dalton represented the weight of an atom of hydrogen by 1, because it is the lightest of bodies. In this he has been followed by the chemists of the Royal Institution, by Mr. Philips, Dr. Henry, and Dr. Turner, and perhaps some others whose names I do not at present recollect.   Dr. Wollaston, in his paper on Chemical Equivalents, represented the atomic weight of oxygen by 1, because it enters into a greater number of combinations than any other substance ; and this plan has been adopted by Berzelius, by myself, and by the greater number, if not the whole, of the chemists on the continent.   Perhaps the advantage which Dr. Wollaston assigned for making the atom of oxygen unity will ultimately disappear : for there is no reason for believing that the other supporters of combustion are not capable of entering into as many compounds as oxygen.   But, from the constitution of the atmosphere, it is obvious that the compounds into which oxygen enters will always be of more importance to us than any others; and in this point of view it may be attended with considerable convenience to have oxygen represented by 1. In the present state of the atomic theory there is another reason for making the atom of oxygen unity, which I think of considerable importance. Chemists are not yet agreed about the atom of hydrogen. Some consider water a compound of 1 atom of oxygen and 2 atoms of hydrogen; others, of 1 atom of oxygen and 1 atom of hydrogen. According to the first view, the atom of hydrogen is only 1-16th of the weight of an atom of oxygen; according to the second, it is 1-8th. If, therefore, we were to represent the atom of hydrogen by 1, the consequence would be, that two tables of atomic weights would be requisite—all the atoms in one being double the weight of the atoms in the other: whereas, if we make the atom of oxygen unity, it will be the atom of hydrogen only that will differ in the two tables. In the one table it will be 0·125, in the other it will be 0·0625: or, reckoning with Berzelius the atom of oxygen = 100, we have that of hydrogen = 12·5 or 6·25, according as we view water to be a compound of 1 atom of oxygen with 1 or 2 atoms of hydrogen.

In the year 1809 Gay-Lussac published in the second volume of the Mémoires d'Arcueil a paper on the union of the gaseous substances with each other. In this paper he shows that the proportions in which the gases unite with each other are of the simplest kind. One volume of one gas either combining with one volume of another, or with two volumes, or with half a volume. The atomic theory of Dalton had been opposed with considerable keenness by Berthollet in his Introduction to the French translation of my System of Chemistry.   Nor was this opposition to be wondered at; because its admission would of course overturn all the opinions which Berthollet had laboured to establish in his Chemical Statics.   The object of Gay-Lussac's paper was to confirm and establish the new atomic theory, by exhibiting it in a new point of view.   Nothing can be more ingenious than his mode of treating the subject, or more complete than the proofs which he brings forward in support of it.   It had been already established that water is formed by the union of one volume of oxygen and two volumes of hydrogen gas.   Gay-Lussac found by experiment, that one volume of muriatic acid gas is just saturated by one volume of ammoniacal gas: the product is sal ammoniac.   Fluoboric acid gas unites in two proportions with ammoniacal gas: the first compound consists of one volume of fluoboric gas, and one volume of ammoniacal; the second, of one volume of the acid gas, and two volumes of the alkaline.   The first forms a neutral salt, the second an alkaline salt.   He showed likewise, that carbonic acid and ammoniacal gas could combine also in two proportions; namely, one volume of the acid gas with one or two volumes of the alkaline gas.

M. Amédée Berthollet had proved that ammonia is a compound of one volume of azotic, and three volumes of hydrogen gas.   Gay-Lussac himself had shown that sulphuric acid is composed of one volume sulphurous acid gas, and a half-volume of oxygen gas. He showed further, that the compounds of azote and oxygen were composed as follows:

	Azote.	Oxygen.
Protoxide of azote	1 volume	$+ \frac{1}{2}$ volume
Deutoxide of azote	1 ,,	$+ 1$
Nitrous acid .	1 ,,	$+ 2$

He showed also, that when the two gases after combining remained in the gaseous state, the diminution of volume was either 0, or $\frac{1}{3}$, or $\frac{1}{2}$.

The constancy of these proportions left no doubt that the combinations of all gaseous bodies were definite. The theory of Dalton applied to them with great facility. We have only to consider a volume of gas to represent an atom, and then we see that in gases one atom of one gas combines either with one, two, or three atoms of another gas, and never with more. There is, indeed, a difficulty occasioned by the way in which we view the composition of water. If water be composed of one atom of oxygen and one atom of hydrogen, then it follows that a volume of oxygen contains twice as many atoms as a volume of hydrogen. Consequently, if a volume of hydrogen gas represent an atom, half a volume of oxygen gas must represent an atom.

Dr. Prout soon after showed that there is an intimate connexion between the atomic weight of a gas and its specific gravity. This indeed is obvious at once. I afterwards showed that the specific gravity of a gas is either equal to its atomic weight multiplied by $1\cdot111\dot{1}$ (the specific gravity of oxygen gas), or by $0\cdot555\dot{5}$ (half the specific gravity of oxygen gas), or by $0\cdot277\dot{7}$ (1-4th of the specific gravity of oxygen gas), these differences depending upon the relative condensation which the gases undergo when their elements unite. The following table exhibits the atoms and specific gravity of these three sets of gases :

I.   Sp. Gr. $=$ Atomic Weight $\times 1\cdot111\dot{1}$

	Atomic weight.		Sp. gravity.
Oxygen gas	1	.	$1\cdot111\dot{1}$
Fluosilicic acid	$3\cdot25$	.	$3\cdot611\dot{1}$

II. Sp. Gr. = Atomic Weight × 0·5555̇.

	Atomic weight.		Sp. gravity.
Hydrogen .	. 0·125	.	. 0·0694̇
Azotic . .	. 1·75	.	. 0·0722̇
Chlorine . .	. 4·5	.	. 2·5
Carbon vapour	. 0·75	.	. 0·4166̇
Phosphorus vapour	2	.	. 1·1111̇
Sulphur vapour	. 2	·	. 1·1111̇
Tellurium vapour	. 4	.	. 2·2222̇
Arsenic vapour	. 4 75	.	. 2·6388̇
Selenium vapour	. 5	.	. 2·7777̇
Bromine vapour	. 10	.	. 5·5555̇
Iodine vapour	. 15·75	.	. 8·75
Steam . .	. 1·125	.	. 0·625
Carbonic oxide gas .	1·75	.	. 0·9722̇
Carbonic acid .	. 2·75	.	. 1·5277̇
Protoxide of azote	. 2·75	.	. 1·5277̇
Nitric acid vapour	. 6·75	.	. 3·75
Sulphurous acid	. 4	.	. 2.2222̇
Sulphuric acid vapour	5	.	. 2·7777̇
Cyanogen . .	. 3·25	.	. 1·8055̇
Fluoboric acid	. 4·25	.	. 2·3611̇
Bisulphuret of carbon	4·75	.	. 2·6388̇
Chloro-carbonic acid	6·25	.	. 3·4722̇

III. Sp. Gr.=Atomic Weight × 0·2777̇.

	Atomic weight.		Sp. gravity.
Ammoniacal gas .	2·125	.	. 0·59027̇
Hydrocyanic acid .	3·375	.	. 0·9375
Deutoxide of azote .	3·75	.	. 1·0416̇
Muriatic acid .	4·625	.	. 1·28472̇
Hydrobromic acid .	10·125	.	. 2·8125
Hydriodic acid .	15·875	.	. 4·40973

When Professor Berzelius, of Stockholm, thought
of writing his Elementary Treatise on Chemistry, the
first volume of which was published in the year
1808, he prepared himself for the task by reading
several chemical works which do not commonly fall
under the eye of those who compose elementary
treatises.   Among other books he read the Stochiometry of Richter, and was much struck with the explanations there given of the composition of salts,
and the precipitation of metals by each other.   It
followed from the researches of Richter, that if we
were in possession of good analyses of certain salts,
we might by means of them calculate with accuracy
the composition of all the rest.   Berzelius formed
immediately the project of analyzing a series of salts
with the most minute attention to accuracy.   While
employed in putting this project in execution, Davy
discovered the constituents of the alkalies and earths,
Mr. Dalton gave to the world his notions respecting the atomic theory, and Gay-Lussac made known
his theory of volumes.   This greatly enlarged his
views as he proceeded, and induced him to embrace
a much wider field than he had originally contem-
plated.   His first analyses were unsatisfactory ; but
by repeating them and varying the methods, he detected errors, improved his processes, and finally obtained results, which agreed exceedingly well with
the theoretical calculations.   These laborious investigations occupied him several years.   The first
outline of his experiments appeared in the 77th
volume of the Annales de Chimie, in 1811, in a
letter addressed by Berzelius to Berthollet.   In this
letter he gives an account of his methods of analyses
together with the composition of forty-seven compound bodies.   He shows that when a metallic
protosulphuret is converted into a sulphate, the
sulphate is neutral ; that an atom of sulphur is twice

as heavy as an atom of oxygen; and that when sulphite of barytes is converted into sulphate, the sulphate is neutral, there being no excess either of acid or base. From these and many other important facts he finally draws this conclusion : " In a compound formed by the union of two oxides, the one which (when decomposed by the galvanic battery) attaches itself to the positive pole (the *acid* for example) contains two, three, four, five, &c., times as much oxygen, as the one which attaches itself to the negative pole (the alkali, earth, or metallic oxide)." Berzelius's essay itself appeared in the third volume of the Afhandlingar, in 1810. It was almost immediately translated into German, and published by Gilbert in his Annalen der Physik. But no English translation has ever appeared, the editors of our periodical works being in general unacquainted with the German and other northern languages In 1815 Berzelius applied the atomic theory to the mineral kingdom, and showed with infinite ingenuity that minerals are chemical compounds in definite or atomic proportions, and by far the greater number of them combinations of acids and bases. He applied the theory also to the vegetable kingdom by analyzing several of the vegetable acids, and showing their atomic constitution. But here a difficulty occurs, which in the present state of our knowledge, we are unable to surmount. There are two acids, the *acetic* and *succinic*, that are composed of exactly the same number, and same kind of atoms, and whose atomic weight is 6·25. The constituents of these two acids are

Atomic weight.

2 atoms hydrogen 0·25
4 ,,   carbon   3
3 ,,   oxygen   3
————
6·25

So that they consist of *nine* atoms. Now as these
two acids are composed of the same number and
the same kind of atoms, one would expect that their
properties should be the same ; but this is not
the case : acetic acid has a strong and aromatic
smell, succinic acid has no smell whatever. Acetic
acid is so soluble in water that it is difficult to
obtain it in crystals, and it cannot be procured
in a separate state free from water ; for the crystals of acetic acid are composed of one atom of
acid and one atom of water united together ; but
succinic acid is not only easily obtained free from
water, but it is not even very soluble in that liquid.
The nature of the salts formed by these two acids
is quite different; the action of heat upon each
is quite different; the specific gravity of each differs.
In short all their properties exhibit a striking contrast. Now how are we to account for this ? Undoubtedly by the different ways in which the atoms
are arranged in each. If the electro-chemical theory of combination be correct, we can only view
atoms as combining two by two. A substance then,
containing nine atoms, such as acetic acid, must
be of a very complex nature. And it is obvious
enough that these nine atoms might arrange themselves in a great variety of binary compounds, and
the way in which these binary compounds unite may,
and doubtless does, produce a considerable effect
upon the nature of the compound formed. Thus, if
we make use of Mr. Dalton's symbols to represent
the atoms of hydrogen, carbon and oxygen, we may
suppose the nine atoms constituting acetic and succinic acid to be arranged thus :

Or thus:

Now, undoubtedly these two arrangements would produce a great change in the nature of the compound.

There is something in the vegetable acids quite different from the acids of the inorganic kingdom, and which would lead to the suspicion that the electro-chemical theory will not apply to them as it does to the others. In the acids of carbon, sulphur, phosphorus, selenium, &c., we find one atom of a positive substance united to one, two, or three of a negative substance: we are not surprised, therefore, to find the acid formed negative also. But in acetic and succinic acids we find every atom of oxygen united with two electro-positive atoms: the wonder then is, that the acid should not only retain its electro-negative properties, but that it should possess considerable power as an acid. In benzoic acid, for every atom of oxygen, there are present no fewer than seven electro-positive atoms.

Berzelius has returned to these analytical experiments repeatedly, so that at last he has brought his results very near the truth indeed. It is to his labours chiefly that the great progress which the atomic theory has made is owing.

In the year 1814 there appeared in the Philosophical Transactions a description of a Synoptical Scale of Chemical Equivalents, by Dr. Wollaston. In this paper we have the equivalents or atomic weights of seventy-three different bodies, deduced chiefly from a sagacious comparison of the previous analytical experiments of others, and almost all of them very near the truth. These numbers are laid down upon a sliding rule, by means of a table of logarithms, and over against them the names of the substances. By means of this rule a great many important questions respecting the substances contained on the scale may be solved. Hence the scale is of great advantage to the practical chemist. It gives, by bare inspection, the constituents of all the salts contained on it, the quantity of any other ingredient necessary to decompose any salt, and the weights of the new constituents that will be formed. The contrivance of this scale, therefore, may be considered as an important addition to the atomic theory. It rendered that theory every where familiar to all those who employed it. To it chiefly we owe, I believe, the currency of that theory in Great Britain; and the prevalence of the mode which Dr. Wollaston introduced, namely, of representing the atom of oxygen by unity, or at least by ten, which comes nearly to the same thing.

Perhaps the reader will excuse me if to the preceding historical details I add a few words to make him acquainted with my own attempts to render the atomic theory more accurate by new and careful analyses. I shall not say any thing respecting the experiments which I undertook to determine the specific gravity of the gases; though they were performed with much care, and at a considerable expense, and though I believe the results obtained approached accuracy as nearly as the present state of chemical apparatus enables us to go. In the year 1819 I began a set of experiments to determine the exact composition of the salts containing the different elementary bodies by means of double decomposition, as was done by Wenzel, conceiving that in that way the results would be very near the truth, while the experiments would be more easily made. My mode was to dissolve, for example, a certain weight of muriate of barytes in distilled water, and then to ascertain by repeated trials what weight of sulphate of soda must be added to precipitate the whole of the barytes without leaving any surplus of sulphuric acid in the liquid. To determine this I put into a watch-glass a few drops of the filtered liquor consisting of the mixture of solutions of the two salts : to this I added a drop of solution of sulphate of soda. If the liquid remained clear it was a proof that it contained no sensible quantity of barytes. To another portion of the liquid, also in a watch-glass, I added a drop of muriate of barytes. If there was no precipitate it was a proof that the liquid contained no sensible quantity of sulphuric acid. If there was a precipitate, on the addition of either of these solutions, it showed that there was an excess of one or other of the salts. I then mixed the two salts in another proportion, and proceeded in this way till I had found two quantities which when mixed exhibited no evidence of the residual liquid containing any sulphuric acid or barytes. I considered these two weights of the salts as the equivalent weights of the salt, or as weights proportional to an integrant particle of each salt. I made no attempt to collect the two new formed salts and to weigh them separately.

I published the result of my numerous experiments in 1825, in a work entitled " An Attempt to establish the First Principles of Chemistry by Experiment." The most valuable part of this book is the account of the salts ; about three hundred of which I subjected to actual analysis. Of these the worst executed are the phosphates ; for with respect to them I was sometimes misled by my method of double decomposition. I was not aware at first, that,

in certain cases, the proportion of acid in these salts
varies, and th e phosphate of soda which I employed
gave me a wrong number for the atomic weight of
phosphoric acid.

CHAPTER VII.

OF THE PRESENT STATE OF CHEMISTRY.

To finish this history it will be now proper to lay before the reader a kind of map of the present state of chemistry, that he may be able to judge how much of the science has been already explored, and how much still remains untrodden ground.

Leaving out of view light, heat, and electricity, respecting the nature of which only conjectures can be formed, we are at present acquainted with fifty-three simple bodies, which naturally divide themselves into three classes; namely, *supporters, acidifiable bases,* and *alkalifiable bases.*

The supporters are oxygen, chlorine, bromine, iodine, and fluorine. They are all in a state of negative electricity: for when compounds containing them are decomposed by the voltaic battery they all attach themselves to the positive pole. They have the property of uniting with every individual belonging to the other two classes. When they combine with the acidifiable bases in certain proportions they constitute *acids;* when with the alkalifiable bases, *alkalies.* In certain proportions they constitute *neutral* bodies, which possess neither the properties of acids nor alkalies.

The acidifiable bases are seventeen in number; namely, hydrogen, azote, carbon, boron, silicon, sul-

phur, selenium, tellurium, phosphorus, arsenic, antimony, chromium, uranium, molybdenum, tungsten, titanium, columbium. These bodies do not form acids
with every supporter, or in every proportion; but
they constitute the bases of all the known acids,
which form a numerous set of bodies, many of
which are still very imperfectly investigated. And
indeed there are a good many of them that may be
considered as unknown. These acidifiable bases are all
electro-positive; but they differ, in this respect, considerably from each other; hydrogen and carbon
being two of the most powerful, while titanium and
columbium have the least energy. Sulphur and selenium, and probably some other bodies belonging to
this class are occasional electro-negative bodies, as
well as the supporters. Hence, when united to other
acidifiable bases, they produce a new class of acids,
analogous to those formed by the supporters. These
have got the name of sulphur acids, selenium acids,
&c. Sulphur forms acids with arsenic, antimony,
molybdenum, and tungsten, and doubtless with
several other bases. To distinguish such acids from
alkaline bases, I have of late made an alteration in
the termination of the old word *sulphuret*, employed
to denote the combination of sulphur with a base.
Thus *sulphide* of arsenic means an acid formed by
the union of sulphur and arsenic; *sulphuret* of copper means an alkaline body formed by the union of
sulphur and copper. The term *sulphide* implies an
acid, the term *sulphuret* a *base*. This mode of
naming has become necessary, as without it many
of these new salts could not be described in an intelligible manner. The same mode will apply to
the acid and alkaline compounds of silenium. Thus
a *selenide* is an acid compound, and a *seleniet* an
alkaline compound in which selenium acts the part
of a supporter or electro-negative body. The same

mode of naming might and doubtless will be extended to all the other similar compounds, as soon as it becomes necessary. In order to form a systematic momenclature it will speedily be requisite to new-model all the old names which denote acids and bases; because unless this is done the names will become too numerous to be remembered. At present we denote the alkaline bodies formed by the union of manganese and oxygen by the name of *oxides of manganese*, and the acid compound of oxygen and the same metal by the name of *manganesic acid*. The word *oxide* applies to every compound of a base and oxygen, whether neutral or alkaline; but when the compound has acid qualities this is denoted by adding the syllable *ic* to the name of the base. This mode of naming answered tolerably well as long as the acids and alkalies were all combinations of oxygen with a base; but now that we know the existence of eight or ten classes of acids and alkalies, consisting of as many supporters, or acidifiable bases united to bases, it is needless to remark how very defective it has become. But this is not the place to dwell longer upon such a subject.

The alkalifiable bases are thirty-one in number; namely, potassium, sodium, lithium, barium, strontium, calcium, magnesium, aluminum, glucinum, yttrium, cerium, zirconium, thorinum, iron, manganese, nickel, cobalt, zinc, cadmium, lead, tin, bismuth, copper, mercury, silver, gold, platinum, palladium, rhodium, iridium, osmium. The compounds which these bodies form with oxygen, and the other supporters, constitute all the alkaline bases or the substances capable of neutralizing the acids.

Some of the acidifiable bases, when united to a certain portion of oxygen, constitute, not acids, but *bases* or *alkalies*. Thus the *green oxides of chro-*

mium and uranium are alkalies; while, on the other hand, there is a compound of oxygen and manganese which possesses acid properties. In such cases it is always the compound containing the least oxygen which is an alkali, and that containing the most oxygen that is an acid.

The opinion at present universally adopted by chemists is, that the ultimate particles of bodies consist of *atoms*, incapable of further division; and these atoms are of a size almost infinitely small. It can be demonstrated that the size of an atom of *lead* does not amount to so much as $\frac{1}{888,492,000,000,000}$ of a cubic inch.

But, notwithstanding this extreme minuteness, each of these atoms possesses a peculiar weight and a peculiar bulk, which distinguish it from the atoms of every other body. We cannot determine the absolute weight of any of them, but merely the relative weights; and this is done by ascertaining the relative proportions in which they unite. When two bodies unite in only one proportion, it is reasonable to conclude that the compound consists of 1 atom of the one body, united to 1 atom of the other. Thus oxide of bismuth is a compound of 1 oxygen and 9 bismuth; and, as the bodies unite in no other proportion, we conclude that an atom of bismuth is nine times as heavy as an atom of oxygen. It is in this way that the atomic weights of the simple bodies have been attempted to be determined. The following table exhibits these weights referred to oxygen as unity, and deduced from the best data at present in our possession :

	Atomic weight.			Atomic weight.
Oxygen .	. 1		Calcium .	. 2·5
Fluorine	. 2·25		Magnesium .	1·5
Chlorine	. 4·5		Aluminum .	1·25
Bromine	. 10		Glucinum .	2·25

	Atomic weight.			Atomic weight.
Iodine .	. 15·75		Yttrium .	. 4·25
Hydrogen .	0·125		Zirconium .	5
Azote . .	1·75		Thorinum .	7·5
Carbon . .	0·75		Iron . . .	3·5
Boron . .	1		Manganese .	3·5
Silicon . .	1		Nickel . .	3·25
Phosphorus	2		Cobalt . .	3·25
Sulphur .	2		Cerium . .	6·25
Selenium .	5		Zinc . . .	4·25
Tellurium .	4		Cadmium .	7
Arsenic . .	4·75		Lead . . .	13
Antimony .	8		Tin . . .	7·25
Chromium .	4		Bismuth . .	9
Uranium .	26		Copper . .	4
Molybdenum	6		Mercury . .	12·5
Tungsten .	12·5		Silver . .	13·75
Titanium .	3·25		Gold . . .	12·5
Columbium	22·75		Platinum .	12
Potassium .	5		Palladium .	6·75
Sodium . .	3		Rhodium .	6·75
Lithium .	0·75		Iridium . .	12·25
Barium . .	8·5		Osmium . .	12·5
Strontium .	5·5			

The atomic weights of these bodies, divided by their specific gravity, ought to give us the comparative size of the atoms. The following table, constructed in this way, exhibits the relative bulks of these atoms which belong to bodies whose specific gravity is known :

	Volume.			Volume.
Carbon . .	1		Platinum ⎰ Palladium ⎱	. 2·6
Nickel ⎰ Cobalt ⎱	. . 1·75		Zinc . . .	. 2·75
Manganese ⎱ Copper ⎰ Iron ⎰	. 2		Rhodium ⎱ Tellurium ⎰ Chromium ⎰	. 3

	Volume.		Volume.
Molybdenum .	3·25	Gold ⎫	
Silica ⎫		Silver ⎬ . 6	
Titanium ⎭ . 3·5		Osmium ⎭	
Cadmium . .	3·75	Oxygen ⎫	
Arsenic ⎫		Hydrogen ⎬ . 9·33	
Phosphorus ⎬ . 4		Azote ⎪	
Antimony ⎭		Chlorine ⎭	
Tungsten ⎫		Uranium . . 13·5	
Bismuth ⎬ . 4·25		Columbium ⎫ 14	
Mercury ⎭		Sodium ⎭	
Tin ⎫		Bromine . . 15·75	
Sulphur ⎭ . 4·66		Iodine . . . 24	
Selenium ⎫ . 5·4		Potassium . 27	
Lead ⎭			

We have no data to enable us to determine the shape of these atoms. The most generally received opinion is, that they are spheres or spheroids; though there are difficulties in the way of admitting such an opinion, in the present state of our knowledge, nearly insurmountable.

The probability is, that all the supporters have the property of uniting with all the bases, in at least three proportions. But by far the greater number of these compounds still remain unknown. The greatest progress has been made in our knowledge of the compounds of oxygen; but even there much remains to be investigated; owing, in a great measure, to the scarcity of several of the bases which prevent chemists from subjecting them to the requisite number of experiments. The compounds of chlorine have also been a good deal investigated; but bromine and iodine have been known for so short a time, that chemists have not yet had leisure to contrive the requisite processes for causing them to unite with bases.

The acids at present known amount to a very

great number. The oxygen acids have been most investigated. They consist of two sets : those consisting of oxygen united to a single base, and those in which it is united to two or more bases. The last set are derived from the animal and vegetable kingdoms : it does not seem likely that the electro-chemical theory of Davy applies to them. They must derive their acid qualities from some electric principle not yet adverted to ; for, from Davy's experiments, there can be little doubt that they are electro-negative, as well as the other acids. The acid compounds of oxygen and a single base are about thirty-two in number. Their names are

Hyponitrous acid	Selenic acid
Nitrous acid ?	Arsenious acid
Nitric acid	Arsenic acid
Carbonic acid	Antimonious acid
Oxalic acid	Antimonic acid
Boracic acid	Oxide of tellurium
Silicic acid	Chromic acid
Hypophosphorous acid	Uranic acid
Phosphorous acid	Molybdic acid
Phosphoric acid	Tungstic acid
Hyposulphurous acid	Titanic acid
Subsulphurous acid	Columbic acid
Sulphurous acid	Manganesic acid
Sulphuric acid	Chloric acid
Hyposulphuric acid	Bromic acid
Selenious acid	Iodic acid.

The acids from the vegetable and animal kingdoms (not reckoning a considerable number which consist of combinations of sulphuric acid with a vegetable or animal body), amount to about forty-three : so that at present we are acquainted with very nearly eighty acids which contain oxygen as an essential constituent.

The other classes of acids have been but imper-

fectly investigated.   Hydrogen enters into combination and forms powerful acids with all the supporters except oxygen.   These have been called
hydracids.   They are

Muriatic acid, or hydrochloric acid
Hydrobromic acid
Hydriodic acid
Hydrofluoric acid, or fluoric acid
Hydrosulphuric acid
Hydroselenic acid
Hydrotelluric acid

These constitute (such of them as can be procured)
some of the most useful and most powerful chemical
reagents in use.   There is also another compound
body, *cyanogen*, similar in its characters to a supporter : it also forms various acids, by uniting to
hydrogen, chlorine, oxygen, sulphur, &c.   Thus
we have

Hydrocyanic acid
Chlorocyanic acid
Cyanic acid
Sulphocyanic acid, &c.

We know, also, fluosilicic acid and fluoboric acids.
If to these we add fulminic acid, and the various
sulphur acids already investigated, we may state,
without risk of any excess, that the number of acids
at present known to chemists, and capable of uniting
to bases, exceeds a hundred.

The number of alkaline bases is not, perhaps, so
great ; but it must even at present exceed seventy ;
and it will certainly be much augmented when chemists turn their attention to the subject.   Now
every base is capable of uniting with almost every
acid,* in all probability in at least three different

* Acids and bases of the same class all unite.   Thus sulphur acids unite with sulphur bases ; oxygen acids with oxygen bases, &c.

proportions: so that the number of *salts* which they are capable of forming cannot be fewer than 21,000. Now scarcely 1000 of these are at present known, or have been investigated with tolerable precision. What a prodigious field of investigation remains to be traversed must be obvious to the most careless reader. In such a number of salts, how many remain unknown that might be applied to useful purposes, either in medicine, or as mordants, or dyes, &c. How much, in all probability, will be added to the resources of mankind by such investigations need not be observed.

The animal and vegetable kingdoms present a still more tempting field of investigation. Animal and vegetable substances may be arranged under three classes, acids, alkalies, and neutrals. The class of acids presents many substances of great utility, either in the arts, or for seasoning food. The alkalies contain almost all the powerful medicines that are drawn from the vegetable kingdom. The neutral bodies are important as articles of food, and are applied, too, to many other purposes of firstrate utility. All these bodies are composed (chiefly, at least) of hydrogen, carbon, oxygen, and azote; substances easily procured abundantly at a cheap rate. Should chemists, in consequence of the knowledge acquired by future investigations, ever arrive at the knowledge of the mode of forming these principles from their elements at a cheap rate, the prodigious change which such a discovery would make upon the state of society must be at once evident. Mankind would be, in some measure, independent of climate and situation; every thing could be produced at pleasure in every part of the earth; and the inhabitants of the warmer regions would no longer be the exclusive possessors of comforts and conveniences to which those in less favoured regions of the

earth are strangers.   Let the science advance for
another century with the same rapidity that it has
done during the last fifty years, and it will produce
effects upon society of which the present race can
form no adequate idea.   Even already some of these
effects are beginning to develop themselves;—our
streets are now illuminated with gas drawn from the
bowels of the earth; and the failure of the Green-
land fishery during an unfortunate season like the
last, no longer fills us with dismay.   What a change
has been produced in the country by the introduction of steam-boats! and what a still greater improvement is at present in progress, when steam-
carriages and railroads are gradually taking the
place of horses and common roads.   Distances will
soon be reduced to one-half of what they are at present; while the diminished force and increased rate
of conveyance will contribute essentially to lower
the rest of our manufactures, and enable us to enter
into a successful competition with other nations.

I must say a few words upon the application of
chemistry to physiology before concluding this imperfect sketch of the present state of the science.
The only functions of the living body upon which
chemistry is calculated to throw light, are the processes of digestion, assimilation, and secretion.   The
nervous system is regulated by laws seemingly quite
unconnected with chemistry and mechanics, and, in
the present state of our knowledge, perfectly in-
scrutable.   Even in the processes of digestion, assimilation, and secretion, the nervous influence is
important and essential.   Hence even of these functions our notions are necessarily very imperfect; but
the application of chemistry supplies us with some
data at least, which are too important to be altogether neglected.

The food of man consists of solids and liquids,

and the quantity of each taken by different individuals is so various, that no general average can be struck. I think that the drink will, in most cases, exceed the solid food in nearly the proportion of 4 to 3; but the solid food itself contains not less than 7-10ths of its weight of water. In reality, then, the quantity of liquid taken into the stomach is to that of solid matter as 10 to 1. The food is introduced into the mouth, comminuted by the teeth, and mixed up with the saliva into a kind of pulp.

The saliva is a liquid expressly secreted for this purpose, and the quantity certainly does not fall short of ten ounces in the twenty-four hours: indeed I believe it exceeds that amount: it is a liquid almost as colourless as water, slightly viscid, and without taste or smell: it contains about $\frac{3}{1000}$ of its weight of a peculiar matter, which is transparent and soluble in water: it has suspended in it about $\frac{1\cdot4}{1000}$ of its weight of mucus; and in solution, about $\frac{2\cdot8}{1000}$ of common salt and soda: the rest is water.

From the mouth the food passes into the stomach, where it is changed to a kind of pap called *chyme*. The nature of the food can readily be distinguished after mastication; but when converted into chyme, it loses its characteristic properties. This conversion is produced by the action of the eighth pair of nerves, which are partly distributed on the stomach; for when they are cut, the process is stopped: but if a current of electricity, by means of a small voltaic battery, be made to pass through the stomach, the process goes on as usual. Hence the process is obviously connected with the action of electricity. A current of electricity, by means of the nerves, seems to pass through the food in the stomach, and to decompose the common salt which is always mixed with the food. The muriatic acid is set at liberty, and dissolves the food; for *chyme* seems to be simply a solution of the food in muriatic acid.

The chyme passes through the pyloric orifice of the stomach into the duodenum, the first of the small intestines, where it is mixed with two liquids, the bile, secreted by the liver, and the pancreatic juice, secreted by the pancreas, and both discharged into the duodenum to assist in the further digestion of the food. The chyme is always acid; but after it has been mixed with the bile, the acidity disappears. The characteristic constituent of the bile is a bitter-tasted substance called *picromel*, which has the property of combining with muriatic acid, and forming with it an insoluble compound. The pancreatic juice also contains a peculiar matter, to which chlorine communicates a red colour. The use of the pancreatic juice is not understood.

During the passage of the chyme through the small intestines it is gradually separated into two substances; the *chyle*, which is absorbed by the lacteals, and the excrementitious matter, which is gradually protruded along the great intestines, and at last evacuated. The chyle, in animals that live on vegetable food, is semitransparent, colourless, and without smell; but in those that use animal food it is white, slightly similar to milk, with a tint of pink. When left exposed to the air it coagulates as blood does. The coagulum is *fibrin*. The liquid portion contains *albumen*, and the usual salts that exist in the blood. Thus the chyle contains two of the constituents of blood; namely, *albumen*, which perhaps may be formed in the stomach, and *fibrin*, which is formed in the small intestines. It still wants the third constituent of blood, namely, the *red* globules.

From the lacteals the chyle passes into the thoracic duct; thence into the left subclavian vein, by which it is conveyed to the heart. From the heart it passes into the lungs, during its circulation through which the *red globules* are supposed to be formed, though of this we have no direct evidence.

The lungs are the organs of *breathing*, a function so necessary to hot-blooded animals, that it cannot be suspended, even for a few minutes, without occasioning death. In general, about twenty inspirations, and as many expirations, are made in a minute. The quantity of air which the lungs of an ordinary sized man can contain, when fully distended, is about 300 cubic inches. But the quantity actually drawn in and thrown out, during ordinary inspirations and expirations, amounts to about sixteen cubic inches each time.

In ordinary cases the volume of air is not sensibly altered by respiration; but it undergoes two remarkable changes. A portion of its oxygen is converted into carbonic acid gas, and the air expired is saturated with humidity at the temperature of 98º. The moisture thus given out amounts to about seven ounces troy, or very little short of half an avoirdupois pound. The quantity of carbonic acid formed varies much in different individuals, and also at different times in the day; being a maximum at twelve o'clock at noon, and a minimum at midnight. Perhaps four of carbonic acid, in every 100 cubic inches of air breathed, may be a tolerable approach to the truth; that is to say, that every six respirations produce four cubic inches of carbonic acid. This would amount to 19,200 cubic inches in twenty-four hours. Now the weight of 19,200 cubic inches of carbonic acid gas is 18·98 troy ounces, which contain rather more than five troy ounces of carbon.

These alterations in the air are doubtless con-

nected with corresponding alterations in the blood,
though with respect to the specific nature of these
alterations we are ignorant.   But there are two
purposes which respiration answers, the nature of
which we can understand, and which seem to afford
a reason why it cannot be interrupted without death.
It serves to develop the *animal heat,* which is so
essential to the continuance of life; and it gives
the blood the property of stimulating the heart;
without which it would cease to contract, and put
an end to the circulation of the blood.   This stimulating property is connected with the scarlet colour
which the blood acquires during respiration; for
when the scarlet colour disappears the blood ceases
to stimulate the heart.

   The temperature of the human body in a state
of health is about 98° in this country; but in the
torrid zone it is a little higher.   Now as we are
almost always surrounded by a medium colder than
98°, it is obvious that the human body is constantly
giving out heat; so that if it did not possess the
power of generating heat, it is clear that its temperature would soon sink as low as that of the surrounding atmosphere.

   It is now generally understood that common combustion is nothing else than the union of oxygen
gas with the burning body.   The substances commonly employed as combustibles are composed
chiefly of carbon and hydrogen.   The heat evolved
is proportional to the oxygen gas which unites with
these bodies.   And it has been ascertained that
every 3¾ cubic inches of oxygen which combine with
carbon or hydrogen occasion the evolution of 1° of
heat.

   There are reasons for believing that not only carbon but also hydrogen unite with oxygen in the
lungs, and that therefore both carbonic acid and

water are formed in that organ. And from the late experiments of M. Dupretz it is clear that the heat evolved in a given time, by a hot-blooded animal, is very little short of the heat that would be evolved by the combustion of the same weight of carbon and hydrogen consumed during that time in the lungs. Hence it follows that the heat evolved in the lungs is the consequence of the union of the oxygen of the air with the carbon and hydrogen of the blood, and that the process is perfectly analogous to combustion.

The specific heat of arterial blood is somewhat greater than that of venous blood. Hence the reason why the temperature of the lungs does not become higher by breathing, and why the temperature of the other parts of the body are kept up by the circulation.

The blood seems to be completed in the kidneys. It consists essentially of albumen, fibrin, and the red globules, with a considerable quantity of water, holding in solution certain salts which are found equally in all the animal fluids. It is employed during the circulation in supplying the waste of the system, and in being manufactured into all the different secretions necessary for the various functions of the living body. By these different applications of it we cannot doubt that its nature undergoes very great changes, and that it would soon become unfit for the purposes of the living body were there not an organ expressly destined to withdraw the redundant and useless portions of that liquid, and to restore it to the same state that it was in when it left the lungs. These organs are the *kidneys;* through which all the blood passes, and during its circulation through which the urine is separated from it and withdrawn altogether from the body. These organs are as necessary for the continuance of life as the lungs themselves; accordingly, when they are diseased or destroyed, death very speedily ensues.

The quantity of urine voided daily is very various; though, doubtless, it bears a close relation to that of the drink. It is nearly but not quite equal to the amount of the drink; and is seldom, in persons who enjoy health, less than 2 lbs. avoirdupois in twenty-four hours. Urine is one of the most complex substances in the animal kingdom, containing a much greater number of ingredients than are to be found in the blood from which it is secreted.

The water in urine voided daily amounts to about 1·866 lbs. The blood contains no acid except a little muriatic. But in urine we find sulphuric, phosphoric, and uric acids, and sometimes oxalic and nitric acids, and perhaps also some others. The quantity of sulphuric acid may be about forty-eight grains daily, containing nineteen grains of sulphur. The phosphoric acid about thirty-three grains, containing about fourteen grains of phosphorus. The uric acid may amount to fourteen grains. These acids are in combination with potash, or soda, or ammonia, and also with a very little lime and magnesia. The common salt evacuated daily in the urine amounts to about sixty-two grains. The urea, a peculiar substance found only in the urine, amounts perhaps to as much as 420 grains.

It would appear from these facts that the kidneys possess the property of converting the sulphur and phosphorus, which are known to exist in the blood, into acids, and likewise of forming other acids and urea.

The quantity of water thrown out of the system by the urine and lungs is scarcely equal to the amount of liquid daily consumed along with the food. But there is another organ which has been ascertained to throw out likewise a considerable quantity of moisture, this organ is the skin; and the process is called *perspiration*. From the experiments of Lavoisier and Seguin it appears that the quantity of moisture given out daily by the skin amounts to 54·89 ounces: this added to the quantity evolved from the lungs and the urine considerably exceeds the weight of liquid taken with the food, and leaves no doubt that water as well as carbonic acid must be formed in the lungs during respiration.

Such is an imperfect sketch of the present state of that department of physiology which is most intimately connected with Chemistry. It is amply sufficient, short as it is, to satisfy the most careless observer how little progress has hitherto been made in these investigations; and what an extensive field remains yet to be traversed by future observers.

THE END.

C. WHITING, BEAUFORT HOUSE, STRAND.

HISTORY, PHILOSOPHY AND
SOCIOLOGY OF SCIENCE

Classics, Staples and Precursors

An Arno Press Collection

Aliotta, [Antonio]. **The Idealistic Reaction Against Science.** 1914

Arago, [Dominique François Jean]. **Historical Eloge of James Watt.** 1839

Bavink, Bernhard. **The Natural Sciences.** 1932

Benjamin, Park. **A History of Electricity.** 1898

Bennett, Jesse Lee. **The Diffusion of Science.** 1942

[Bronfenbrenner], Ornstein, Martha. **The Role of Scientific Societies in the Seventeenth Century.** 1928

Bush, Vannevar. **Endless Horizons.** 1946

Campanella, Thomas. **The Defense of Galileo.** 1937

Carmichael, R. D. **The Logic of Discovery.** 1930

Caullery, Maurice. **French Science and its Principal Discoveries Since the Seventeenth Century.** [1934]

Caullery, Maurice. **Universities and Scientific Life in the United States.** 1922

Debates on the Decline of Science. 1975

de Beer, G. R. **Sir Hans Sloane and the British Museum.** 1953

Dissertations on the Progress of Knowledge. [1824]. 2 vols. in one

Euler, [Leonard]. **Letters of Euler.** 1833. 2 vols. in one

Flint, Robert. **Philosophy as Scientia Scientiarum and a History of Classifications of the Sciences.** 1904

Forke, Alfred. **The World-Conception of the Chinese.** 1925

Frank, Philipp. **Modern Science and its Philosophy.** 1949

The Freedom of Science. 1975

George, William H. **The Scientist in Action.** 1936

Goodfield, G. J. **The Growth of Scientific Physiology.** 1960

Graves, Robert Perceval. **Life of Sir William Rowan Hamilton.** 3 vols. 1882

Haldane, J. B. S. **Science and Everyday Life.** 1940

Hall, Daniel, et al. **The Frustration of Science.** 1935

Halley, Edmond. **Correspondence and Papers of Edmond Halley.** 1932

Jones, Bence. **The Royal Institution.** 1871

Kaplan, Norman. **Science and Society.** 1965

Levy, H. **The Universe of Science.** 1933

Marchant, James. **Alfred Russel Wallace.** 1916

McKie, Douglas and Niels H. de V. Heathcote. **The Discovery of Specific and Latent Heats.** 1935

Montagu, M. F. Ashley. **Studies and Essays in the History of Science and Learning.** [1944]

Morgan, John. **A Discourse Upon the Institution of Medical Schools in America.** 1765

Mottelay, Paul Fleury. **Bibliographical History of Electricity and Magnetism Chronologically Arranged.** 1922

Muir, M. M. Pattison. **A History of Chemical Theories and Laws.** 1907

National Council of American-Soviet Friendship. **Science in Soviet Russia: Papers Presented at Congress of American-Soviet Friendship.** 1944

Needham, Joseph. **A History of Embryology.** 1959

Needham, Joseph and Walter Pagel. **Background to Modern Science.** 1940

Osborn, Henry Fairfield. **From the Greeks to Darwin.** 1929

Partington, J[ames] R[iddick]. **Origins and Development of Applied Chemistry.** 1935

Polanyi, M[ichael]. **The Contempt of Freedom.** 1940

Priestley, Joseph. **Disquisitions Relating to Matter and Spirit.** 1777

Ray, John. **The Correspondence of John Ray.** 1848

Richet, Charles. **The Natural History of a Savant.** 1927

Schuster, Arthur. **The Progress of Physics During 33 Years (1875-1908).** 1911

Science, Internationalism and War. 1975

Selye, Hans. **From Dream to Discovery: On Being a Scientist.** 1964

Singer, Charles. **Studies in the History and Method of Science.** 1917/1921. 2 vols. in one

Smith, Edward. **The Life of Sir Joseph Banks.** 1911

Snow, A. J. **Matter and Gravity in Newton's Physical Philosophy.** 1926

Somerville, Mary. **On the Connexion of the Physical Sciences.** 1846

Thomson, J. J. **Recollections and Reflections.** 1936

Thomson, Thomas. **The History of Chemistry.** 1830/31

Underwood, E. Ashworth. **Science, Medicine and History.** 2 vols. 1953

Visher, Stephen Sargent. **Scientists Starred 1903-1943 in American Men of Science.** 1947

Von Humboldt, Alexander. **Views of Nature: Or Contemplations on the Sublime Phenomena of Creation.** 1850

Von Meyer, Ernst. **A History of Chemistry from Earliest Times to the Present Day.** 1891

Walker, Helen M. **Studies in the History of Statistical Method.** 1929

Watson, David Lindsay. **Scientists Are Human.** 1938

Weld, Charles Richard. **A History of the Royal Society.** 1848. 2 vols. in one

Wilson, George. **The Life of the Honorable Henry Cavendish.** 1851